Lecture Notes in Physics

Editorial Board

R. Beig, Wien, Austria
B.-G. Englert, Ismaning, Germany
U. Frisch, Nice, France
P. Hänggi, Augsburg, Germany
K. Hepp, Zürich, Switzerland
W. Hillebrandt, Garching, Germany
D. Imboden, Zürich, Switzerland
R. L. Jaffe, Cambridge, MA, USA
R. Lipowsky, Golm, Germany
H. v. Löhneysen, Karlsruhe, Germany
I. Ojima, Kyoto, Japan
D. Sornette, Nice, France, and Los Angeles, CA, USA
S. Theisen, Golm, Germany
W. Weise, Trento, Italy, and Garching, Germany
J. Wess, München, Germany
J. Zittartz, Köln, Germany

Springer
Berlin
Heidelberg
New York
Hong Kong
London
Milan
Paris
Tokyo

Physics and Astronomy   ONLINE LIBRARY

http://www.springer.de/phys/

Editorial Policy

The series *Lecture Notes in Physics* (LNP), founded in 1969, reports new developments in physics research and teaching -- quickly, informally but with a high quality. Manuscripts to be considered for publication are topical volumes consisting of a limited number of contributions, carefully edited and closely related to each other. Each contribution should contain at least partly original and previously unpublished material, be written in a clear, pedagogical style and aimed at a broader readership, especially graduate students and nonspecialist researchers wishing to familiarize themselves with the topic concerned. For this reason, traditional proceedings cannot be considered for this series though volumes to appear in this series are often based on material presented at conferences, workshops and schools (in exceptional cases the original papers and/or those not included in the printed book may be added on an accompanying CD ROM, together with the abstracts of posters and other material suitable for publication, e.g. large tables, colour pictures, program codes, etc.).

Acceptance

A project can only be accepted tentatively for publication, by both the editorial board and the publisher, following thorough examination of the material submitted. The book proposal sent to the publisher should consist at least of a preliminary table of contents outlining the structure of the book together with abstracts of all contributions to be included.
Final acceptance is issued by the series editor in charge, in consultation with the publisher, only after receiving the complete manuscript. Final acceptance, possibly requiring minor corrections, usually follows the tentative acceptance unless the final manuscript differs significantly from expectations (project outline). In particular, the series editors are entitled to reject individual contributions if they do not meet the high quality standards of this series. The final manuscript must be camera-ready, and should include both an informative introduction and a sufficiently detailed subject index.

Contractual Aspects

Publication in LNP is free of charge. There is no formal contract, no royalties are paid, and no bulk orders are required, although special discounts are offered in this case. The volume editors receive jointly 30 free copies for their personal use and are entitled, as are the contributing authors, to purchase Springer books at a reduced rate. The publisher secures the copyright for each volume. As a rule, no reprints of individual contributions can be supplied.

Manuscript Submission

The manuscript in its final and approved version must be submitted in camera-ready form. The corresponding electronic source files are also required for the production process, in particular the online version. Technical assistance in compiling the final manuscript can be provided by the publisher's production editor(s), especially with regard to the publisher's own Latex macro package which has been specially designed for this series.

Online Version/ LNP Homepage

LNP homepage (list of available titles, aims and scope, editorial contacts etc.):
http://www.springer.de/phys/books/lnpp/
LNP online (abstracts, full-texts, subscriptions etc.):
http://link.springer.de/series/lnpp/

Michael Beyer (Ed.)

CP Violation in Particle, Nuclear and Astrophysics

 Springer

Editor

Michael Beyer
Universität Rostock
Fachbereich Physik
18051 Rostock, Germany

Cover picture: **with courtesy of M. Beyer, University of Rostock, Germany**

Library of Congress Cataloging-in-Publication Data applied for.

Beyer, Michael:
CP violation in particle, nuclear and astrophysics / Michael Beyer. - Berlin
; Heidelberg ; New York ; Barcelona ; Hong Kong ; London ; Milan ; Paris ;
Tokyo : Springer, 2002
   (Lecture notes in physics ; Vol. 591)
   (Physics and astronomy online library)
   ISBN 3-540-43705-3

ISSN 0075-8450
ISBN 3-540-43705-3 Springer-Verlag Berlin Heidelberg New York

This work is subject to copyright. All rights are reserved, whether the whole or part of the material is concerned, specifically the rights of translation, reprinting, reuse of illustrations, recitation, broadcasting, reproduction on microfilm or in any other way, and storage in data banks. Duplication of this publication or parts thereof is permitted only under the provisions of the German Copyright Law of September 9, 1965, in its current version, and permission for use must always be obtained from Springer-Verlag. Violations are liable for prosecution under the German Copyright Law.

Springer-Verlag Berlin Heidelberg New York
a member of BertelsmannSpringer Science+Business Media GmbH

http://www.springer.de

© Springer-Verlag Berlin Heidelberg 2002
Printed in Germany

The use of general descriptive names, registered names, trademarks, etc. in this publication does not imply, even in the absence of a specific statement, that such names are exempt from the relevant protective laws and regulations and therefore free for general use.

Typesetting: Print-ready by the authors/editor
Camera-data conversion by Steingraeber Satztechnik GmbH Heidelberg
Cover design: *design & production*, Heidelberg

Printed on acid-free paper
SPIN: 10880127   54/3141/du - 5 4 3 2 1 0

Preface

The exciting experiments of the BABAR and BELLE collaborations have now proven violation of CP symmetry in the neutral B system. This has renewed strong interest in the physics of CP violation. Novel experimental techniques and new highly intense neutron sources are now becoming available to further test the related time reversal symmetry. They will substantially lower the current limit on the neutron electric dipole moment and hence open up new tests of theoretical concepts beyond the Standard Model. These are strongly required to explain the decisive excess of matter versus antimatter in our Universe.

There is a definite need to communicate these exciting developments to younger scientists, and therefore we organized a summer school in October 2000 on "CP Violation and Related Topics", which was held in Prerow, a small Baltic Sea resort. These Lecture Notes were inspired by the vivid interest of the participants, and I am grateful to the authors, who faced the unexpected and delivered all the material for an up-to-date introduction to this broad field.

It is a great pleasure for me to warmly thank the Co-organizers of the summer school, Henning Schröder, Thomas Mannel, Klaus R. Schubert and my colleague Roland Waldi.

Also I would like to express my sincere thanks to the Volkswagen-Stiftung for their financial support of this inspiring summer school.

Rostock, July 2002 *Michael Beyer*

Table of Contents

Introduction
Michael Beyer 1
References 3

CP, T, and CPT Symmetries
Ernest M. Henley 4
1   Introduction, Parity, Charge Conjugation 4
     1.1   Parity 4
     1.2   Charge Conjugation 7
2   The CP Transformation 9
3   Time Reversal 12
     3.1   Time Reversal 12
     3.2   Consequences of Time Reversal Invariance 14
     3.3   The Strong CP Problem 19
4   CPT, PCT, TPC, ... 22
5   Summary 24
References 24

CP Violation in the K^0 System
Konrad Kleinknecht 27
1   The Big Bang and the Expanding Universe 27
2   Symmetries 28
3   Phenomenology and Models of CP Violation 30
4   Theoretical Estimates for the Parameter ϵ' of Direct CP Violation . 32
5   Experiments on Direct CP Violation 33
     5.1   Early Experiments: The Observation of Direct CP Violation . 33
     5.2   The NA48 Detector 34
     5.3   The KTeV Detector 36
     5.4   Data Analysis NA48: Confirmation of Direct CP Violation ... 37
     5.5   Analysis of KTeV Data: Confirmation of Direct CP Violation 39
6   Conclusion 41
References 43

Flavour Oscillation and CP Violation of B Mesons

Roland Waldi 43
1 Introduction 43
    1.1 The Experiments 44
2 Quark Mixing and Particle Antiparticle Oscillations 45
    2.1 The Unitary CKM Matrix 45
    2.2 Oscillation Phenomenology.............................. 57
    2.3 Experimental Determination
        of the Mixing Parameters of B Mesons 81
3 CP Violation 96
    3.1 CP Eigenstates Versus Mass Eigenstates 98
    3.2 CP Violating Interference Effects in B Decays 102
    3.3 Direct CP Violation..................................... 103
    3.4 CP Violation in the Oscillation 106
    3.5 CP Violation in Common Final States of B^0 and $\bar{B}^0$ 110
    3.6 Measurement of Time-Dependent Asymmetries
        of Neutral B Mesons 146
    3.7 Experimental Data on $B \to J/\psi K_S$ 183
    3.8 Experimental Data on $B \to \pi\pi$ 188
4 Outlook 189
References 190

CP Asymmetries in Neutral Kaon and Beon Decays

Klaus R. Schubert 196
References 205

Time Reversal Invariance in Nuclear Physics: From Neutrons to Stochastic Systems

Christopher R. Gould, Edward David Davis 206
1 Introduction 206
2 Overview of Electric Dipole Moment
   and Transmission Tests of Time Reversal Invariance 207
    2.1 Electric Dipole Moments................................ 207
    2.2 Neutron Transmission Tests of Symmetry Violation 210
3 Total Cross Section and Forward Elastic Scattering
   Amplitude for Arbitrary States of Polarization 214
    3.1 Generalization of the Optical Theorem
        to Include Spin Degrees of Freedom 214
    3.2 Statistical Density Matrix
        for the Projectile Target Spin Space 217
    3.3 Partial Wave Expansions
        on the Energy Momentum Shell 219
    3.4 Decomposition of the Total Cross Section 223
    3.5 Decomposition of the Forward Elastic Scattering Amplitude.. 227
References 235

CP Violation and Baryogenesis
Werner Bernreuther 237
1 Introduction 237
2 Some Basics of Cosmology 238
   2.1 The Standard Model of Cosmology 238
   2.2 Equilibrium Thermodynamics 241
   2.3 Departures from Thermal Equilibrium 243
3 The Baryon Asymmetry of the Universe 244
   3.1 Heuristic Considerations 244
   3.2 The Sakharov Conditions 246
4 CP and B Violation in the Standard Model 249
5 Electroweak Baryogenesis 255
   5.1 Why the SM Fails 259
   5.2 EW Phase Transition in SM Extensions 262
   5.3 CP Violation in SM Extensions 263
   5.4 Electroweak Baryogenesis 269
   5.5 Role of the KM Phase 275
6 Out-of-Equilibrium Decay of Super-Heavy Particle(s) 276
   6.1 GUT Baryogenesis 277
   6.2 Baryogenesis Through Leptogenesis 281
7 Summary 284
References 290

Physics Beyond the Standard Model
Gian Francesco Giudice 294
1 Introduction 294
2 Grand Unified Theories 295
   2.1 SU(5) 295
   2.2 Experimental Tests of GUTs 296
   2.3 SO(10) and Neutrino Masses 298
3 The Hierarchy Problem 299
4 Supersymmetry 300
   4.1 Supersymmetric Unification 302
   4.2 Electroweak Symmetry Breaking 303
   4.3 Higgs Sector 304
   4.4 Supersymmetry and Experiments 306
   4.5 The Flavour Problem 307
   4.6 Recent Developments in Supersymmetry 308
   4.7 More on Experimental Consequences 312
5 Technicolour 313
6 Extra Dimensions 316
   6.1 Opening New Problems 319
   6.2 Experimental Tests of Extra Dimensions 321
7 Conclusions 324
References 324

Subject Index 329

List of Contributors

Werner Bernreuther
Institut für Theoretische Physik,
RWTH Aachen,
52056 Aachen, Germany
breuther@physik.rwth-aachen.de

Michael Beyer
Fachbereich Physik,
Universität Rostock,
18051 Rostock, Germany
michael.beyer
    @physik.uni-rostock.de

Edward David Davis
Physics Department,
Kuwait University,
P.O. Box 5969,
Safat, Kuwait
davis@kuc01.kuniv.edu.kw

Gian Francesco Giudice
Theoretical Physics Division,
CERN,
1211 Geneva 23,
Switzerland
Gian.Giudice@cern.ch

Christopher R. Gould
North Carolina State University,
Raleigh NC 27695, USA,
and
Triangle Universities Nuclear
Laboratory,
Durham NC 27708, USA
chris_gould@ncsu.edu

Ernest M. Henley
Department of Physics,
and Institute for Nuclear Theory,
Box 351560,
University of Washington,
Seattle, WA 98195-1560, USA
henley
    @nuclthy.phys.washington.edu

Konrad Kleinknecht
Institut für Physik,
Johannes Gutenberg-Universität,
Staudinger Weg 7,
55099 Mainz, Germany
kleinknecht
    @dipmza.physik.uni-mainz.de

Klaus R. Schubert
Institut für Kern-
und Teilchenphysik,
Technische Universität Dresden,
01062 Dresden, Germany
schubert@physik.uni-dresden.de

Roland Waldi
Fachbereich Physik,
Universität Rostock,
18051 Rostock, Germany
roland.waldi
    @physik.uni-rostock.de

Introduction

Michael Beyer

Fachbereich Physik, Universität Rostock, 18051 Rostock, Germany

Symmetries or approximate symmetries play an essential role in physics. Symmetries are related to invariants of the system, i.e. conservation laws and integrals. Therefore a symmetry that holds leads to a substantial reduction of work in the dynamical description. The case of a broken symmetry is equally important. Reasons for a symmetry to be violated can be many-fold, e.g., new forces, new particles, properties/structure of the vacuum state, phase freedom, etc. In its turn the investigation of symmetries provide an excellent tool to find "new physics".

The discrete symmetries *particle-antiparticle conjugation* C, *parity* P, and *time reversal* T are outstanding in this context. Starting with rather irrational statements of belief or disbelief, e.g., about parity as a valid symmetry in the early times of quantum mechanics [1], it is now proven that these symmetries are violated. Presently, the study of CP symmetry that is the combination of parity and particle anti-particle conjugation is one of the most exciting activity in physics. It touches upon different areas of physics, among them are elementary particle physics including the standard model and the physics "beyond", as well as atomic, nuclear, and astrophysics. In addition, the mathematical description of CP symmetry and also T symmetry is based on "not everyday math" such as anti-unitarity. The basis of discrete symmetries and the consequences of their violation are given in a general introduction by Dr. Henley.

Violation of CP symmetry has been discovered by the celebrated experiment of Christenson, Cronin, Fitch, and Turlay [2] in 1964. The details and the present day status of CP violation in the K system are given in Dr. Kleinknecht's lecture. This CP violation can be accommodated by a third generation of quarks introduced into the Cabibbo mixing matrix by Kobayashi and Maskawa [3] in 1973 that by now is part of the standard model of elementary particle physics. As a consequence of the independent phase in the CKM matrix one can expect CP violation in other (heavy) mesons as well.

An important discovery along the way of studying CP violation in a different than the K system has been the large mixing in the neutral B system by the ARGUS collaboration [4] in 1987. Only last year, 38 years after its first discovery, two collaborations, BABAR at SLAC and Belle at KEK, have independently and clearly established CP violation in the B system. The experiments use the difference in the time dependent decay rates of the "golden decay" of B^0 or $\bar{B}^0$ to J/ψ and K_S^0. The asymmetry is parameterized as an angle β that is directly related to the CKM matrix. The presently latest values are $\sin 2\beta = 0.75 \pm 0.09 \pm 0.04$ by the BABAR collaboration [5] and $\sin 2\beta = 0.82 \pm 0.12 \pm 0.05$ by the Belle collaboration [6]. All the details are given in Dr. Waldi's lecture.

Because of the different physics aspects, the different background, and the long time elapsed between the discovery of CP violation in the K system and the B system, differences in language and phrasing have occurred when describing the various phenomena in both systems. On the basis of a unified description a detailed comparison between CP violation in the K system (kaon) and the B system ("beon") is given in Dr. Schubert's lecture.

Although well hidden, the violation of CP symmetry has dramatic consequences, namely the matter-antimatter asymmetry in the present day universe. Presently, there is no evidence how the matter-antimatter asymmetry could be established *without* violation of CP symmetry. However, a satisfying explanation of matter-antimatter asymmetry in the universe is still lacking. An elementary introduction and an overview of CP violation in the context of matter-antimatter asymmetry is given in Dr. Bernreuther's lecture.

The search for violation of CP symmetry and T symmetry in other than the K system is ongoing. An outstanding experiment is the search for a neutron electric dipole moment (edm). Both P and T must be violated to establish a finite value for the dipole moment. In addition the net effect is flavor diagonal. In the past decades experimental limits have been improved by orders of magnitude. The present upper limit [7] $|d_n| < 6.3 \times 10^{-26} ecm$ is still five to six orders of magnitude above the standard model result induced by the CKM matrix. In the next few years substantial progress can be expected at this front, because of new experimental techniques using polarized ^{3}He as a reference to reduce the systematic uncertainties by approximately two orders of magnitude and/or new highly intense neutron sources. More on this and other nuclear physics, namely transmission experiments, is given in the lecture by Drs. Gould and Davis.

Additionally, the current limit on the neutron edm induces a limit on the parity and time reversal violating θ-term of quantum chromodynamics (QCD), $\bar\theta \lesssim 10^{-9}$. This small value $\bar\theta$ that is a difference of the effect of the QCD θ-term and the mass matrix is known as strong CP problem. Dr. Henley's lecture further touches upon this issue.

It is clear that CP violation provides an excellent tool to discriminate various extensions of the standard model and/or more fundamental theories that include the standard model as a limiting case. To date the number of possible nontrivial phases of possible new particles that might lead to CP violation seems not yet exhausted. A view on what might be expected as new fundamental theories is given in Dr. Giudice's lecture.

Editorial Remarks on Conventions

The representation of CP symmetry for mesons, $\Phi(x)$, is given by successive application of P and C symmetry operations. The representation of C can be used to define antiparticle states

$$\Phi^C(x) = \eta'_C \Phi^\dagger(x) \equiv \eta_C \Phi^{\text{anti}}(x)$$

with an arbitrary phase η_C. The representation of parity for mesons is

$$\Phi^P(x) = \eta_P \Phi(P^{-1}x)$$

where $P^{-1}x = Px = (x^0, -x^1, -x^2, -x^3)$ represents parity in Minkowski space and acting on the coordinate vector $x = (x^\mu)$. The combination leads to

$$\Phi^{CP}(x) = \eta_C \eta_P \Phi^{\text{anti}}(P^{-1}x).$$

In case of K^0 and its antiparticle $\bar{K}^0$ the arbitrary phase has been chosen $\eta_C \eta_P = +1$ in Dr. Henley's and $\eta_C \eta_P = -1$ in Dr. Kleinknecht's lecture. In Dr. Schubert's and Dr. Waldi's lectures no explicit phase convention has been chosen.

In the K system various notations of ϵ's are used. The same relations in all lectures are those between the ϵ's and the observables η_{+-} and η_{00}. Note, however, that intermediate formulas may differ, depending on additional conventions used.

References

1. A. Pais, *Inward bound* (Oxford University Press, New York, 1986)
2. J.H. Christenson, J.W. Cronin, V.L. Fitch, R. Turlay, Phys. Rev. Lett. 13 (1964), 138.
3. M. Kobayashi and T. Maskawa, Progr. Theor. Phys. **49**,652 (1973).
4. H. Albrecht et al., Phys. Lett. **B192** (1987) 245.
5. BABAR Collab., SLAC-PUB-9153 (2002).
6. BELLE Collab., presented at the XXXVII Recontres de Moriond, March 2002.
7. P. G. Harris *et al.*, Phys. Rev. Lett. **82** (1999) 904.

CP, T, and CPT Symmetries

Ernest M. Henley

Department of Physics, and Institute for Nuclear Theory University of Washington, Seattle, WA 98195, USA

Abstract. The parity (P), charge conjugation (C) symmetries are briefly reviewed. CP and time reversal (T) transformations and conservation laws are studied; the strong CP problem is discussed. Experimental measurements of the violation of these symmetries are presented. Finally, the CPT theorem and its consequences are discussed.

1 Introduction, Parity, Charge Conjugation

In order to discuss CP and its consequences it is necessary to first present parity (P) and charge conjugation (C) separately [1,2]. I will do so briefly as well as introduce some of the current research. I will then introduce the CP symmetry and its breaking in the K^0 system. Time reversal is a non-unitary transformation and this causes some difficulties. I will discuss the operation and the symmetry under time reversal and some tests of it that have been carried out. Lastly I will present the CPT "theorem" and discuss its consequences and tests thereof. I will use $\hbar = c = 1$.

1.1 Parity

It was 46 years ago that Lee and Yang discovered that there was no evidence for parity (P) conservation in the weak interactions and suggested experiments to test it [3]. They realized that a pseudoscalar observable was required.

What is the parity transformation? It is a reflection of all coordinates in the origin so that $\boldsymbol{r} \to -\boldsymbol{r}$. It is also called left-right or mirror reflection because a reflection in a plane followed by a rotation of 180° corresponds to the parity transformation. Symmetry under a parity transformation says that you cannot tell whether you are examining the real or mirror world. Let me illustrate with a half-charged sphere which is rotating about the z-axis so that there is the required preferred direction in space. The object has both an electric dipole moment, $\boldsymbol{d}$ and a magnetic dipole moment, $\boldsymbol{\mu}$ in the z-direction by dint of the rotating charge distribution, shown in Fig. 1. If a mirror is placed in the midplane then the object looks as shown on the right hand side of Fig. 1. The dipole moments are parallel to the angular momentum in the real world, but the electric dipole moment $\boldsymbol{d}$ is antiparallel to it in the

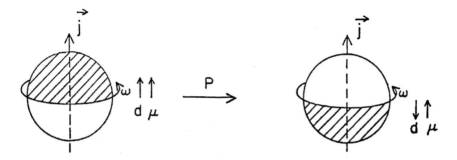

Fig. 1. Parity transformation of electric and magnetic dipole moments

mirror world. Thus, for an object with an electric dipole moment you can tell the difference between the two worlds. It follows that if P is a conserved quantum number, i.e., $[P, H] = 0$, where H is the energy operator for the system, that an object, or a quantum mechanical state, may not possess an electric dipole moment (if the state is not a degenerate one). The experiment suggested by Lee and Yang and carried out by Dr. Wu and collaborators [4] was the preferential emission of electrons antiparallel to the spin of ^{60}Co. The weak interactions are now known to violate P by 100 % because they are purely left-handed. The theory of electroweak interactions makes full use of this feature [5]. By contrast, there are a large number of experiments which show that the strong and electromagnetic interactions conserve P [6].

Table 1. Transformation of some operators under the parity operation

	r, t	p, E	j, ρ J	A, A^0
$P \Downarrow$	$-r, t$	$-p, E$	$-j, \rho$ J	$-A, A^0$

Formally, we have the transformations shown in Table 1. The transformation operator is Hermitian, $P = P^\dagger$, so that the eigenvalues can be chosen to be real. If we carry out two parity transformations in succession for a bosonic system we are back to our original picture. Therefore, state functions can be either even ($\eta_P = 1$) or odd ($\eta_P = -1$) under the parity transformation; if they are of mixed parity, then P is not conserved. Thus, we have

$$P\phi(\boldsymbol{r}, t) = \eta_P \phi(-\boldsymbol{r}, t) \ . \tag{1}$$

Particles generally have an intrinsic parity, which you can think of as due to their internal structure. For bosons the intrinsic parity can be measured; for instance that of the pion is negative as deduced from the observation of the reaction $\pi^- d \to n\, n$ with pions at rest [2]. In this reaction, the two neutrons must be in the $^3P_1$ state to conserve angular momentum for a pion captured at rest. For a photon, the intrinsic parity has to be negative in view of the transformation of the vector potential, $\boldsymbol{A}$. The gauge boson responsible for the strong interaction also has $\eta_P = -1$.

For fermions, it actually takes four parity transformations to return to the original picture, because a rotation of 4π is required. Thus the eigenvalue, η_P can be chosen to be ± 1 as well as $\pm i$; the former is used for most fermions, but not for Majorana neutrinos. Relativistically, the situation is complicated by the spinors and the parity transformation is

$$P\psi(\boldsymbol{r},t) = \eta_P \gamma^0 \psi(-\boldsymbol{r},t) \ . \tag{2}$$

It follows from this transformation that

$$Pu = \eta_P \gamma^0 u = \eta_P u \ , \ Pv = \eta_P \gamma^0 v = -\eta_P v \ , \tag{3}$$

where u and v are the positive and negative energy spinors. Therefore, a particle and its antiparticle have opposite parities. Since P is multiplicative, the parity of a system of two particles is the product of the intrinsic parities multiplied by $(-1)^\ell$, where ℓ is the relative orbital angular momentum. The angular momentum transformation follows from the behavior of the spherical harmonics, $PY_\ell^m(\theta,\phi) = Y_\ell^m(\pi-\theta,\phi+\pi) = (-1)^\ell Y_\ell^m(\theta,\phi)$. If we have two bosons of the same family, e.g., $\pi^+\pi^-$ in a state of relative orbital angular momentum ℓ, then the parity of the system is $(-1)^\ell$. On the other hand for a fermionic particle-antiparticle system the parity is $(-1)^{\ell+1}$.

The study of parity nonconservation due to the weak interactions is ongoing. Thus, there are continued searches for an electric dipole moment (edm) of the neutron and of atoms. However, as I will show later an edm requires time reversal violation as well as parity nonconservation. Another experiment of note is a recent finding of the anapole moment of a Cs atom [7]. The anapole moment is a toroidal dipole moment and corresponds to an axial coupling of a photon to (in this case) a hadron. This parity-violating coupling of photons to the nucleus of Cs comes from a weak *nuclear* force. The experiment measures an interference of the weak and electromagnetic amplitudes present in the atom and uses the difference of hyperfine levels to see the effect. It has been observed in a measurement made to an accuracy of $1/2\%$, even though the entire parity-violating effect is miniscule. There are also ongoing studies of the interference of the strong and weak forces by examining the dependence of a cross section on the direction of longitudinally polarized hadrons. The effect sought here is of order 10^{-7} to 10^{-8} and measurements in proton proton scattering have been carried out to better than 10% [8]. In slow neutron scattering off heavy nuclear targets one can take advantage of the very small spacings between states of opposite parity in a compound nucleus to get effect of several percent, an enhancement of $\geq 10^5$ [9].

It is noteworthy that once the theory is understood, the small PNC effects present in electromagnetic and strong interactions can be used to probe structure effects. Thus, in another ongoing experiment PNC in electron scattering is used to observe strangeness in the nucleon [10]; there are also proposals to measure the neutron radii of nuclei [11] because the weak interaction of electrons with neutrons is more than a factor of ten larger for neutrons than protons. Because strange quarks have a weak isospin of 1/2 and a strong one of 0, and because SU(3) is required when strange quarks are included, the interference of the weak and electromagnetic interactions produces measurable effects of strangeness; for example the weak axial current for the nucleon has only isospin 1 in the absence of strangeness but can have isospin 0 if strangeness is present. The experiment, carried out at Bates at MIT, is also ongoing at the Jefferson Lab. Indications are that the strangeness vector matrix element in the proton is small and positive, opposite in sign to what most theorists had predicted, e.g., based on a proton virtually dissociating into a kaon and strange baryon.

1.2 Charge Conjugation

Charge conjugation (C) is a misnomer; it should be called particle-antiparticle conjugation. Of course, charge (if any) is reversed in this process, but so are all other additive quantum numbers, e.g. baryon number; on the other hand linear and angular momenta are unchanged. Thus, if we call q_{gen} a generalized charge, we have $C|q_{\text{gen}}> = -|q_{\text{gen}}>$. If C is conserved then $[C, H] = 0$, but only if $q_{\text{gen}} = 0$ can there be an eigenvalue of C since $[C, Q_{\text{gen}}] \neq 0$. Formally, we have [2]

$$C|q_{\text{gen}}, \boldsymbol{p}, s> = |-q_{\text{gen}}, \boldsymbol{p}, s>, \quad (4)$$

where $\boldsymbol{p}$ is the linear momentum and s is the spin of the particle. The application of C twice leads back to the original system, so that $C^2 = 1$; it follows that the eigenvalues of the operator C are $\eta_C = \pm 1$, and the eigenvalues are called charge parity. Of course, only neutral systems, with $q_{\text{gen}} = 0$ can be eigenvectors of C. For a neutral system of bosons such as $\pi^+\pi^-$, the eigenvalue is the product of the intrinsic values of η_C multiplied by $(-1)^\ell$. Since C interchanges the $+$ and $-$ charges, it corresponds to a parity transformation, i.e.,

$$C|\pi^+\pi^-> = |\pi^-\pi^+> = (-1)^\ell |\pi^+\pi^->, \quad (5)$$
$$C|\pi^0\pi^0> = |\pi^0\pi^0>. \quad (6)$$

Since the charge and currents change sign under the C transformation, the electromagnetic vector potential also changes sign, and it follows that the photon has $\eta_C = -1$. Since a neutral pion decays into two photons we can immediately surmise that $\eta_C(\pi^0) = 1$. It follows that the neutral pion cannot decay into 3 photons if C is a valid symmetry and this decay has not been seen (fractional decay rate $\leq 3 \times 10^{-8}$) [12].

For a fermion or a system of fermions, the spinors make the operation slightly more cumbersome. In order to do it properly, second quantization is needed, but I will stick to first quantization here. In that case we need a 4×4 matrix operator for the spinors, so that the Dirac equation for the electron and positron are

$$(\not{p} \mp e\not{A} - m)\psi = 0 . \tag{7}$$

The matrix $\mathcal{C} = i\gamma^2\gamma^0 = -\mathcal{C}^{-1} = -\mathcal{C}^\dagger = -\mathcal{C}^T$ does the job, with $\mathcal{C}^T$ being the transpose matrix and the matrix notation of Bjorken and Drell [13] has been used. We also have $\psi_C = \mathcal{C}\ \bar{\psi}^T$ and $\mathcal{C}\gamma_\mu^T \mathcal{C}^{-1} = -\gamma_\mu$. It follows that for the free spinors $\mathcal{C}\bar{u}^T = \eta_C v$, and $\mathcal{C}\bar{v}^T = \eta_C u$.

Thus, for a system of a fermion and its antiparticle we obtain

$$C|f\bar{f}> = (-1)^{\ell+s}|f\bar{f}> , \tag{8}$$

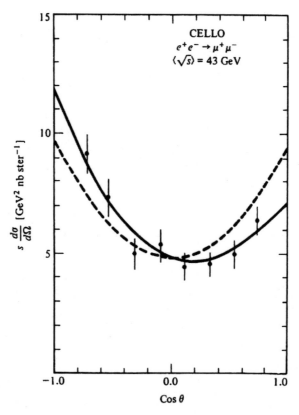

Fig. 2. The angular distribution for $e^+e^- \to \mu^+\mu^-$ at a center-of-mass energy of 43 GeV [from Cello Collaboration, Phys. Lett.**191B**, 209 (1987)]. Dashed line symmetric distribution.

where s is the spin of the system. This follows from the total antisymmetry under the exchange of space, spin, and q_{gen}. The space exchange corresponds to a parity transformation and gives $-(-1)^\ell$; the spin exchange gives $(-1)^{s+1}$, so that we have a factor of $-(-1)^{\ell+s+1}$. Again there have been recent studies of C. The weak interactions violate C as much as they do P since $C|\nu_L\rangle = |\bar{\nu}_L\rangle$ and the antineutrino is right-handed. This was already known in the late 1950's, e.g., from measurements of the electron polarization in beta decay and $\pi^\pm$ decays [14]. The weak interaction effects on stronger interactions can be shown by looking at the forwards-backwards asymmetry in a reaction such as $e^+ e^- \to \mu^+ \mu^-$; the observed asymmetry is illustrated in Fig. 2 at 43 GeV [15]; it comes from the interference of weak and electromagnetic forces.

2 The CP Transformation

CP is the product of the charge conjugation and parity transformations. We now have all the properties we need to discuss CP conservation and its breaking. CP conservation implies no difference between the real world and the mirror world for antiparticles. When it was found that both P and C were broken by the weak interactions, theorists suggested that CP should still be conserved. In that case there must exist an antiparticle to every particle with the same spin, but with opposite handedness and additive quantum numbers. This belief in CP conservation may have been based on Mach's principle that the laws of physics should not depend on the choice of geometrical coordinate system. If CP is conserved, it follows that for a boson

$$CP\phi = \eta_P \eta_C \phi^\dagger , \qquad (9)$$

$$CP\pi^+ = -C\pi^+ = -\pi^- , \qquad (10)$$

$$CP\pi^- = -C\pi^- = -\pi^+ , \qquad (11)$$

$$CP\pi^0 = -C\pi^0 = -\pi^0 . \qquad (12)$$

Table 2. CP transformation of neutral system of 2 and 3 pions

system	P	C	CP
$\pi^+\pi^-$	$(-1)^\ell$	$(-1)^\ell$	$+1$
$\pi^0\pi^0$	1	1	1
$\pi^+\pi^-\pi^0$	$(-1)^{\ell+L+1}$	$(-1)^\ell$	$(-1)^{L+1}$
$\pi^0\pi^0\pi^0$	-1	1	-1

For a neutral (uncharged) system of two or three pions the properties under a CP transformation also follow from those under C and P. They are shown in Table 2. Here L is the orbital angular momentum of the π^0 relative to the center-of-mass of the $\pi^+\pi^-$ and ℓ is that of the π^+ relative to the π^-.

For a neutral system of a fermion and its antiparticle, we have

$$CP(f\bar{f}) = (-1)^{\ell+1}C(f\bar{f}) = (-1)^{2\ell+1+s}(f\bar{f}) = (-1)^{s+1}(f\bar{f}), \quad (13)$$

where $(f\bar{f})$ stands for any fermion and its antifermion system.

CP is most useful for neutral systems and for particles that are neutral, but are not their own antiparticles. This fits the K^0, the D^0 and the B^0 systems. Most CP experiments have been related to the K^0 system, where CP nonconservation was first discovered, but the B-factories are also beginning to produce data.

Since the K^0 is a pseudoscalar meson, we have $CPK^0 = -\bar{K}^0$ and $CP\bar{K}^0 = -K^0$. If CP is a good symmetry, the appropriate weak interaction eigenstates are not K^0 and $\bar{K}^0$, which are useful for the strong interaction where strangeness is conserved, but rather [2]

$$K_1^0 = \frac{K^0 - \bar{K}^0}{\sqrt{2}}$$
$$K_2^0 = \frac{K^0 + \bar{K}^0}{\sqrt{2}} \quad (14)$$

with $CPK_1^0 = K_1^0$ and $CPK_2^0 = -K_2^0$. Because the kaons decay via weak interactions, it is the K_1 and K_2 which have definite lifetimes and not the K^0 and $\bar{K}^0$. From Table 2 we see that the K_1 can decay into $\pi^+\pi^-$ and $\pi^0\pi^0$. But the K_2 cannot do so and can only decay into 3 pions, since the spin of the kaons is zero. Due to the phase space reduction, alone, we expect the decay into 3 pions to be much slower than that into 2 pions. The longer lifetime of the K_2 allows many interesting phenomena to occur. I will cite but two examples. If we produce a K^0 in a strong interaction, its K_1 component decays rapidly and only a K_2 will be left after a short time. This K_2 has a component corresponding to a $\bar{K}^0$, with the opposite strangeness of the particle produced originally. This presence can be tested because it is the $\bar{K}^0$ which can be captured on a proton or neutron to give a pion, e.g. $\bar{K}^0 p \to \pi^+ \Lambda^0$. Also, if you pass the K_2 beam through matter, then it will regenerate K_1's because the $\bar{K}^0$ component of the K_2 is strongly absorbed by nuclei.

In 1964 Christenson, Cronin, Fitch, and Turley [16] found that the much longer lived K_2^0 did decay into two pions in about 2×10^{-3} of cases; thus CP is *not* conserved exactly in the weak interactions and we have to define new weak interaction eigenstates, called K_S and K_L, where S and L stand for short and long

$$K_S = \frac{K_1 + \epsilon K_2}{\sqrt{1+\mid \epsilon \mid^2}}$$
$$K_L = \frac{K_2 + \epsilon K_1}{\sqrt{1+\mid \epsilon \mid^2}} \quad (15)$$
$$< K_L | K_S > = \frac{\epsilon^* + \epsilon}{1+\mid \epsilon \mid^2}. \quad (16)$$

The lifetime of K_S is 0.89×10^{-10} sec and that of K_L is 5.2×10^{-8} sec. The masses of K_S and K_L also differ due to 2nd order weak interactions; the difference is $m_{K_L} - m_{K_S} = (3.489 \pm .008) \times 10^{-12}$ MeV [12]. It is not only the mass eigenstates that may violate CP, but also the decay amplitudes themselves. The amplitudes for isospin $I = 0$ and 2 can be written as

$$A(K^0 \to \pi\pi) = A_I e^{i\delta_I}$$
$$A(\bar{K}^0 \to \pi\pi) = A_I^* e^{i\delta_I}, \qquad (17)$$

where δ_I are the strong final state isopin 0 and 2 pion-pion phase shifts. CPT has been assumed and the amplitudes A_I acquire a phase if CP symmetry is violated. The measure of CP violation is usually expressed in terms of

$$|\eta_{+-}| e^{i\phi_{+-}} = \frac{A(K_L \to \pi^+\pi^-)}{A(K_S \to \pi^+\pi^-)}$$
$$|\eta_{00}| e^{i\phi_{00}} = \frac{A(K_L \to \pi^0\pi^0)}{A(K_S \to \pi^0\pi^0)} \qquad (18)$$

The final state can be written in terms of isospin, e.g.,

$$|\pi^+\pi^-> = \sqrt{\frac{1}{3}}|I = 2> + \sqrt{\frac{2}{3}}|I = 0>.$$

The freedom to choose one arbitrary phase allows one to set A_0 real. One then defines

$$\epsilon' = \sqrt{\frac{1}{2}} \frac{\mathrm{Im} A_2}{A_0} \qquad (19)$$

The relationship between the ηs and the ϵs is

$$\eta_{+-} = \epsilon + \epsilon', \quad \eta_{00} = \epsilon - 2\epsilon' \qquad (20)$$

Measurements give $|\eta_{+-}| \simeq |\eta_{00}| \simeq 2.28 \times 10^{-3}$, $\phi_{+-} \simeq \phi_{00} \simeq 45°$. The fact that $\epsilon' \ll \epsilon$ follows because $|A_2| \ll |A_0|$; it has only recently been measured and it is found $\epsilon'/\epsilon = (22 \pm 3) \times 10^{-4}$ [17]. I am sure that Dr. Kleinknecht will discuss these matters in much more detail.

There are a number of theories that incorporate CP violation. There are, in my biased view, two "natural" ones. The first one was proposed by Lincoln Wolfenstein [18] and makes use of the weakness of CP nonconservation to have a superweak interaction. In this model, there would be no CP violation in the decay amplitude of the neutral kaons ($\epsilon' = 0$); since a nonvanishing ϵ' has been found, this theory has to be ruled out. The other one makes use of the fact that there are three families of quarks to introduce a non-vanishing phase. The quarks as used in QCD are mass eigenstates but not weak interaction eigenstates. The mass eigenstates respect P, C, and CP. Kobayashi and Maskawa [19] proposed a matrix (now called the CKM matrix, after Cabibbo and KM) to connect these eigenstates. By convention, the up, charmed and top quarks are unmixed, but the down, strange and bottom quarks are mixed by a unitary matrix, the CKM matrix

$$\begin{pmatrix} d' \\ s' \\ b' \end{pmatrix} = \begin{pmatrix} V_{ud} & V_{us} & V_{ub} \\ V_{cd} & V_{cs} & V_{cb} \\ V_{td} & V_{ts} & V_{tb} \end{pmatrix} \begin{pmatrix} d \\ s \\ b \end{pmatrix} \quad (21)$$

The diagonal elements of the KM matrix are close to 1. The matrix element V_{us} is known as the sine of the Cabibbo angle $\simeq 0.22 \equiv \lambda$. Wolfenstein [12] has shown that one can use $V_{ub} \sim \lambda^3 A(\rho - i\eta), V_{cb} \sim \lambda^2 A, V_{td} \sim \lambda^3 A(1 - \rho - i\eta)$. This provides an order of magnitude for the various matrix elements. A measurable phase can be introduced because we have three families. Because the matrix is unitary, we have for instance

$$V_{ud}^2 + V_{us}^2 + V_{ub}^2 = 1 \; . \quad (22)$$

Superallowed Fermi beta decays ($J^P = 0^+ \to 0^+$ transitions within an isospin multiplet) allow one to obtain V_{ud} very accurately, $V_{ud} = 0.9736 \pm 0.0006$; V_{us} is known from hyperon and kaon decays to be 0.2205 ± 0.0018, and V_{ub} is sufficiently small that its square can be neglected. The best value found for the sum above is 0.9965 ± 0.0015. [20] Unitarity does not seem to be completely satisfied. Is this a real effect or are the uncertainties due to isospin symmetry-breaking and radiative corrections underestimated? If the effect is real, then it would indicate physics beyond the standard model.

There are, of course, other models, such as left-right symmetric models, multiple Higgs models, R parity violation in supersymmetric models, and others [25]. The problem is that CP nonconservation has so far been found in a single system and it is difficult to test or base a theory on this uniqueness. Hopefully, the B factories will fill in the voids since the B system differs from the K system; See Dr. Waldi's lecture on this subject.

3 Time Reversal

3.1 Time Reversal

Classically, we know that Newton's equation of motion is time reversal invariant ($t \to -t$) in the absence of friction since it is second order in time. By contrast, the Schrödinger equation is a diffusion equation and is first order in time. Can one incorporate time reversal invariance (TRI) in quantum mechanics? [2] The answer is yes and Wigner showed that it requires an antiunitary transformation. Under time reversal a wavefunction $\psi(\mathbf{r}, t) \to \psi_T(\mathbf{r}, -t)$. If $\psi_T = \psi(-t)$, then the Schrödinger equation for ψ_T is

$$i\frac{\partial \psi_T}{\partial t} = -i\frac{\partial \psi(-t)}{\partial(-t)} = H\psi_T \quad (23)$$

is not the same equation. This can be corrected by defining $\psi_T = \psi^*(\boldsymbol{r}, -t)$. The complex conjugate Schrödinger equation is

$$-i\frac{\partial \psi^*(t)}{\partial t} = H^*\psi^*(t) \ . \tag{24}$$

If we let $t \to t' = -t$, we obtain

$$-i\frac{\partial \psi^*(-t)}{\partial(-t)} = i\frac{\partial \psi^*(-t)}{\partial(t)} = i\frac{\partial \psi_T(-t)}{\partial t} = H^*\psi_T \ . \tag{25}$$

If H is real, we can write $T\psi(t) = \psi_T(-t) = K\psi(-t) = \psi^*(-t)$, $KHK = H^* = H$ and we have the usual Schrödinger equation for ψ_T. Here K is a complex conjugation operator. Thus, if we have a state $|\psi\rangle = |a\phi + b\phi'\rangle$, then $T|\psi\rangle = |a^*T\phi + b^*T\phi'\rangle$. Under the time reversal transformation, momentum $\boldsymbol{p}$ should change sign, since it is first order in time, and so should the angular momentum $\boldsymbol{J}$. In quantum mechanics, $\boldsymbol{p} = -i\boldsymbol{\nabla}$ changes sign because of the operation of K. For a plane wave, f.i., we have

$$T\psi(\boldsymbol{r},t) = Te^{i(\boldsymbol{p}\cdot\boldsymbol{r}-Et)} = \psi^*(\boldsymbol{r},-t) = e^{-i(\boldsymbol{p}\cdot\boldsymbol{r}+Et)} = e^{i(-\boldsymbol{p}\cdot\boldsymbol{r}-Et)} \tag{26}$$

which also shows that $\boldsymbol{p}$ changes sign under the time reversal transformation. For the spherical harmonics it follows that

$$TY_\ell^m = KY_\ell^m = Y_\ell^{m*} = (-1)^m Y_\ell^{-m} \ . \tag{27}$$

For many purposes, and as we will see, it is useful to introduce spherical harmonics with a different phase when discussing time reversal, namely

$$\begin{aligned}\mathcal{Y}_\ell^m &= i^\ell Y_\ell^m \ , \\ T\mathcal{Y}_\ell^m &= (-1)^{\ell-m}\mathcal{Y}_\ell^{-m}\end{aligned} \tag{28}$$

It is the complex conjugation which makes TRI more difficult and the time reversal operation non-unitary.

For the case of a fermion, or more generally, we have

$$T\psi(t) = \psi_T(-t) = U_T K\psi(-t) = U_T \psi^*(-t) \tag{29}$$

where U_T is a unitary transformation. We can see the non unitarity of T by examining the scalar product of two states

$$\langle T\psi(t)|T\phi(t)\rangle = \langle U_T\psi^*(-t)|U_T\phi^*(-t)\rangle = \langle \psi^*(-t)|\phi^*(-t)\rangle$$
$$= \langle \psi(-t)|\phi(-t)\rangle^* = \langle \phi(-t)|\psi(-t)\rangle \ ; \tag{30}$$
$$\langle \psi_T|\phi_T\rangle = \langle \psi|\phi\rangle^* = \langle \phi|\psi\rangle \ . \tag{31}$$

For an operator A, such as the momentum, the time reversal transformation is

$$A_T = TAT^{-1} = U_T A^* U_T^\dagger \ . \tag{32}$$

For the matrix element of the operator A we have [2]

$$< T\psi(t)|TAT^{-1}|T\phi(t) > = < U_T\psi^*(-t)|U_T A^* U_T^\dagger|U_T\phi^*(-t) >$$
$$= < \psi(-t)|A|\phi(-t) >^* = < \phi(-t)|A^\dagger|\psi(-t) > . \quad (33)$$

If A is a Hermitian observable, $A_H = A_H^\dagger$, such as the Hamiltonian, we have

$$< \psi_T|A_{H,T}|\phi_T > = < \phi|A_H|\psi > . \quad (34)$$

I have not yet used time reversal symmetry. If $A_T = A$ (e.g., $H_T = H$), then an invariance is implied. If, furthermore, ψ and ϕ are eigenstates of T, e.g., $T\phi = \phi$, then it follows that $< \phi|A|\phi >$ is purely real (pure imaginary if $A_T = -A$, e.g., $\boldsymbol{p}_T = -\boldsymbol{p}$).

For a non-relativistic fermion, $U_T = i\sigma_y$; this follows because σ_y is imaginary and anticommutes with σ_x and σ_z. Relativistically, we can take $U_T = i\gamma_1\gamma_3$ in the Bjorken-Drell metric [13]. With this transformation we obtain for the electromagnetic charge and current and the vector potentials

$$Tj^0T^{-1} = j^0, \quad T\boldsymbol{j}T^{-1} = -\boldsymbol{j} , \quad (35)$$
$$TA^0T^{-1} = A^0, \quad T\boldsymbol{A}T^{-1} = -\boldsymbol{A} . \quad (36)$$

We also find

$$T|\boldsymbol{p}, \frac{1}{2}, m> = (-1)^{\frac{1}{2}-m}|-\boldsymbol{p}, \frac{1}{2}, -m> , \quad (37)$$

where m is the magnetic quantum number of the spin. The usefulness of the phase introduced in the spherical harmonic $\mathcal{Y}_\ell^m$ now becomes apparent because we can readily generalize this transformation by substituting s for the spin $\frac{1}{2}$,

$$T|\boldsymbol{p}, s, m> = (-1)^{s-m}|-\boldsymbol{p}, s, -m> . \quad (38)$$

3.2 Consequences of Time Reversal Invariance

Time reversal invariance implies that [2]

$$THT^{-1} = H . \quad (39)$$

We note, first, that under the application of T twice we revert to the original state. Thus, we have from the last equation

$$T^2|\psi> = e^{i\alpha}|\psi> = (-1)^{2s}|\psi> . \quad (40)$$

A bosonic system, $|b>$ has eigenvalues of $+1$ and a fermionic one $|f>$ of -1. For any observable A_H, we have $T^2 A_H T^{-2} = A_{H,TT} = A_H$. It then follows that A_H cannot connect bosons and fermions, or

$$< f|A_H|b > = 0 . \quad (41)$$

Consider, next a system with an odd number of fermions, $|f>$ and its time reversed state $|\psi> = T|f>$. We have with Eq.(31)

$$<f|\psi> = <T^2 f|Tf> = -<f|Tf> \,. \tag{42}$$

It follows that $|f>$ and $T|f> = |\psi>$ are orthogonal. If the Hamiltonian of the system is time reversal invariant, $[T, H] = 0$, then $|f>$ and $T|f>$ have the same energy and the state is doubly degenerate; this is known as Kramer's theorem [2]. It applies to all odd A nuclei, where A is the number of nucleons. The degeneracy can only be lifted by a field that does not commute with T, e.g., a magnetic field, but not an electric field.

In general, tests of TRI require a comparison of two reaction rates; the exception is a process that is sufficiently weak that perturbation with a Hermitian interaction Hamiltonian can be used and *there are no strong interactions among the final or initial state particles*. Consider an electromagnetic or weak transition from a state $|i> = |a, j_i, m_i>$ to a state $|f> = |b, j_f, m_f>$, where a and b stand for all other quantum numbers than the angular momenta, with a Hamiltonian H' responsible for the transition. If $H'_T = H'$, it follows from Eq. (34) that

$$<f_T|H'|i_T> = (-1)^{j_i+j_f-m_i-m_f} <b, j_f, -m_f|H'|a, j_i, -m_i> = <f|H'|i>^* \,. \tag{43}$$

If H' is a scalar or pseudoscalar operator, then $j_i = j_f$ and $m_i = m_f$. Since $2(j-m)$ is even, we find that the matrix element must be real. Thus, the matrix element of a Hermitian and rotationally invariant operator must be real.

You can generalize this reality to electromagnetic transition multipoles, see Eq. (30)

$$T\, T_q^{(k)}\, T^{-1} = \eta_k (-1)^{k-q} T_{-q}^{(k)} \,, \tag{44}$$

where k is the tensor order, q the spherical component, and $\eta_k = \pm 1$. After some algebra [2] and use of symmetries of Clebsch-Gordon coefficients, we obtain

$$<f_T|T\, T_q^{(k)}\, T^{-1}|i_T> = \eta_k <f|T_q^{(k)}|i> = <f|T_q^{(k)}|i>^* \,. \tag{45}$$

Thus, the matrix elements of the tensor operators are either real or imaginary, but it is always possible to choose phases such that $\eta_k = +1$ and all matrix elements are then real. Most experiments that have been carried out make use of a search for a phase different from $0°$ or $180°$ in an interference of an E2 (quadrupole) and M1 (magnetic dipole) transition matrix elements. Such a phase would indicate a time reversal invariance violation. Experimentally, one looks for a term proportional to $\boldsymbol{k} \cdot \boldsymbol{j_f}\, \boldsymbol{k} \cdot \boldsymbol{j_f} \times \boldsymbol{j_i}$, where $\boldsymbol{j_i}(\boldsymbol{j_f})$ is the polarization of the initial (final) state, and $\boldsymbol{k}$ is the momentum of the photon emitted in the transition. The alignment of the final state is detected by a subsequent electromagnetic transition, as illustrated in Fig. 3.

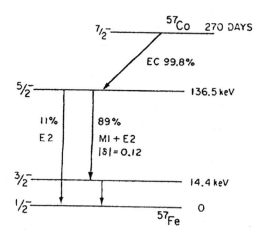

Fig. 3. Decay scheme for ^{57}Co, showing the mixed decay and subsequent gamma-ray transition

The most significant limit has perhaps been found in the study of ^{57}Fe, with a decay scheme shown in Fig. 3. The sine of the interference angle between the E2 and M1 matrix elements is found to be $\sin\eta = (2.9 \pm 6.6) \times 10^{-4}$ [21]. There have been a slew of other similar experiments, but no violation of TRI has been found. One has to be careful in that any interaction of the photon with atomic shell electrons spoils the hermiticity of the operator and thus the test. These effects are small; they can be estimated and corrections made. This was done in the above experiment.

A similar development holds for beta decay, but here parity is not conserved. One can therefore search for a phase in the interference of a Fermi and a Gamow-Teller transition matrix element. A T-odd term would be of the form $D\boldsymbol{j} \cdot \boldsymbol{p}_e \times \boldsymbol{p}_\nu$ and $R\boldsymbol{J} \cdot \boldsymbol{s}_e \times \boldsymbol{p}_e$, where $\boldsymbol{j}$ refers to the polarization of the parent nucleus, and the other terms to the momenta of the subscripted particle and polarization (s_e) of the electron. Measurements of D and R are also limited by final state interactions of the emitted electron and nucleus, which spoils Hermiticity of H' by introducing an imaginary part to the amplitude. You then can get effects which mimic a TRI violation but is really just a final state interaction. To-date, the accuracy ($\sim 10^{-3}$) is not sufficient to probe these effects. No violation of TRI has been found [21].

Time reversal invariance can also be tested in reactions by means of the principle of reciprocity. We define the scattering matrix S by

$$S|in> = |out>, \tag{46}$$
$$SS^\dagger = S^\dagger S = 1, \tag{47}$$
$$S_T = TST^{-1} = S^\dagger \tag{48}$$

The rate for reaching a final state $|f>$ from an initial one $|i>$ is

$$\mathcal{R}_{fi} = \text{const.} \, |<f|S|i>|^2 \rho_f \equiv \text{const.} \, |S_{fi}|^2 \rho_f \,, \tag{49}$$

where ρ_f is the final phase space density. If TRI holds, we have

$$\frac{\mathcal{R}_{fi}}{\mathcal{R}_{i_T f_T}} = \frac{\rho_f}{\rho_i} \,. \tag{50}$$

This is the principle of reciprocity, with

$$<i_T|S|f_T> = S_{i_T f_T} = (-1)^{j_i+j_f-m_i-m_f}$$
$$\times <a_T, -\boldsymbol{p}_i, j_i, -m_i|S|b_T, -\boldsymbol{p}_f, j_f, -m_f>\,. \tag{51}$$

Here a and b stand for all other quantum numbers. The time reversed states have both their momenta and spins reversed. If parity conservation holds we can change $-\boldsymbol{p}$ to $\boldsymbol{p}$ in both initial and final states. If we average over the spins (do not measure polarizations), we obtain the principle of detailed balance

$$\sum_{m_i,m_f} |<b, \boldsymbol{p}_f.j_f, m_f|S|a, \boldsymbol{p}_i, j_i, m_i>|^2 =$$
$$\sum_{m_i,m_f} |<a_T, \boldsymbol{p}_i.j_i, m_i|S|b_T, \boldsymbol{p}_f, j_f, m_f>|^2\,. \tag{52}$$

The best detailed balance experiment is the reaction $d+^{27}\text{Al} \leftrightarrow ^4\text{He}+^{24}\text{Mg}$ and shows no asymmetry at the level of $\leq 5 \times 10^{-4}$ [22].

A clever null-type experiment was suggested by Stodolsky [23]. It can be carried out in either a parity conserving or parity-violating mode. It is the forward elastic scattering of slow polarized neutrons from transversally polarized nuclei (e.g., Fe) as shown in Fig. 4. Although it is a null experiment and appears to require a weak interaction with a Hermitian interaction H', the search for terms (in forward elastic scattering) $A\boldsymbol{\sigma} \cdot \boldsymbol{s} \times \boldsymbol{p}$, which is P-odd and T-odd or $B\boldsymbol{\sigma} \cdot \boldsymbol{s} \times \boldsymbol{p}\,\boldsymbol{p} \cdot \boldsymbol{s}$, which is P-even and T-odd, actually makes use of reciprocity. This is illustrated in Fig. 4. From this figure it is seen that the time reversed situation for elastic forward scattering is identical to the initial one, except for the reversal of the target spin ($\boldsymbol{s}$). Thus, if time reversal is a valid symmetry, the terms A or B, above, are not allowed. The test is being examined on a nucleus such as ^{139}La, where a large ($\sim 10^5$) enhancement was found for parity nonconservation due to the mixing of very closely spaced compound nucleus resonances.

Perhaps the most sensitive test of TRI is the search for an electric dipole moment (d_E) in a neutral system such as a neutron or atom. Earlier, we showed that parity non-conservation is required to have a nonvanishing d_E. As pointed out there, it is *also* necessary to have TRI violated. The argument is similar to that for parity, and is illustrated in Fig. 5. When time is reversed,

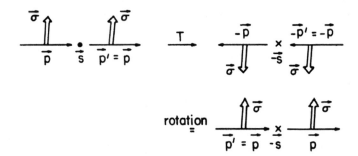

Fig. 4. Reciprocity test of TRI with polarized neutron(σ) transmission through a polarized(s) target

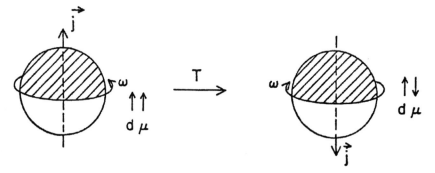

Fig. 5. Behavior of electric and magnetic dipole moments under time reversal; the angular momentum defines a direction in space

the angular momentum changes sign, but the electric dipole moment does not; hence if d_E does not vanish, we can tell the difference between a movie running forwards and backwards, e.g., TRI is violated.

Ever more accurate experiments have attempted to find a non-vanishing d_E and the best present upper limits are [24]

$$d_E(n) \leq 6.3 \times 10^{-26} \text{ecm},\tag{53}$$

$$d_E(\text{Hg}^{199}) \leq 8 \times 10^{-28} \text{ecm}.\tag{54}$$

^{199}Hg has paired electrons but the spin of the nucleus is 1/2. The numbers for d_E are very small. You might expect $d_E \sim e \times$ neutron size $\times$ relative strength of weak interaction $\times$ reduction due to need of T violation $\sim e \times 10^{-13}$cm $\times 10^{-6} \ldots 10^{-7} \times 10^{-3} \sim 10^{-22} \ldots 10^{-23}$ecm. However, the CKM matrix must be invoked twice for an expectation value in a nucleon state since flavor is conserved; this leads to a further reduction of $10^{-6} \cdot 10^{-7}$ or an expected $d_E \sim 10^{-28} \ldots 10^{-30}$ecm. Calculations based on the CKM matrix give an even smaller number $\sim 10^{-31} \ldots 10^{-33}$ ecm, [25] but present experiments have already ruled out several models of TRI violation.

It is noteworthy that QCD allows a T-odd interaction term

$$H' = \frac{g_s^2}{32\pi^2} \theta \, G_{\mu\nu}^a \tilde{G}^{a,\mu\nu} \,, \tag{55}$$

where g_s is the strong interaction coupling constant θ is a parameter, G is the gluon field, $\tilde{G}$ is its dual (see below) and a is a color index. The dipole moment measurements limit $\theta \leq 10^{-9}$, a very small value. This is called the strong CP problem because we do not understand why θ is so small. I will discuss this problem in more detail in a short while.

Last year I could still say that no direct test of TRI has shown any violation, but this is no longer true. At CERN, the CPLEAR team has measured the [26] asymmetry

$$a = \frac{\mathcal{R}(\bar{K}^0 \to e^+\pi^-\nu) - \mathcal{R}(K^0 \to e^-\pi^+\bar{\nu})}{\mathcal{R}(\bar{K}^0 \to e^+\pi^-\nu) + \mathcal{R}(K^0 \to e^-\pi^+\bar{\nu})}) = (6.6 \pm 1.3 \pm 1.0) \times 10^{-3} \,, \tag{56}$$

where the first error is statistical and the second one systematic. Why is this a test of TRI [27]? Only if TRI holds will

$$\mathcal{R}[\bar{K}^0(t=0) \to K^0(t=T)] = \mathcal{R}[K^0(t=0) \to \bar{K}^0(t=T)] \,. \tag{57}$$

The test does assume CPT invariance, which I will discuss shortly. However, Ellis and Mavromatos [27] show that the test cannot be interpreted as a maintenance of T invariance and a violation of the CPT theorem.

3.3 The Strong CP Problem

Mention has been made above of the strong CP problem. Let me restate the problem here and go into some detail. For a review see [28].

The most general QCD Lagrangian density, consistent with Lorentz invariance, hermiticity, and gauge and chiral invariances is (color indices are omitted, except where required)

$$\mathcal{L} = \bar{\psi} i \slashed{D} \psi - \frac{1}{4} G_{\mu\nu} G^{\mu\nu} + \frac{g_s^2}{32\pi^2} \theta G_{\mu\nu} \tilde{G}^{\mu\nu} = \mathcal{L}_{\text{QCD}} + \mathcal{L}_\theta \,, \tag{58}$$

where θ is an angle parameter, g_s is the strong coupling constant,

$$\slashed{D} \equiv (\partial_\mu + i g_s A_\mu^a \lambda^a/2) \gamma^\mu,$$
$$\tilde{G}_{\mu\nu} \equiv \frac{1}{2} \epsilon_{\mu\nu\alpha\beta} G^{\alpha\beta},$$
$$G_{\mu,\nu} \equiv [\partial_\mu + i g_s A_\mu, \partial_\nu + i g_s A_\nu] \,, \tag{59}$$

with A_μ the gluon field and a is a color index. The θ term violates both P and T and is therefore CP-odd, but it is gauge invariant and renormalizable.

It is a total derivative and therefore does not contribute classically or in perturbation theory, but only to non-perturbative effects. It arises because there are an infinite number of nontrivial degenerate QCD vacua with gluon fields in their lowest state [28]. One can also write higher dimensional terms, but they are expected to be small. The angle variable θ is used because the S-matrix has a periodicity of 2π in θ. The problem is complicated by the possible presence of a complex mass term

$$m_q \bar{\psi} e^{i\theta_m \gamma_5} \psi , \qquad (60)$$

which has a P- and T-odd part $im_q \sin\theta_m \bar{\psi}\gamma_5\psi$. A chiral rotation can shift some of the θ term into the θ_m mass term or vice versa, but you cannot rotate both away simultaneously. If either θ or θ_m were alone, then it could be rotated away by a chiral rotation,

$$q_i \to \exp(i\gamma_5 \alpha_i/2) \; q_i$$
$$\theta \to \theta - 2\alpha_i . \qquad (61)$$

Thus, it is only $\tilde{\theta} \equiv (\theta + \theta_m)$, which is measurable and meaningful. For low energy calculations it is simpler to shift all of the $\tilde{\theta}$ to the θ_m term [28,29].

There are a number of ways of avoiding the strong CP problem by making $\tilde{\theta} = 0$. The first one occurs if one of the quark masses (presumably m_u) is zero; for 3 quarks, the CP-violating term is

$$\mathcal{L}_{\cancel{CP}} = i\tilde{\theta}\beta(\bar{u}\gamma_5 u + \bar{d}\gamma_5 d + \bar{s}\gamma_5 s) , \qquad (62)$$

$$\beta = (\sum_{i=1}^{3} 1/m_i)^{-1} = \frac{m_u m_d m_s}{m_u m_d + m_u m_s + m_d m_s} .$$

However, we know from decays and masses of hadrons that m_u is not zero.

There are a number of consequences of the strong CP problem. Two of the best known ones are the decay $\eta \to \pi\pi$ and the electric dipole moment of the neutron and of atoms. The absence of the η decay to two pions limits $\tilde{\theta}$ to about 10^{-3}, but the absence of a neutron electric dipole moment places a more stringent limit on $\tilde{\theta}$. The first calculations made use of the dominance of the coupling of the photon to the pion in a pion loop (see Fig. 6) to compute the neutron electric dipole moment [30], with a CP-violating coupling $ig_{\pi NN}\bar{\psi}\tau\psi \cdot \pi$, with $g'_{\pi NN} = 0.028\,\tilde{\theta}$. This gives

$$d_E(n) = \frac{eg_{\pi NN} g'_{\pi NN}}{4\pi^2 m_N} \ln \frac{m_N}{m_\pi}$$

with $g_{\pi NN}$ being the strong pion nucleon coupling. There have also been calculations with various bag models [28], with QCD sum rules [29,31] and by other means [28]. They all give approximately the same result, $d_E(n) \simeq -(3\ldots 6) \times 10^{-16}\tilde{\theta}\; e$, so that $\tilde{\theta} \leq 2 \times 10^{-10}$. We thus have the strong CP problem as to why $\tilde{\theta}$ is so small.

Fig. 6. Feynman diagrams contributing to $d_E(n)$. The cross corresponds to a CP-violating πNN vertex, $ig'_{\pi NN}\bar{N}\tau N \cdot \pi$

A host of suggestions have been made from time to time for solving the strong CP problem [28]. The most natural one and the generally accepted one is that proposed by Peccei and Quinn [32]. In essence, they introduce $\tilde{\theta}$ as a dynamical field instead of a parameter. Thus, the different θ fields distinguish different vacuum states of the same theory but with different vacuum energies [32,34]. The one with the lowest energy, the true vacuum is that with $\tilde{\theta} = 0$. The dynamical field corresponds to the pseudoscalar axion a and $\tilde{\theta}$ is then written as a/f_a, where f_a is the coupling of the axion. The theory has a new U(1) symmetry, and requires two Higgs fields which couple to the quarks. I will not go into details here, except to point out that the new U(1) symmetry requires a Goldstone boson, the axion. It acquires mass due the breaking of the symmetry by the QCD axial anomaly or by instantons. Details can be found in [28,32,34]. There have been many searches for axions [33]. Originally, the axion was proposed by Weinberg and Wilczek [35] on the basis of a broken chiral symmetry, but this axion has not been seen. The decay $a \to \gamma\gamma$ was sought and the relation of lifetime and mass is shown in Fig. 7 from [33].

There is also a modified Peccei-Quinn symmetry which can be introduced and is free of the QCD anomaly. In this case the masses of the up and down quarks are zero to begin with but the symmetry is spontaneously broken by quark condensates $<\bar{q}q>$; this gives rise to both pion and axion masses. The axion is then expected to have an extremely small mass with very weak couplings; it has therefore been called an "invisible axion." Ongoing searches for such an invisible axion are summarized in Fig. 8 [33]. The mass and coupling f_a of the axion are also limited by cosmology [28]. If, for instance, f_a is too small then the axion would dominate the energy density of the present universe. The limits are 10^{-5} eV $\leq m_a \leq 10^{-1}$ eV and 10^7 GeV $\leq f_a \leq 10^{12}$ GeV.

The fact that the strong CP problem may be solved by the Peccei-Quinn mechanism does not mean that weak CP goes away. There can be higher order terms than those discussed here and supersymmetric theories can give rise to a weak CP violation of the θ type. However, the $\tilde{\theta}$ term does not

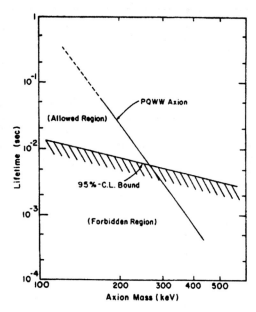

Fig. 7. Limits on axion lifetime versus axion mass from the decay $a \to \gamma\gamma$ (PQWW refers to Peccei-Quinn, Weinberg, Wilczek)

contribute to the CP violation observed in the neutral kaon system since it is flavor diagonal.

4 CPT, PCT, TPC, ...

$\Theta = CPT = CTP = TCP = ...$ is the combined operation of C, P, and T taken in any order, with

$$\Theta|q_{\text{gen}}, \boldsymbol{p}, s, m \rangle = \eta_\Theta| - q_{\text{gen}}, \boldsymbol{p}, s, -m \rangle \tag{63}$$

$$\Theta H(t, \boldsymbol{r})\Theta^{-1} = H(-t, -\boldsymbol{r}), \tag{64}$$

where the last equality holds if Θ invariance holds. Since T is an antiunitary operator, so is Θ. We have similar equations as Eqs. (32)-(39) for the operator Θ. Unlike T, CP, and P the CPT theorem, first observed by Schwinger and Lüders [36] and generalized by Pauli and Bell [37] can be proven on very general grounds. The proof requires (1) Lorentz invariance, (2) locality, (3) the usual connection between spin and statistics, and (4) a Hermitian Hamiltonian [2,38]. Until a few years ago I could have said that all accepted theories satisfy these conditions, but there are now string theories which predict a small violation of CPT due to non-locality. This has reawakened an interest in tests of the theorem.

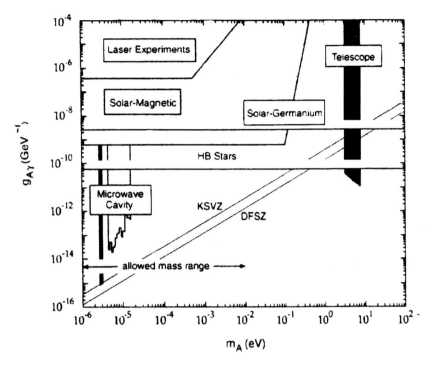

Fig. 8. Excluded region for the coupling of an axion to photons, $g_{a\gamma\gamma}$ as a function of the axion mass m_A (The two lines labeled KSVZ and DFSZ refer to two models.)

There are several interesting consequences of the CPT theorem. Since the mass of an object is independent of spin orientation we have

$$m = <\psi|H|\psi> = \bar{m} = <\Theta\psi|H|\Theta\psi> \ . \tag{65}$$

A very good test is the limit on the mass difference of the K^0 and $\bar{K}^0$ obtained from the measured mass difference of the K_1 and K_2,

$$\Delta m = \frac{|m_{K^0} - m_{\bar{K}^0}|}{m_{K^0}} \le 10^{-18} \ . \tag{66}$$

Sanda [39] believes that this value is not meaningful because the denominator is so large, but it is quite usual to compare differences to sums of the same two quantities. The CPT theorem also predicts equality in magnitude of the magnetic dipole moment of particle-antiparticle pairs. Because $g-2$ has been measured very accurately for the electron and muon and their antiparticles, one finds [12]

$$\frac{g(e^+) - g(e-)}{g(\text{average})} = (-0.5 \pm 2.1) \times 10^{-12} \ , \tag{67}$$

$$\frac{g(\mu^+) - g(\mu^-)}{g(\text{average})} = (-2.6 \pm 1.6) \times 10^{-8} \tag{68}$$

For a weak interaction decay due to a Hermitian interaction, the CPT theorem predicts $\tau = \bar{\tau}$. This follows from Eq.(65), which applies equally to lifetimes with H replaced by H_{weak}. For μ^+ and μ^-, the ratio of lifetimes is $\frac{\tau(\mu^+)}{\tau(\mu^-)} = 1.00002 \pm 0.00008$. The CPT theorem does not predict equality of partial lifetimes, that is the decay to any one particular channel, especially if strong interactions are present in the final state.

There are further consequences but no test has shown any violation of CPT. For instance, some quite recent tests, (1) with the muonium hyperfine structure, which is said to provide a limit of 4×10^{-21} [40] on a CPT violation and (2) the frequency difference of co-located ^{129}Xe and ^{3}He Zeeman masers which shows no effect at a level of 10^{-31} GeV. Both these tests are based on a search for a sidereal dependence on the frequencies involved.

It is quite clear that if there is a violation of CPT, it is bound to be very very small.

5 Summary

I have reviewed the separate transformations of parity and charge conjugation, discussed CP invariance and T invariance in detail and have shown some examples of the violation of these symmetries in the weak interaction sector of the standard model. Along the way I discussed the strong CP problem. Finally, I gave some information on the combined CPT transformation and its invariance. Although string theories predict some violation of CPT, none has been found to a very high accuracy.

References

1. H. Frauenfelder and E.M. Henley, *Subatomic Physics*, 2nd edn. (Prentice Hall, Englewood Cliffs, NJ, 1991)
2. H. Frauenfelder and E.M. Henley, *Nuclear and Particle Physics* (W.A. Benjamin, Reading, MA, 1975)
3. T.D. Lee and C.N. Yang, Phys. Rev. **104**, 254 (1956)
4. C.S. Wu et al., Phys. Rev. **105**, 1413 (1957)
5. See e.g., E. D. Commins and P.M. Bucksbaum *Weak Interactions of Leptons and Quarks* (Cambridge Univ. Press, London, 1983)
6. E. G. Adelberger and W.C. Haxton, Ann. Rev. Nucl. Part. Sci. **35**, 501 (1985)
7. C.S. Wood et al., Science **275**, 1759 (1997)
8. W. Haeberli and B.R. Holstein in *Symmetries and Fundamental Interactions in Nuclei* ed. by W.C. Haxton and E. M. Henley (World Sci., Singapore, 1995) p. 17
9. C.R. Gould et al., Int. J. Mod. Phys. **A5**, 2181 (1990)
10. D.T. Spayde et al., Phys. Rev. Lett. **84**, 1106 (2000)

11. C.J. Horowitz et al., LANL Archiv. nucl-th 9912038 (1999); S. J. Pollock and M. Welliver (private communication)
12. Review of Particle Physics, Particle Data Group, Europ. Phys. J. **C15**, 1 (2000)
13. J.D. Bjorken and S.D. Drell, *Relativistic Quantum Mechanics* (McGraw-Hill, NY, 1964)
14. H. Frauenfelder and A. Rossi in *Methods in Experimental Physics*, ed. by L.C. Yuan and C.S. Wu (Academic Press, NY, 1963)
15. H.G. Behrend, Cello Collabor. Phys. Lett. **191B**,209 (1987)
16. J.H. Christenson, J.W. Cronin, W.L. Fitch, and R. Turlay, Phys. Rev. Lett. **13**, 138 (1964)
17. G.D. Barr et al., Phys. Lett. **B317**, 233 (1993); A. Alavi-Harati, KTeV Collabor., Phys. Rev. Lett. **83**, 22 (1999)
18. Lincoln Wolfenstein, Phys. Rev. Lett **13**, 562 (1964) and Comm. Nucl. Part. Phys. **21**, 275 (1994)
19. M. Kobayashi and T. Maskawa, Progr. Theor. Phys. **49**,652 (1973); N. Cabibbo, Phys. Rev. Lett. **10**, 531 (1963)
20. I.S. Towner and J.C. Hardy in *Symmetries and Fundamental Interactions in Nuclei*, loc.cit. p.183
21. F. Boehm in *Symmetriesand Fundamental Interactions in Nuclei*, loc.cit. p.67
22. E. Blanke et al., Phys. Rev. Lett. **51**,355 (1983)
23. L. Stodolski in *Tests of Time Reversal Invariance in Neutron Physics*, ed. by N. Roberson, C.R. Gould, and J.D. Bowman (World Sci., Singapore, 1987), p. 12
24. P.G. Harris et al., Phys. Rev. Lett. **82**, 904 (1999); J.P. Jacobs et al., Phys. Rev. **A52**, 3521 (1995)
25. L. Wolfenstein, Ann. Rev. Nucl. Part. Sci. **36**, 137 (1986); X.G. He, B.H.J. McKellar, and S, Pakvasa, Int. J. Mod. Phys. **A4** 5 (1986), N.F. Ramsay, Ann. Rev. Nucl. Part. Sci., **40**, 1 (1990)
26. A. Angelopoulos et al., CPLEAR Collabor. Phys. Lett. **B444**, 43 (1998)
27. J. Ellis and N.E. Navromatos, Phys. Rep. **320**, 341 (1999)
28. H-Y Cheng, Phys. Rep. **158**, 1 (1988)
29. C-T Chan, Ph.D. thesis , Univ. of Washington, 1996 and C-T Chan, E.M. Henley, and T. Meissner, LANL arch. hep-ph/9905317 (1999)
30. V. Baluni, Phys. Rev. **D19**, 2227 (1979); R. J Crewther et al., Phys. Lett. **B88**, 123 (1979), erratum ibid **B91**, 487 (1980)
31. M. Pospelov and A. Ritz,, Phys. Rev. Lett. **83**, 2526 (1999) and Nucl. Phys. **B573**, 177 (2000)
32. R.D. Peccei and H.R. Quinn, Phys. Rev. Lett. **38**, 1440 (1977); Phys. Rev **D16**, 179 (1977)
33. L.J. Rosenberg, and K.A. van Bibber, Phys. Rep. **325** ,1 (2000)
34. J.E. Kim, Phys. Rep. **150**, 1 (1987)
35. S. Weinberg, Phys. Rev. Lett. **40**, 223 (1978); F. Wilczek, Phys. Rev. Lett. **40**, 279 (1978)
36. J. Schwinger, Phys. Rev. **82**, 914 (1951) and **91**, 713 (1951); G. Lueders, Z. Phys. **133**, 325 (1952)
37. W. Pauli in *Niels Bohr and the Development of Physics* (Pergamon, London, 1955); J.S. Bell, Proc. Roy. Soc., London,**A231**, 479 (1955)
38. J.J. Sakurai, *Invariance Principles and Elementary Particles*, (Princeton Univ. Press, Princeton, NJ, (1964)

39. A.I. Sanda, LANL Archiv. hep-ph 9902353 (1999)
40. R. Bluhm, V.A. Kostalecky, and C.D. Lane, Phys. Rev. Lett. **84**, 1098 (2000)
41. D. Bear et al., LANL Archiv. phys-0007049 (2000)

CP Violation in the K^0 System

Konrad Kleinknecht

Institut für Physik, Johannes Gutenberg-Universität
55099 Mainz, Staudinger Weg 7, Deutschland

Abstract. A small matter-antimatter asymmetry of the weak force was experimentally established. This CP violation may be related to the small excess of matter from the big bang. The nature of CP violation in the K^0 system has been clarified after 35 years of experimentation: it is due to a small part of the weak interaction ("illiweak interaction"). A non-trivial phase in the weak quark mixing matrix generates "direct CP violation" in the weak Hamiltonian. The experiments demonstrating direct CP violation are discussed.

1 The Big Bang and the Expanding Universe

In our universe, there is no indication of antimatter from the big bang. If there exists antimatter, it must show up through the γ radiation which is produced when antimatter hits matter and the two annihilate to γ radiation.

In the big bang, matter and antimatter were made at extremely high temperatures and in equal quantities because the forces which are responsible for their production are completely symmetric with respect to matter and antimatter. In this hot fireball, creation and annihilation of particles and antiparticles led to an equilibrium of approximately equal numbers of particles, antiparticles, and photons. The temperature corresponds to the average energy of the particles. The expanding fireball cools down, and below a certain temperature the creation of particle-antiparticle pairs stops while annihilation goes on. If all forces were symmetric with respect to matter and antimatter, there would be no matter left over in the cold phase but only the photons shifted to the infared regime. These photons were indeed discovered 1965 by Penzias and Wilson as the cosmic microwave background radiation. Their wave-length spectrum perfectly resembles the Planck radiation of a black body at a temperature of 2.73 degrees Kelvin (Fig. 1). This is the echo of the hot photons from the big bang. Today the density of this radiation is 500000 photons per liter.

The big problem however is the complete absence of primordial antimatter in our universe and the small density of matter as compared to photons: about 6×10^{-5} nucleons per liter. The ratio of nucleons over photons is about 10^{-10}, while is was of order one in the early phases of the universe.

A possible explanation for this phenomenon was given by Sacharov and Kuzmin. They stated that this small surplus of matter is possible only if

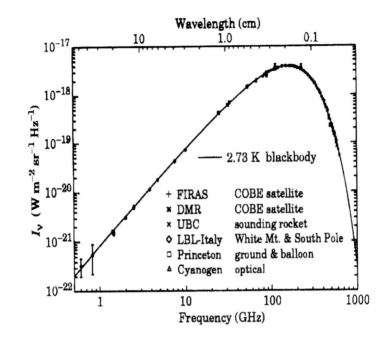

Fig. 1. Frequency distribution of the cosmic microwave background variation, as measured by the COBE satellite

- one force violates matter-antimatter symmetry
- baryon number is violated as well, and
- the expansion goes through phases when there is no thermodynamic equilibrium.

2 Symmetries

The idea of Sacharov postulated that one of the known forces or a new force violates the symmetry between matter and antimatter and thus produces a small surplus of matter. All the remaining matter annihilates with the corresponding amount of antimatter, and at the end we are left with the surplus (of 10^{-10}) of matter and the red-shifted photons. Such a symmetry violation goes against principles which were cherished for centuries.

Symmetries and conservation laws have long played an important role in physics. The simplest examples of macroscopic relevance are the conservation of energy and momentum, which are due to the invariance of forces under translation in time and space, respectively. This was demonstrated by Emmy Noether.

In the domain of quantum phenomena, there are also conservation laws corresponding to discrete transformations. One of these is reflection in space ("parity operation") P. Invariance of laws of nature under P means that the mirror image of an experiment yields the same result in its reflected frame of reference as the original experiment in the original frame of reference. This means that "left" and "right" cannot be defined in an absolute sense.

Similarly, the particle-antiparticle conjugation C transforms each particle into its antiparticle, whereby all additive quantum numbers change their sign. C invariance of laws again means that experiments in a world consisting mainly of antiparticles will give identical results to the ones in our world with the one exception that all names of particles are "anti" relative to ours. Here again it will be not possible to define in an absolute way whether a particle consists of antimatter or matter: an antiatom composed of antinucleons and positrons emits the same spectral lines as the corresponding atom.

A third transformation of this kind is time reversal T, which reverses momenta and angular momenta. This corresponds formally to an inversion of the direction of time. According to the CPT theorem of Lüders & Pauli [1,2] there is a connection between these three transformations such that under rather weak assumptions in a local field theory all processes are invariant under the combined operation C P T.

For a long time it was assumed that all elementary processes are also invariant under the application of each of the three operations C, P, and T separately. However, the work of Lee & Yang [3] questioned the assumption, and the subsequent experiments demonstrated the violation of P and C invariance in weak decays of nuclei and of pions and muons. This violation can be visualized by the longitudinal polarization of neutrinos emerging from a weak vertex: they are left-handed when they are particles and right-handed when antiparticles. Application of P or C to a neutrino leads to an unphysical state (Fig. 2).

The combined operation CP, however, transforms a left-handed neutrino into a right-handed antineutrino, thus connecting two physical states. CP invariance therefore was considered [4] to be replacing the separate P and C invariance of weak interactions.

A unique testing ground for CP invariance in the microworld are elementary particles called neutral K mesons. They have the unique property that the particle K^0 and its antiparticle $\overline{K^0}$ differ in one quantum number, called strangeness, but still these two particles can mix through a transition mediated by the weak interaction. One consequence of this postulated CP invariance for the neutral K mesons was predicted by Gell-Mann & Pais[5] : there should be a long-lived partner to the known $V^0(K_1^0)$ particle of short lifetime (10^{-10} sec). According to this proposal these two physical particles are mixtures of two strangeness eigenstates, $K^0(S=+1)$ and $\overline{K^0}(S=-1)$ produced in strong interactions. Weak interactions do not conserve strangeness and the physical particles should be eigenstates of CP if the weak interactions are CP invariant. These eigenstates are (with $\overline{K^0} = CP\ K^0$)

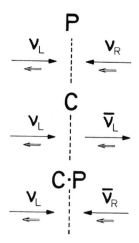

Fig. 2. The mirror image of a left-handed neutrino under P, C, and CP mirror operations

$$CPK_1 = CP[(K^0 + \overline{K^0})/\sqrt{2}] = (\overline{K^0} + K^0)/\sqrt{2} = K_1$$
$$CPK_2 = CP[(K^0 - \overline{K^0})/\sqrt{2}] = (\overline{K^0} - K^0)/\sqrt{2} = -K_2.$$

Because of $CP\,(\pi^+\pi^-) = (\pi^+\pi^-)$ for π mesons in a state with angular momentum zero, the decay into $\pi^+\pi^-$ is allowed for the K_1 but forbidden for the K_2; hence the longer lifetime of K_2, which was indeed confirmed when the K_2 was discovered.

In 1964, however, Christenson, Cronin, Fitch, and Turlay [6] discovered that the long-lived neutral K meson also decays to $\pi^+\pi^-$ with a branching ratio of $\sim 2 \times 10^{-3}$. From then on the long-lived state was called K_L because it was no longer identical to the CP eigenstate K_2; similary, the short-lived state was called K_S.

3 Phenomenology and Models of CP Violation

The phenomenon of CP violation in decays of neutral K^0 mesons is now with us for 38 years [6]. The first ten years of intense experimentation after the discovery of the decay $K_L \to \pi^+\pi^-$ were devoted to the observation of other manifestations of the phenomenon, like the decay [7] $K_L \to \pi^0\pi^0$ and the charge asymmetry in the decays [8] $K_L \to \pi^\pm e^\mp \nu$ and $K_L \to \pi^\pm \mu^\mp \nu$, and to precision experiments on the moduli and phases of the CP violating amplitudes [9]. These experimental results excluded a large number of theoretical models proposed to explain CP violation, such that at the time of the London 1974 conference [10] essentially two classes of models survived: The superweak model postulating a new, very weak, CP violating interaction [11] with $\Delta S = 2$, and milliweak models invoking a small (10^{-3}) part of the

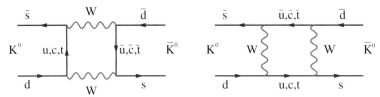

Fig. 3. Box diagram for $K^0 - \bar{K}^0$ mixing connected to CP violating parameter ϵ.

normal $\Delta S = 1$ weak interaction as the source of CP violation. In this case, there is a direct decay of a Kaon state with $CP = -1$ into a two-pion state with $CP = +1$ through a milliweak Hamiltonian. This is called "direct CP violation", as opposed to CP violation by $K^0/\bar{K}^0$ mixing. The key question then became: which of these models is describing the phenomenon? Can one devise experiments distinguishing between those models?

In this context it was very important that a specific milliweak model within the standard model was proposed by Kobayashi and Maskawa [12] in 1973. At the time of the discovery of CP violation, only 3 quarks were known, and there was no possibility of explaining CP violation as a genuine phenomenon of weak interactions. This situation remained unchanged with the fourth quark because the 2×2 weak quark mixing matrix has only one free parameter, the Cabibbo angle, and no non-trivial complex phase. However, as remarked by Kobayashi and Maskawa, the picture changes if six quarks are present. Then the 3×3 mixing matrix naturally contains a phase, apart from three mixing angles. It is then possible to construct CP violating weak amplitudes from "box-diagrams" of the form shown in Fig. 3.

A necessary consequence of this model of CP violation is the non-equality of the relative decay rates for $K_L \to \pi^+\pi^-$ and $K_L \to \pi^0\pi^0$. This "direct CP violation" is due to "Penguin diagrams" of the form given in Fig. 4.

For a quantitative discussion, we use the conventional notations. Let $\bar{K}^0 = CPK^0$, then the eigenstates of CP are:

$$K_1 = (K^0 + \bar{K}^0)/\sqrt{2} = +CPK_1$$
$$K_2 = (K^0 - \bar{K}^0)/\sqrt{2} = -CPK_2$$

The physical long-lived (K_L) and short-lived (K_S) states are then

$$K_S = (K_1 + \epsilon_S K_2)/(1 + |\epsilon_S|^2)^{1/2}$$
$$K_L = (K_2 + \epsilon_L K_1)/(1 + |\epsilon_L|^2)^{1/2}$$

With CPT invariance, $\epsilon_S = \epsilon_L = \epsilon$. The admixture parameter

$$\epsilon = \frac{\mathrm{Im}\,\Gamma_{12}/2 + i\,\mathrm{Im}\,M_{12}}{i\,(\Gamma_S - \Gamma_L)/2 - (m_S - m_L)}$$

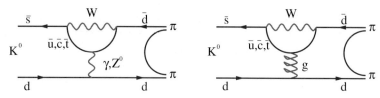

Fig. 4. Penguin diagrams for $K^0 \to 2\pi$ decay with direct CP violation (amplitude ϵ').

is given in terms of the $K^0/\bar{K}^0$ mass matrix M and decay matrix Γ. The phase of ϵ, assuming CPT, is

$$\mathrm{Arg}(\epsilon) = \arctan(2\Delta m/\Gamma_S) = 43.7° \pm 0.2°.$$

The experimentally observable quantities are

$$|\eta_{+-}|e^{i\phi_{+-}} = \eta_{+-} = <\pi^+\pi^-|\mathcal{T}|K_L>/<\pi^+\pi^-|\mathcal{T}|K_S>,$$
$$|\eta_{00}|e^{i\phi_{00}} = \eta_{00} = <\pi^0\pi^0|\mathcal{T}|K_L>/<\pi^0\pi^0|\mathcal{T}|K_S>.$$

It can be shown that these amplitude ratios consist of the contribution from CP violation in the $K^0/\bar{K}^0$ mixing (box diagrams in the KM model), called ϵ above, and another one from CP violation in the weak $K \to 2\pi$ amplitudes (penguin diagrams in the KM model), called ϵ':

$$\eta_{+-} = \epsilon + \epsilon'$$
$$\eta_{00} = \epsilon - 2\epsilon'$$

In this way η_{+-}, η_{00} and $3\epsilon'$ form a triangle in the complex plane, the Wu-Yang triangle. The CP violating decay amplitude ϵ' is due to interference of $\Delta I = 1/2 (A_0)$ and $\Delta I = 3/2 (A_2)$ amplitudes:

$$\epsilon' = \frac{i\,\mathrm{Im} A_2}{2A_0} \exp[i(\delta_2 - \delta_0)]$$

and its phase is given by the $\pi\pi$ phase shifts in the $I=0$ and $I=2$ states, δ_0 and δ_2 (CPT assumed):

$$\mathrm{Arg}(\epsilon') = (\delta_2 - \delta_0) + \pi/2$$

which experimentally is $(61 \pm 9)°$ (Ref. [13]) or $(45 \pm 9)°$ (Ref. [14]). The two models discussed above differ significantly: the superweak model predicts vanishing direct CP violation in weak decays, $\epsilon' = 0$, and therefore $\eta_{+-} = \eta_{00} = \epsilon$, while in milliweak models one expects $\epsilon' \neq 0$.

4 Theoretical Estimates for the Parameter ϵ' of Direct CP Violation

The prediction for ϵ' within the weak quark-mixing model of Kobayashi and Maskawa [12] (KM model) can be estimated if one infers the magnitude of the mixing angles from other experiments and if the hadronic matrix elements for box graphs and penguin graphs are calculated. Typical values for $|\epsilon'/\epsilon|$ are in the range $+(0.2\ldots 3) \times 10^{-3}$ for three generations of quarks. A measurement of this quantity to this level of precision therefore becomes the "experimentum crucis" for our understanding of CP violation. Since the phase of ϵ' is close to the one of ϵ, and since $|\epsilon'/\epsilon| \ll 1$ in good approximation, we get:

$$\epsilon'/\epsilon = \mathrm{Re}(\epsilon'/\epsilon) = \left(1 - |\eta_{00}/\eta_{+-}|^2\right)/6$$

A measurement of the double ratio

$$R = \frac{|\eta_{00}|^2}{|\eta_{+-}|^2} = \frac{\Gamma(K_L \to 2\pi^0)/\Gamma(K_L \to \pi^+\pi^-)}{\Gamma(K_S \to 2\pi^0)/\Gamma(K_S \to \pi^+\pi^-)}$$

to a precision of better than 0.3% is therefore required to distinguish between the two remaining models.

Various methods are used to calculate the value of $\mathrm{Re}(\epsilon'/\epsilon)$. Due to the difficulties in calculating hadronic matrix elements, which involve long distance effects, the task turns out to be very difficult. The following results have been obtained recently:

1. The Dortmund group uses the $1/N_C$ expansion and Chiral Pertubation Theory. They quote a range of $1.5 \times 10^{-4} < \epsilon'/\epsilon < 31.6 \times 10^{-4}$ (Ref. [15]) from scanning the complete range of input parameters.
2. The Munich group uses a phenomenological approach in which as many parameters as possible are taken from experiment. Their result (Ref. [16]) is $1.5 \times 10^{-4} < \epsilon'/\epsilon < 28.8 \times 10^{-4}$ from a scanning of the input parameters and $\epsilon'/\epsilon = (7.7^{+6.0}_{-3.5}) \times 10^{-4}$ using a Monte Carlo method to determine the error.
3. The Rome group uses lattice calculation results for the input parameters. Their result is $\epsilon'/\epsilon = (4.7^{+6.7}_{-5.9}) \times 10^{-4}$ (Ref. [17]).
4. The Trieste group uses a chiral quark model to calculate ϵ'/ϵ. Their result is $7 \times 10^{-4} < \epsilon'/\epsilon < 31 \times 10^{-4}$ (Ref. [18]) from scanning.

It is hoped that reliable hadronic matrix elements will be obtained in the near future by lattice gauge theory calculations.

5 Experiments on Direct CP Violation

5.1 Early Experiments: The Observation of Direct CP Violation

The first observation of direct CP violation was made by a collaboration of physicists at the European laboratory for particle physics CERN at Geneva, Switzerland in 1988. Their experiment, called NA31, was based on the concurrent detection of $2\pi^0$ and $\pi^+\pi^-$ decays. Collinear beams of K_S and K_L

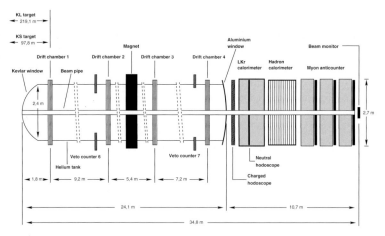

Fig. 5. Layout of the main detector components of the NA48 experiment.

were employed alternately. Kaons with energies around 100 GeV were produced by a 450 GeV proton beam from the proton accelerator SPS at CERN. The energies of the decay products were measured in a combination of a high-resolution Liquid-Argon electromagnetic calorimeter and an iron-scintillator hadronic calorimeter. In the first exposures, about 100000 decays of the type $K_L \to \pi^0\pi^0$ and 295000 decays of $K_L \to \pi^*\pi^-$ were observed, and the result for the CP parameter was $\text{Re}(\epsilon'/\epsilon) = (33 \pm 11) \times 10^{-4}$ (Ref. [19]). In further improved experimentation, the number of observed $K_L \to \pi^0\pi^0$ decays was increased to 4×10^5.

A similar sensitivity was achieved by an experiment at Fermilab near Chicago, called E 731. In 1992/93 the experiments NA31 at CERN and E731 at FNAL presented final results. The CERN result [20] $(23.0 \pm 6.5) \times 10^{-4}$ shows with more than 3 Standard deviations a clear evidence for direct CP violation whereas the Fermilab result [21] with $(7.4 \pm 5.9) \times 10^{-4}$ is consistent with zero. While the CERN experiment had observed direct CP violation, the Fermilab experiment did not concur. As a consequence of this disagreement, two new experiments were constructed in order to verify the earlier result. They were called NA48 at CERN and kTeV at Fermilab.

5.2 The NA48 Detector

The new CERN experiment NA48 (Fig. 5) was designed

- to measure all four decay modes concurrently,
- to register data at a rate 10 times higher than NA31,
- to achieve an improved energy resolution for photons (liquid Krypton calorimeter) and for pions (magnetic spectrometer).

In the design of the NA48 detector the cancellation of systematic uncertainties in the double ratio is exploited as much as possible. Important properties of the experiment are 1. two almost collinear beams which lead to an almost identical illumination of the detector and 2. the lifetime weighting of the events defined as K_L events.

The K_L target is located 126 m upstream of the beginning of the decay region. As the decay lengths at the average kaon momentum of 110 GeV/c are $\lambda_S = 5.9$ m and $\lambda_L = 3\,400$ m respectively, the neutral beam derived from this target is dominated by K_L. The K_S target is located 6 m upstream of the decay region so that this beam is dominated by K_S. The two beams are almost collinear: The K_S target is situated 7.2 cm above the center of the K_L beam. The relative angle of the beams is 0.6 mrad so that they converge at the position of the electromagnetic calorimeter.

The beginning of the K_S decay region is defined by an anticounter (AKS). This detector is used to veto Kaon decays occuring upstream of the counter. The position of the AKS also defines the global Kaon energy scale as the energy is directly correlated to the distance scale. The decay region itself is contained in a 90 m long evacuated tank.

The identification of K_S decays is done by a detector (tagger) consisting of an array of scintillators situated in the proton beam directed to the K_S target. If a proton signal is detected within a time window of ± 2 ns with respect to a decay, the event is defined as K_S event. The absence of a proton defines a K_L event.

A magnetic spectrometer is used to reconstruct $K_{S,L} \to \pi^+\pi^-$ decays. The spectrometer consists of a dipole magnet with "momentum kick" of 265 MeV/c and four drift chambers which have a spatial resolution of ~ 90 μm. This leads to a mass resolution of 2.5 MeV/c^2. A hodoscope consisting of two planes of plastic scintillator provides the time of a charged event with a resolution of about 200 ps.

A quasi-homogeneous liquid krypton electro-magnetic calorimeter (LKR) is used to identify the four photons from a $\pi^0\pi^0$ event. Liquid krypton has a radiation length of $X_0 = 4.7$ cm which allows one to build a compact calorimeter with high energy resolution ($\Delta E/E = 1.35\,\%$ measured for electrons coming from a $K_L \to \pi e \nu$ (K_{e3}) decay) and very good time resolution (< 300 ps) and very good linearity. It consists of 13212 2×2 cm^2 cells pointing to the average K_S decay position. The transverse spatial resolution is better than 1.3 mm.

The electromagnetic calorimeter is complemented by an iron-scintillator sandwich calorimeter with a depth of 6.8 nuclear interaction lenghts which measures the remaining energy of hadrons for use in the trigger for charged events.

A muon veto detector, consisting of three planes of scintillator shielded by 80 cm of iron, is used to identify muons to veto $K_L \to \pi\mu\nu(K_{\mu 3})$ events.

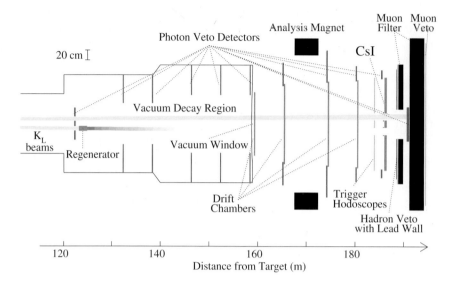

Fig. 6. Layout of the main detector components of the KTeV experiment

5.3 The KTeV Detector

The main elements of the KTeV detector (fig. 6) are similar to those in NA48 since both experiments work in a similar environment. The main difference is the way in which K_S mesons are produced.

KTeV uses two parallel well separated kaon beams derived from a single target. In one beam, the K_L mesons from the target pass through a collimator and then decay in an evacuated decay region of 30 m length. In the other beam the K_L mesons traverse a slab of matter ("regenerator"). During this passage, the K^0 and $\overline{K^0}$ components of the K_L are affected differently by interactions with matter. Therefore the wave emerging behind the regenerator is a slightly different superpostion of these two components. In the forward direction, the energing wave contains a K_S wave coherent with the outgoing K_L wave. This is the regenerated K_S beam used in the experiment. The regenerator alternates between both beams every minute in order to keep the detector illumination identical for the K_S and the K_L components.

The decay region of the regenerator beam is defined by a lead-scintillator counter at the downstream end of the regenerator. The decay region of the vacuum beam starts at a mask anticounter.

Similar to NA48 the KTeV spectrometer consists of four drift chambers; the magnet provides a momentum kick of $p_t = 412$ MeV/c, leading to a mass resolution of $\sigma_{m(\pi^+\pi^-)} = 1.6$ MeV/c^2. For triggering of charged events a scintillator hodoscope is used.

The electromagnetic calorimeter (CsI) consists of 3100 pure Cesium-Iodide crystals with a radiation length of $X_0 = 1.85$ cm. In the inner region the size

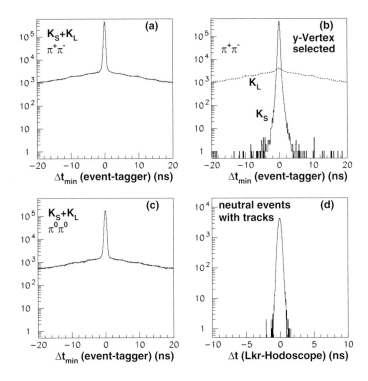

Fig. 7. (a),(c): Minimal difference between tagger time and event time ($\Delta t_{\min}$). (b) $\Delta t_{\min}$ for charged K_L and K_S events. (d) Comparison between charged and neutral event time. For this measurement decays, selected by the neutral trigger, with tracks are used (γ conversion and Dalitz decays $K_S \to \pi^0 \pi_D^0 \to \gamma\gamma\gamma e^+ e^-$).

is 2.5×2.5 cm² and in the outer region 5.0×5.0 cm². Two beam holes of 15×15 cm² allow the two kaon beams to pass to the beam dump. The energy resolution at large photon energies above 10 GeV is 0.75% as measured with K_{e3} decays.

In addition 10 lead-scintillator "photon veto" counters are used to detect particles escaping the decay volume. The background is further reduced by a muon veto counter consisting of 4 m of steel and a hodoscope.

5.4 Data Analysis NA48: Confirmation of Direct CP Violation

To identify events coming from the K_S target a coincidence window of ± 2 ns between the proton signal in the tagger and the event time is chosen (see Fig. 7a,c). Due to inefficiencies in the tagger and in the proton reconstruction a fraction α_{SL} of true K_S events are misidentified as K_L events. On the other hand there is a constant background of protons in the tagger which have not led to a good K_S event. If those protons accidentally coincide with a true K_L

event, this event is misidentified as a K_S decay. This fraction α_{LS} depends only on the proton rate in the tagger and the width of the coincidence window.

Both effects, α_{SL}^{+-} and α_{LS}^{+-}, can be measured (see figure 7b) in the charged mode as K_S and K_L can be distinguished by the vertical position of the decay vertex. The results are $\alpha_{SL}^{+-} = (1.63 \pm 0.03) \times 10^{-4}$ and $\alpha_{LS}^{+-} = (10.649 \pm 0.008)\%$. This means that about 11% of true K_L events are misidentified as K_S events, however, this quantity is precisely measured to the 10^{-4} level. What is important for the measurement of R is the difference between the charged and the neutral decay modes $\Delta\alpha_{LS} = \alpha_{LS}^{00} - \alpha_{LS}^{+-}$. Proton rates in the sidebands of the tagging window are measured in both modes to measure $\Delta\alpha_{LS}$. The result is $\Delta\alpha_{LS} = (4.3 \pm 1.8) \times 10^{-4}$. Several methods have been used to measure $\Delta\alpha_{SL}$, leading to the conclusion that there is no measurable difference between the mistaggings within an uncertainty of $\pm 0.5 \times 10^{-4}$.

Another important correction is the background subtraction. Decays $K_L \to \pi e \nu$ and $K_L \to \pi \mu \nu$ can be misidentified as $K \to \pi^+ \pi^-$ decays, as the ν is undetectable. However, since the ν carries away momentum and energy, these events can be identified by their high transverse momentum p'_t and their reconstructed invariant mass. The remaining background can be measured by extrapolating the shape of the background in the $m - p'^2_t$-plane into the signal region. In this way the charged background fraction leads to an overall correction on R of $(16.9 \pm 3.0) \times 10^{-4}$.

A similar extrapolation can be done in the neutral decay mode. Here the background comes from $K_L \to 3\pi^0$ decays, where two γs are not detected. This leads to a misreconstruction of the invariant π^0 masses. In this case the background leads to a correction of R by $(-5.9 \pm 2.0) \times 10^{-4}$.

The number of signal events after these corrections are summarised in table 1.

Table 1. Event numbers after tagging correction (only NA48) and background subtraction.

Event statistics ($\times 10^6$)					
	NA48	KTeV		NA48	KTeV
$K_S \to \pi^+\pi^-$	22.221	4.52	$K_L \to \pi^+\pi^-$	14.453	2.61
$K_S \to \pi^0\pi^0$	5.209	1.42	$K_L \to \pi^0\pi^0$	3.290	0.86

The efficiency of the triggers used to record neutral and charged events have been determined. Independent triggers are used which accept a downscaled fraction of events. In the neutral decay mode the efficiency is measured to be 0.99920 ± 0.00009 without measurable difference between K_S and K_L decays. The $\pi^+\pi^-$ trigger efficiency is measured to be $(98.319 \pm 0.038)\%$ for K_L and $(98.353 \pm 0.022\%)$ for K_S decays. Here a small difference between the trigger efficiency in K_S and K_L decays is found. This leads to a correction to the double ratio of $(-4.5 \pm 4.7) \times 10^{-4}$. The error on this measurement is dominated by the total number of events registered with the independent trigger. This error is one of the main contributions to the systematic error of the measurement of R.

The distance D from the LKR to the decay vertex is reconstructed using the position of the four γ clusters. From the kinematics of the decay one obtains

$$D = \frac{1}{M_K}\sqrt{\sum_{i,j} E_i\, E_j\, r_{ij}^2},$$

where E_i is the energy of cluster i and r_{ij} the distance between cluster i and cluster j. This formula directly relates the distance scale to the energy scale. It is therefore possible to fix the global energy scale with the measurement of the known AKS position in the neutral decay mode. In addition more checks on the energy scale and the linearity of the energy measurement can be performed, such as the measurement of the invariant π^0 mass and the use of the known position of an added thin CH_2 target (a π^- beam produces $\pi^0 \to 2\gamma$). The comparison of all methods gives an uncertainty of $\pm 3\times 10^{-4}$ in the global energy scale.

Another systematic problem is the minimization of acceptance corrections by weighting of the K_L events. The difference in the lifetime between K_S and K_L events produce a different illumination of the detector: There are more K_L events decaying closer to the detector and they are therefore also measured at smaller radii closer to the beampipe. NA48 weights the K_L events according to the measured lifetime such that the distribution of the z-position of the decay vertex of K_S and K_L events and the detector acceptances become equal. Using this method the influence of detector inhomogenities is minimised and the analysis becomes nearly independent of acceptance calculations by Monte Carlo methods. In fact the acceptance correction due to small detector differences is quite small. The price to pay for the gain in systematics is the loss in statistics.

Although the acceptance is almost equal there are nevertheless small differences in the beam geometry and detector illumination between decays coming from the K_S and the K_L target. These remaining differences are corrected for with Monte Carlo methods. Using the Monte Carlo to calculate the double ratio R the deviation from the input value 1 is $(26.7 \pm 5.7)\times 10^{-4}$.

Summing up the systematic uncertainties of all different sources in quadrature, they amount to a total 12.4×10^{-4} in R, and to 2.1×10^{-4} in ϵ'/ϵ.

The result of NA48 using the data sample from 1997, 1998 and 1999 is [23], [24]

$$\epsilon'/\epsilon = (\ 15.3\ \pm\ 2.6\ ) \times 10^{-4}. \qquad (1)$$

5.5 Analysis of KTeV Data: Confirmation of Direct CP Violation

KTeV has similar physical backgrounds as NA48, but in addition two-pion decays from K_S mesons produced by incoherent regeneration in the regen-

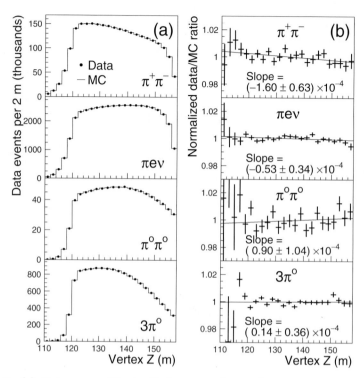

Fig. 8. (a) Data versus Monte Carlo comparisons of z-vertex distributions. (b) Linear fits to the data/MC ratio of (a).

erator have to be subtracted. Typical numbers are: a fraction of 6.9×10^{-4} charged background from K_{e3} and $K_{\mu 3}$ decays and a fraction of 27×10^{-4} neutral background from $3\pi^0$ decays. The background levels in the regenerator beam are 107×10^{-4} in the neutral mode (this gives rise to a large systematic error) and 7.2×10^{-4} in the charged mode. The event numbers after background subtraction.

The main difference in the analysis techniques of the two experiments is the treatment of the acceptance correction. KTeV is not using event weighting but uses Monte Carlo studies to correct for detector differences in the K_S and K_L decays. K_{e3} and $3\pi^0$ decays are used to model the detector and the agreement between data and Monte Carlo is good (see Fig. 8). The acceptance correction to R calculated by Monte Carlo simulation is then $(231 \pm 13) \times 10^{-4}$. This can be compared to the size of the total effect of ϵ'/ϵ on R, 168×10^{-4}, as measured by KTeV. The main source of systematic uncertainty is a slight disagreement between data and Monte Carlo comparison in the $\pi^+\pi^-$ decay mode in the vacuum beam. A slope of $(-1.60 \pm 0.63) \times 10^{-4}$ m^{-1} has been found which is applied as a systematic error.

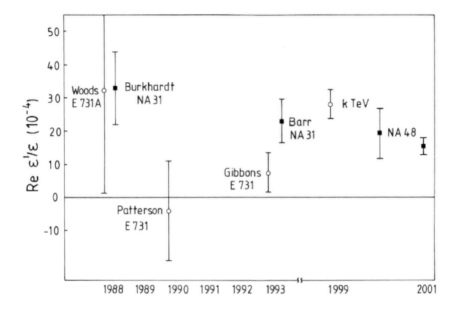

Fig. 9. Measurements of the parameter ϵ'/ϵ of direct CP violation.

The energy scale is determined using the known position of the regenerator edge. The comparison of the measured position of the vacuum window with the real position gives an uncertainty of the global energy scale of 4.2×10^{-4} on R.

The result is obtained by a fit of $\mathrm{Re}(\epsilon'/\epsilon)$, the regeneration amplitude and phase to the event numbers per energy bin. The result is published in 1999 was [22]

$$\epsilon'/\epsilon = (\ 28.0\ \pm\ 3.0\ \mathrm{(stat)}\ \pm\ 2.8\ \mathrm{(sys)}\ ) \times 10^{-4} \qquad (2)$$

but this was subsequently corrected due to an error in background subtraction:

$$\epsilon'/\epsilon = (\ 23.2\ \pm\ 3.0\ \mathrm{(stat)}\ \pm\ 3.2\ \mathrm{(sys)}\ \pm\ 0.7\ \mathrm{(MC)}\ ) \times 10^{-4} \qquad (3)$$

6 Conclusion

The two experiments kTeV and NA48 have definitively confirmed the original observation of the NA31 team that direct CP violation exists. The results of all published results on ϵ'/ϵ are shown in fig. 9. Therefore, CP violation as observed in the K meson system is a part of the weak interaction, due to weak quark mixing. Exotic new interactions like the superweak interaction are not needed. With more data, both experiments will reach a precision of

$\mathcal{O}(2\times 10^{-4})$. We therefore have a very precise experimental result on ϵ'/ϵ. The theoretical calculations of ϵ'/ϵ within the Standard Model however, are still not very precise. This does not change the main conclusion of the experiments, that ϵ' is different from zero and positive, i.e. direct CP violation exists.

However, this milliweak CP violation is probably not large enough to explain the observed matter-antimatter asymmetry from the big bang. An additional, stronger CP violation is needed for this. There are speculations that this might be due to CP violation in the lepton sector in the early universe. Very heavy Majorana neutrinos could play a role in the formation of an antimatter-matter asymmetry at this time [25].

References

1. Lüders, G. 1954. Kgl. Danske Videnskab. Selskab, Matfys. Medd. 28(5) : 1
2. Pauli, W. 1955. In Niels Bohr and the Development of Physics, ed. W. Pauli, p. 30. Oxford: Pergamon, ed. W. Pauli
3. Lee, T. D.. Yang, C. N. 1956, Phys. Rev. 104 : 254
4. Landau, L. D. 1957. Nucl. Phys. 3 : 127
5. Gell-Mann, M., Pais, A. 1955. Phys. Rev. 97 : 1387
6. J.H. Christenson, J.W. Cronin, V.L. Fitch, R. Turlay, Phys. Rev. Lett. 13 (1964), 138.
7. J.M. Gaillard et al. (1967), Phys. Rev. Lett. 18, 20
   J.W. Cronin et al. (1967), Phys. Rev. Lett. 18, 25.
8. S. Bennett et al. (1967), Phys. Rev. Lett. 19, 993
   D. Dorfan et al. (1967), Phys. Rev. Lett. 19, 987.
9. Reviewed in K. Kleinknecht (1976), Ann. Rev. Nucl. Sci. 26, 1.
10. K. Kleinknecht (1974), Proc. 17th Int. Conf. High Energy Phys., London, ed. J.R. Smith (Chilton, Didcot, UK), p.III-23.
11. L. Wolfenstein (1964), Phys. Rev. Lett. 13, 562.
12. M. Kobayashi, T. Maskawa (1973), Prog. Theor. Phys. 49, 652,
    N. Cabibbo (1963), Phys. Rev. Lett. 10, 531.
13. N.N. Biswas et al. (1981), Phys. Rev. Lett. 47, 1378 and priv. comm. to W. Ochs.
14. W. Ochs, priv. comm. from analysis of CERN-Munich data.
15. T. Hambye, G.O. Köhler, E.A. Paschos, P.H. Soldan, hep-ph/9906434.
16. A.J. Buras, M. Gorbahn, S.Jäger, M. Jamin, M.E. Lautenbacher, L. Silvestrini, hep-ph/9904408, hep-ph/9908395.
17. G. Martinelli, Presentation at Kaon99, Chicago, http://hep.uchicago.edu/kaon99/talks/martinelli/
18. S. Bertolini, M. Fabbrichesi, J.O. Eeg, hep-ph/9802405.
19. H. Burkhardt et al., Phys. Lett. B 206 (1988), 169.
20. G.D. Barr et al., Phys. Lett. B317 (1993), 233.
21. L.K. Gibbons et al., Phys. Rev. Lett. 70 (1993), 1203.
22. A. Alavi-Harati et al., Phys. Rev. Lett. 83 (1999), 22.
23. V. Fanti et al., Phys. Lett. B, 465 (1999) 335.
24. A. Lai et al., A precise measurement of the direct CP violation parameter Re(ϵ'/ϵ), Europ. Phys. Journal C22 (2001)231.
25. W. Buchmüller, Ann. Phys. (Leipzig) 10 (2001) 95.

Flavour Oscillation and CP Violation of B Mesons

Roland Waldi
Universität Rostock, D-18051 Rostock, Germany

Abstract. The phenomena of meson-anti-meson flavour oscillation and CP violation are presented with emphasis on their description within the Standard Model. The experimental methods used in their investigation are discussed, and the presently available results on B mesons are shown.

1 Introduction

Our understanding of physics in general and particle physics in particular has been mainly put forward by the discovery of **symmetries**. It is remarkable, that most of the symmetries discovered have, however, finally turned out to be only "almost-symmetries", i. e. to be more or less broken.

Mirror symmetry (parity P) is broken by weak interaction, which makes a maximal distinction between fermions of left and right chirality. First ideas of this unexpected behaviour emerged as a solution of the "$\Theta\ \tau$ puzzle", the fact that the kaon decays both to $P = +1$ and $P = -1$ eigenstates [1], and a direct observation as left-right-asymmetry in weak beta decays followed soon [2]. It is most pronounced in the massless neutrinos, which are produced in weak interactions only with lefthanded helicity, or righthanded in the case of antineutrinos, thus violating the charge-conjugation symmetry (C) at the same time.

The product of both discrete symmetries, CP, is almost intact, and seemed to be conserved even in weak interaction processes. A small violation has first been observed in 1964 [3] in K^0 decays, which was for 37 years the only system known not to respect CP symmetry completely. The explanation of this violation in the Standard Model will be briefly discussed in the next chapter. This is not the only possible description, but the one with no additional assumptions. At the same time, the Standard Model predicts CP violating effects [4] in the decay of beauty mesons (B^0, B_s, B^+), which are large in some rare decay channels. These have been observed for the first time in 2001 [5,6].

There are a few common notions about CP violation that are wrong, but still wide-spread prejudice. The most persistent is:
- The mass eigenstates K^0_S, K^0_L or B_H, B_L are (at least approximate) CP eigenstates.

As will be discussed in more detail below, the transformation of these states under CP is a mere convention, and CP is not an observable for any of these states. While a suitable convention for K_S^0, K_L^0 can be very handy in shortening arguments about CP eigenvalues of final states involving any of these, a similar convention for B mesons can be very much in the way of understanding CP violating phenomena there.

- CP violation in the Standard Model is a small effect, typically $\mathcal{O}(10^{-3})$ or less.

The truth is, that asymmetries up to 60% have been observed in $K^0/\overline{K}^0 \to \pi^+\pi^-$, and up to 100% asymmetries are expected in the $B_s/\overline{B}_s$ system. However, it is a rare effect, suppressed by either small branching fractions or its occurrence after several mean lifetimes or both.

- CP violation within the Standard Model occurs only with "mixing mesons", i.e. $K^0/\overline{K}^0$, $B^0/\overline{B}^0$, $B_s/\overline{B}_s$, and possibly $D^0/\overline{D}^0$.

This one has, furtunately, mostly vanished, since many people nowadays look at CP violation in B^+/B^- decays, or at $\Xi/\overline{\Xi}$ and other baryon decays. The combination of the latter two statements, however, is true: CP violating asymmetries in other than the "mixing mesons" are small within the Standard Model.

In the following sections I will try to derive the Standard Model description of oscillation and CP violation in a formalism that can be applied to all four mesons, and give the translations between the preferred descriptions for K mesons and that for B mesons. I will also try to point the reader to other conventions in terminology or the choice of signs.

1.1 The Experiments

After the discovery of the Υ states, B meson properties have been investigated since the mid-80s at e^+e^- storage rings operating at the $\Upsilon(4S)$ resonance: DORIS II at DESY (Gemany), with the experiment ARGUS, and CESR at Cornell (USA) with the detector CLEO. ARGUS had collected 0.25 Million $B\overline{B}$ events when it stopped in 1992, CLEO has accumulated 10 Million $B\overline{B}$ events by the end of 2000. These experiments have the advantage to investigate events with nothing else but two B mesons, which are even almost at rest since the mass of the $\Upsilon(4S)$ is only 20 MeV above the $B\overline{B}$ threshold.

This source is also exploited by the asymmetric e^+e^- colliders PEP-II at SLAC (USA) and KEK-B at KEK (Japan), where the experiments BABAR and BELLE have started taking data in 1999. They both have reached record luminosities above $3 \cdot 10^{33}/\mathrm{cm}^2/s$, and had collected 23 and 11 Million $B\overline{B}$ pairs by the end of 2000, respectively. Their data samples are increasing rapidly, and they are aiming both for over 100 Million at the end of 2002. These B factories have different electron and positron energies to produce the $\Upsilon(4S)$ with a boost of $\beta\gamma = 0.55$ (BABAR) and 0.42 (BELLE) in order to measure the difference of the lifetime of the two B mesons. This is an essential information for the observation of time-dependent CP asymmetries, as will be discussed below in Chap. 3.

In the 1990s the four experiments ALEPH, DELPHI, L3 and OPAL at the LEP storage ring at CERN (Switzerland) started investigating $b\bar{b}$ jets from Z^0 decays. They have each a sample of almost one Million $b\bar{b}$ events. They were joined by SLD which accumulated polarized Z^0 events at the linear collider SLC at SLAC (USA).

Hadronic production of $b\bar{b}$ jets in addition to the fragments of the original particles are the source of B mesons at the $p\bar{p}$ storage ring Tevatron at Fermilab (USA). Hadronic production of $b\bar{b}X$ at high energies is orders of magnitude higher than any other source, but the samples of triggered and detected events being only a small fraction. The exploitation of these vast amounts of data is a challenge, which has been met in the past by the CDF detector which was the first experiment to collect enough $B^0 \to J/\psi K^0_S$ decays for a meaningful exclusive CP violation analysis. Both CDF and D0 will start collecting new data this year. Hadronic production will also be the source of B mesons at the planned experiments ATLAS, CMS and the dedicated experiment LHCb at the LHC pp storage ring, and the BTeV experiment at the Tevatron. These experiments will ultimately deliver enough $b\bar{b}X$ events for high precision measurements of CP violation parameters that can be expected about ten years from now.

2 Quark Mixing and Particle Antiparticle Oscillations

Mesons are neither particles nor antiparticles in a strict sense, since they are composed of a quark and an antiquark. This implies the existence of mesons with vacuum quantum numbers (e.g. f_0). More important is the existence of pairs of charge-conjugate mesons, which can be transformed into each other via flavour changing weak interaction transitions. These are $K^0/\overline{K}^0$ ($\bar{s}d/s\bar{d}$), $D^0/\overline{D}^0$ ($c\bar{u}/\bar{c}u$), $B^0/\overline{B}^0$ ($\bar{b}d/b\bar{d}$), and $B_s/\overline{B}_s$ ($\bar{b}s/b\bar{s}$).

2.1 The Unitary CKM Matrix

The charged current weak interactions responsible for flavour changes are described in the Standard Model by the couplings $gW^\mu J^{cc}_\mu$ of the W boson to the current

$$J^{cc}_\mu = \begin{pmatrix} \bar{\nu}_e \\ \bar{\nu}_\mu \\ \bar{\nu}_\tau \end{pmatrix} \gamma_\mu \frac{1-\gamma_5}{2} \begin{pmatrix} e \\ \mu \\ \tau \end{pmatrix} + \sum_{r,g,b} \begin{pmatrix} \bar{u} \\ \bar{c} \\ \bar{t} \end{pmatrix} \gamma_\mu \frac{1-\gamma_5}{2} \mathbf{V} \cdot \begin{pmatrix} d \\ s \\ b \end{pmatrix} \quad (1)$$

with a non-trivial transformation matrix $\mathbf{V}$ in the quark sector, the Cabibbo–Kobayashi–Maskawa (CKM) Matrix [7,8]:

$$\mathbf{V} = \begin{pmatrix} V_{ud} & V_{us} & V_{ub} \\ V_{cd} & V_{cs} & V_{cb} \\ V_{td} & V_{ts} & V_{tb} \end{pmatrix}$$

A coupling via a scalar boson would allow a general 3×3 coupling matrix. However, local gauge invariance which is realized via the gauge bosons $W^\pm$ requires that one universal coupling constant connects the triplet of up-type quarks with the triplet of down-type quarks. The only complication permitted is a unitary transformation to another basis of states, which is accomplished by the CKM matrix.

If there were more than three quark families, this would not hold for the 3×3 submatrix. However, this possibility is unlikely, given the limit on neutrino flavours from LEP experiments, who find $n_\nu = 2.994 \pm 0.012$ [9] for neutrinos with mass much below the Z^0 mass. Thus, if a fourth generation exists, it must incorporate a massive neutrino which is many orders of magnitude above the masses indicated by neutrino oscillation experiments, and more than a factor 1000 heavier than the experimental upper limit for the mass of the tau neutrino.

From the 9 real parameters of a general unitary matrix, 5 can be absorbed in 1 global phase, 2 relative phases between u, c, t and 2 relative phases between d, s, b which are all subject to convention and in principle unobservable. If two quarks within one of these two groups were degenerate in mass, even the sixth phase could be removed by redefining the basis in their two-dimensional subspace.

Rephasing may be accomplished by applying a phase factor to every row and column:

$$V_{jk} \to \mathrm{e}^{\mathrm{i}(\phi_j - \phi_k)} V_{jk} \qquad (2)$$

Note that $j = u, c, t$, $k = d, s, b$, and the six numbers $\phi_u, \phi_c, \phi_t, \phi_d, \phi_s, \phi_b$ represent only five independent phases in the CKM matrix, since different sets of $\{\phi_j, \phi_k\}$ yield the same result. Any product where each row and column enters once as V_{ij} and once via a complex conjugate V^*_{kl} like $V_{ij} V_{kl} V^*_{il} V^*_{kj}$ is **invariant** under the transformation (2). This implies that observable phases must always correspond to similar products of CKM matrix elements with equal numbers of V and V^* factors and appropriate combination of indices.

Removing as much unphysical phases as possible, the CKM matrix is described by **4 real parameters**, where only one is a phase parameter, while the other three are rotation angles in flavour space. The physical phase is not one unique number due to the arbitrary choice of the unphysical phases. Unambiguous representations of this phase as the angles of unitarity triangles will be discussed below. The standard parametrization [9] (first proposed in [10], notation follows [11]) uses a choice of phases, that leave V_{ud} and V_{cb} real:

$$\begin{aligned}
\mathbf{V} &= \begin{pmatrix} 1 & 0 & 0 \\ 0 & c_{23} & s_{23} \\ 0 & -s_{23} & c_{23} \end{pmatrix} \begin{pmatrix} c_{13} & 0 & s_{13}\mathrm{e}^{-\mathrm{i}\delta_{13}} \\ 0 & 1 & 0 \\ -s_{13}\mathrm{e}^{\mathrm{i}\delta_{13}} & 0 & c_{13} \end{pmatrix} \begin{pmatrix} c_{12} & s_{12} & 0 \\ -s_{12} & c_{12} & 0 \\ 0 & 0 & 1 \end{pmatrix} \\
&= \begin{pmatrix} c_{12}c_{13} & s_{12}c_{13} & s_{13}\mathrm{e}^{-\mathrm{i}\delta_{13}} \\ -s_{12}c_{23}-c_{12}s_{13}s_{23}\mathrm{e}^{\mathrm{i}\delta_{13}} & c_{12}c_{23}-s_{12}s_{13}s_{23}\mathrm{e}^{\mathrm{i}\delta_{13}} & c_{13}s_{23} \\ s_{12}s_{23}-c_{12}s_{13}c_{23}\mathrm{e}^{\mathrm{i}\delta_{13}} & -c_{12}s_{23}-s_{12}s_{13}c_{23}\mathrm{e}^{\mathrm{i}\delta_{13}} & c_{13}c_{23} \end{pmatrix} \qquad (3)
\end{aligned}$$

with $c_{ij} = \cos\theta_{ij}$, $s_{ij} = \sin\theta_{ij}$, and $s_{ij} > 0$, $c_{ij} > 0$ ($0 \leq \theta_{ij} \leq \pi/2$). The angle $\theta_C = \theta_{12}$ is the Cabibbo-angle [7].

A convenient substitution[1] is $s_{12} = \lambda$, $s_{23} = A\lambda^2$, $s_{13}\sin\delta_{13} = A\lambda^3\eta$, and $s_{13}\cos\delta_{13} = A\lambda^3\rho$ [12], which reflects the apparent hierarchy in the size of mixing angles via orders of a parameter λ. This leads to

$$\mathbf{V} = \begin{pmatrix} 1 & 0 & 0 \\ 0 & \sqrt{1-A^2\lambda^4} & A\lambda^2 \\ 0 & -A\lambda^2 & \sqrt{1-A^2\lambda^4} \end{pmatrix} \cdot$$

$$\cdot \begin{pmatrix} \sqrt{1-A^2\lambda^6(\rho^2+\eta^2)} & 0 & A\lambda^3(\rho-i\eta) \\ 0 & 1 & 0 \\ -A\lambda^3(\rho+i\eta) & 0 & \sqrt{1-A^2\lambda^6(\rho^2+\eta^2)} \end{pmatrix} \cdot$$

$$\cdot \begin{pmatrix} \sqrt{1-\lambda^2} & \lambda & 0 \\ -\lambda & \sqrt{1-\lambda^2} & 0 \\ 0 & 0 & 1 \end{pmatrix}$$

$$= \begin{pmatrix} 1 - \frac{\lambda^2}{2} - \frac{\lambda^4}{8} & \lambda & A\lambda^3(\rho-i\eta) \\ -\lambda - A^2\lambda^5(\rho+i\eta-\frac{1}{2}) & 1 - \frac{\lambda^2}{2} - (\frac{1}{8}+\frac{A}{2})\lambda^4 & A\lambda^2 \\ A\lambda^3[1-(\rho+i\eta)(1-\frac{\lambda^2}{2})] & -A\lambda^2 - A\lambda^4(\rho+i\eta-\frac{1}{2}) & 1 - \frac{1}{2}A^2\lambda^4 \end{pmatrix}$$

$$+ \mathcal{O}(\lambda^6) \tag{4}$$

and agrees to $\mathcal{O}(\lambda^3)$ with the Wolfenstein approximation [13]:

$$\mathbf{V} = \begin{pmatrix} 1 - \frac{\lambda^2}{2} & \lambda & A\lambda^3(\rho - i\eta + \frac{i}{2}\eta\lambda^2) \\ -\lambda & 1 - \frac{\lambda^2}{2} - i\eta A^2\lambda^4 & A\lambda^2(1+i\eta\lambda^2) \\ A\lambda^3(1-\rho-i\eta) & -A\lambda^2 & 1 \end{pmatrix} \tag{5}$$

$$\approx \begin{pmatrix} 1 - \frac{\lambda^2}{2} & \lambda & A\lambda^3(\rho - i\eta) \\ -\lambda & 1 - \frac{\lambda^2}{2} & A\lambda^2 \\ A\lambda^3(1-\rho-i\eta) & -A\lambda^2 & 1 \end{pmatrix} \tag{6}$$

Equation (4) is more convenient [14] in higher orders than the original proposal of Wolfenstein, or an exact parametrization [15] using the Wolfenstein parameters.

2.1.1 Unitarity Triangles

If nature provides us with just these three families of fermions, unitarity requires the following 12 conditions to be fulfilled:

---

[1] An equivalent choice is $\lambda = s_{12}c_{13}$ which leads to the same parametrization to $\mathcal{O}(\lambda^5)$.

rows 1×1, uu	$\|V_{ud}\|^2 + \|V_{us}\|^2 + \|V_{ub}\|^2 = 1$	(7a)
rows 2×2, cc	$\|V_{cd}\|^2 + \|V_{cs}\|^2 + \|V_{cb}\|^2 = 1$	(7b)
rows 3×3, tt	$\|V_{td}\|^2 + \|V_{ts}\|^2 + \|V_{tb}\|^2 = 1$	(7c)
columns 1×1, dd	$\|V_{ud}\|^2 + \|V_{cd}\|^2 + \|V_{td}\|^2 = 1$	(7d)
columns 2×2, ss	$\|V_{us}\|^2 + \|V_{cs}\|^2 + \|V_{ts}\|^2 = 1$	(7e)
columns 3×3, bb	$\|V_{ub}\|^2 + \|V_{cb}\|^2 + \|V_{tb}\|^2 = 1$	(7f)
rows 1×2, cu	$V_{ud}^* V_{cd} + V_{us}^* V_{cs} + V_{ub}^* V_{cb} = 0$	(7g)
rows 1×3, tu	$V_{ud}^* V_{td} + V_{us}^* V_{ts} + V_{ub}^* V_{tb} = 0$	(7h)
rows 2×3, tc	$V_{cd}^* V_{td} + V_{cs}^* V_{ts} + V_{cb}^* V_{tb} = 0$	(7i)
columns 1×2, sd	$V_{ud} V_{us}^* + V_{cd} V_{cs}^* + V_{td} V_{ts}^* = 0$	(7j)
columns 1×3, bd	$V_{ud} V_{ub}^* + V_{cd} V_{cb}^* + V_{td} V_{tb}^* = 0$	(7k)
columns 2×3, bs	$V_{us} V_{ub}^* + V_{cs} V_{cb}^* + V_{ts} V_{tb}^* = 0$	(7l)

An arbitrary phase for the whole matrix cancels in $\mathbf{V^+V}$. A phase common to all elements in a line (column), corresponding to arbitrary phases between u, c, t (d, s, b) will vanish in eqns. 7j–l (7g–i) and become a common factor in eqns. 7g–i (7j–l).

Dividing (7k) by $A\lambda^3 \approx -V_{cd}V_{cb}^*$ yields the unitary triangle[2] as shown in Fig. 1a. In the Wolfenstein approximation, it corresponds to

$$(\rho + \mathrm{i}\eta) - 1 + (1 - \rho - \mathrm{i}\eta) = 0 \qquad (8)$$

A second one from (7h) is shown in Fig. 1b. Dividing by $A\lambda^3 \approx -V_{us}^* V_{ts}$ and using the approximation $V_{ud} \approx 1$ gives the same triangle (8). A closer look, however, reveals slightly different lengths and angles to $\mathcal{O}(\lambda^2)$.

The angles[3] of the unitarity triangles bd and tu (7k and h) in Fig. 1 are defined by[4]

$$\mathrm{e}^{\mathrm{i}\alpha} = -\frac{V_{td} V_{ub} V_{ud}^* V_{tb}^*}{|V_{td} V_{ub} V_{ud} V_{tb}|}$$

$$\mathrm{e}^{\mathrm{i}\beta} = -\frac{V_{td}^* V_{cb}^* V_{cd} V_{tb}}{|V_{td} V_{cb} V_{cd} V_{tb}|} \approx \mathrm{e}^{\mathrm{i}\beta'} = -\frac{V_{td}^* V_{us}^* V_{ts} V_{ud}}{|V_{td} V_{us} V_{ts} V_{ud}|}$$

$$\mathrm{e}^{\mathrm{i}\gamma} = -\frac{V_{ub}^* V_{cd}^* V_{cb} V_{ud}}{|V_{ub} V_{cd} V_{cb} V_{ud}|} \approx \mathrm{e}^{\mathrm{i}\gamma'} = -\frac{V_{ub}^* V_{ts}^* V_{us} V_{tb}}{|V_{ub} V_{ts} V_{us} V_{tb}|}$$

These are rephasing invariant expressions, hence the angles resemble physical quantities independent of the CKM parametrization. It was first emphasized

---

[2] this geometric interpretation has been pointed out by Bjorken $\sim$ 1986; its first documentation in printed form is in ref. 16 and more general in ref. 17.
[3] Another naming convention is $\phi_1 = \beta$, $\phi_2 = \alpha$ and $\phi_3 = \gamma$.
[4] in the complex plane, the angle $\alpha - \beta$ between two vectors $A = a\mathrm{e}^{\mathrm{i}\alpha}$ and $B = b\mathrm{e}^{\mathrm{i}\beta}$ is given by $\mathrm{e}^{\mathrm{i}(\alpha-\beta)} = AB^*/|AB|$ and $\sin(\alpha - \beta) = \mathrm{Im}(AB^*)/|AB| = (AB^* - A^*B)/(2\mathrm{i}|AB|)$.

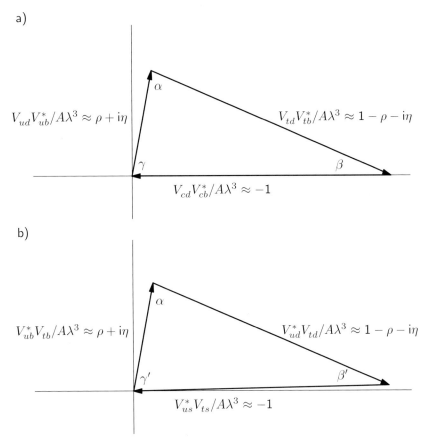

Fig. 1. Unitarity triangles bd and tu in the complex plane, corresponding to **a**: (7k) and **b**: (7h), respectively. Up to corrections of $\mathcal{O}(\lambda^4)$ the top points are (ρ, η) in **(b)**, but $(\bar{\rho} = [1 - \frac{\lambda^2}{2}]\rho, \bar{\eta} = [1 - \frac{\lambda^2}{2}]\eta)$ in **(a)**, and the rightmost points are $(1, 0)$ in **(a)**, but $(1 - \lambda^2[\frac{1}{2} - \rho], \lambda^2\eta)$ in **(b)**. The angles are related via $\gamma - \gamma' = \beta' - \beta \approx \lambda^2 \eta$. Changing the phase convention for the CKM matrix will rotate the triangles in the complex plain, but their shape is invariant under those transformations.

by Jarlskog [18], that CP violation can be described via a rephasing invariant quantity

$$J = \pm \operatorname{Im} V_{ij} V_{kl} V_{il}^* V_{kj}^* \approx A^2 \lambda^6 \eta$$

which is up to a sign independent of i, j, k, l, provided $i \neq k$, $j \neq l$.

$$\begin{aligned}
J &= \operatorname{Im}(V_{ud}V_{cs}V_{us}^*V_{cd}^*) &&= -\operatorname{Im}(V_{ud}V_{cb}V_{ub}^*V_{cd}^*) &&= -\operatorname{Im}(V_{ud}V_{ts}V_{us}^*V_{td}^*) \\
&= \operatorname{Im}(V_{ud}V_{tb}V_{ub}^*V_{td}^*) &&= -\operatorname{Im}(V_{us}V_{cd}V_{ud}^*V_{cs}^*) &&= \operatorname{Im}(V_{us}V_{cb}V_{ub}^*V_{cs}^*) \\
&= \operatorname{Im}(V_{us}V_{td}V_{ud}^*V_{ts}^*) &&= -\operatorname{Im}(V_{us}V_{tb}V_{ub}^*V_{ts}^*) &&= \operatorname{Im}(V_{ub}V_{cd}V_{ud}^*V_{cb}^*) \\
&= -\operatorname{Im}(V_{ub}V_{cs}V_{us}^*V_{cb}^*) &&= -\operatorname{Im}(V_{ub}V_{td}V_{ud}^*V_{tb}^*) &&= \operatorname{Im}(V_{ub}V_{ts}V_{us}^*V_{tb}^*)
\end{aligned}$$

$$\begin{aligned}&= \mathrm{Im}(V_{cd}V_{ts}V_{cs}^*V_{td}^*) &&= -\mathrm{Im}(V_{cd}V_{tb}V_{cb}^*V_{td}^*) &&= -\mathrm{Im}(V_{cs}V_{td}V_{cd}^*V_{ts}^*)\\&= \mathrm{Im}(V_{cs}V_{tb}V_{cb}^*V_{ts}^*) && -\mathrm{Im}(V_{cb}V_{ud}V_{cd}^*V_{tb}^*) &&= -\mathrm{Im}(V_{cb}V_{ts}V_{cs}^*V_{tb}^*)\end{aligned}$$

These terms are all products of the type $\mathrm{Im}\, AB^* = |A||B|\,\mathrm{Im}\, e^{i(\arg A - \arg B)} = |A||B|\sin(\arg A - \arg B)$, which is twice the area of a triangle in the complex plain with sides A and B. The A and B here are sides of a unitarity triangle. The equality of these terms is easily seen, e.g. for the last line replacing d with s is equivalent to applying the unitarity condition (7i)

$$V_{td}V_{cd}^* = -V_{ts}V_{cs}^* - V_{tb}V_{cb}^*$$

which yields

$$\mathrm{Im}(V_{cb}V_{td}V_{cd}^*V_{tb}^*) = -\mathrm{Im}(V_{cb}V_{ts}V_{cs}^*V_{tb}^*) - \mathrm{Im}(V_{cb}V_{tb}V_{cb}^*V_{tb}^*)$$

and the last argument is real, i.e. $\mathrm{Im}(V_{cb}V_{tb}V_{cb}^*V_{tb}^*) = \mathrm{Im}\,|V_{cb}|^2|V_{tb}|^2 = 0$. Hence the areas of all six unitarity triangles defined by (7g–l) are equal and have the value $J/2$. This corresponds to an area $\approx \eta/2$ for the ones in Fig. 1, since their sides have been reduced by the factor $A\lambda^3$. As will be shown below, CP violating observables are typically proportional to the sine of the angles in unitarity triangles, like

$$\sin\gamma = \mathrm{Im}\, e^{i\gamma} = -\frac{\mathrm{Im}(V_{ub}^*V_{cd}^*V_{cb}V_{ud})}{|V_{ub}V_{cd}V_{cb}V_{ud}|} = -\frac{J}{|V_{ub}V_{cd}V_{cb}V_{ud}|}$$

and vanish for $J = 0$, i.e. if all triangles collapse into lines. If the non-trivial phase in the CKM matrix is 0 or π, the parameter η is 0 and hence $J = 0$. This would also be the case if two quarks of a given charge had the same mass, since then a rotation between these two flavours could be chosen that removes the phase factors, as can be seen in (3) where $\theta_{13} = 0$ would remove all terms with the phase δ_{13}.

All six unitarity triangles are shown approximately to scale in Fig. 2. Their angles – which are still largely unknown – can be determined using the standard parametrization (3) in a rewritten form

$$\mathbf{V} = \begin{pmatrix} |V_{ud}| & |V_{us}| & |V_{ub}|e^{-i\tilde\gamma}\\ -|V_{cd}|e^{i\phi_4} & |V_{cs}|e^{-i\phi_6} & |V_{cb}|\\ |V_{td}|e^{-i\tilde\beta} & -|V_{ts}|e^{i\phi_2} & |V_{tb}| \end{pmatrix} \qquad (9)$$

with $\tilde\gamma \equiv \delta_{13}$. Here, absolute values and phases are given as separate factors. The angles $\phi_2 \approx \eta\lambda^2$, $\phi_4 \approx \eta A^2\lambda^4$, and $\phi_6 \approx \eta A^2\lambda^6$ are all positive and very small and their subscript indicates the order in λ of their magnitude. The unitarity triangles in Fig. 1 have angles

$$\begin{aligned}\beta &= \tilde\beta + \phi_4, & \beta' &= \tilde\beta + \phi_2 = \beta + \phi_2 - \phi_4\\ \gamma &= \tilde\gamma - \phi_4, & \gamma' &= \tilde\gamma - \phi_2 = \gamma - \phi_2 + \phi_4\\ \alpha &= \pi - \tilde\beta - \tilde\gamma = \pi - \beta - \gamma = \pi - \beta' - \gamma'\end{aligned}$$

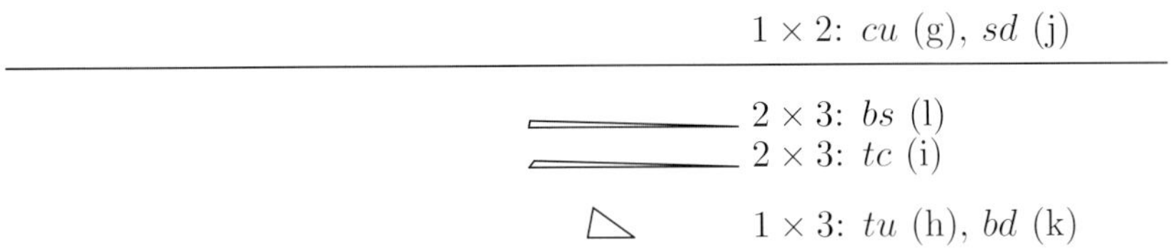

```
                              1 × 2: cu (g), sd (j)

                    ─────────  2 × 3: bs (l)
                    ─────────  2 × 3: tc (i)

                       △       1 × 3: tu (h), bd (k)
```

Fig. 2. The three types of unitarity triangles, approximately to scale (assuming some allowed values for ρ, η that differ from those assumed in Fig. 1). Their areas are equal.

In the Wolfenstein approximation, the unitarity relations read (all terms given to order λ^3 or, if this is still 0, [in brackets] to leading order)

$$-\lambda + \tfrac{1}{2}\lambda^3 + \lambda - \tfrac{1}{2}\lambda^3 + \quad [A^2\lambda^5(\rho + i\eta)] = 0 \quad (7\text{g}')$$
$$A\lambda^3(1 - \rho - i\eta) - \quad A\lambda^3 + \quad A\lambda^3(\rho + i\eta) = 0 \quad (7\text{h}')$$
$$[-A\lambda^4(1 - \rho - i\eta)] - \quad A\lambda^2 + \quad A\lambda^2 = 0 \quad (7\text{i}')$$
$$\lambda - \tfrac{1}{2}\lambda^3 - \lambda + \tfrac{1}{2}\lambda^3 - [A^2\lambda^5(1 - \rho - i\eta)] = 0 \quad (7\text{j}')$$
$$A\lambda^3(\rho + i\eta) - \quad A\lambda^3 + \quad A\lambda^3(1 - \rho - i\eta) = 0 \quad (7\text{k}')$$
$$[A\lambda^4(\rho + i\eta)] + \quad A\lambda^2 - \quad A\lambda^2 = 0 \quad (7\text{l}')$$

and define three pairs of unitarity triangles, 6 in total:

- (7h′) and (7k′) are the ones shown in Fig. 1 with three sides of similar length, all of order $A\lambda^3$. This is **"the unitarity triangle"**. The other ones are quite flat, and it will require very high precision to prove experimentally that they are not degenerate to a line. They are all shown to approximate scale in Fig. 2.

- (7i′) and (7l′) have two sides of length $A\lambda^2$ and one much shorter of order $A\lambda^4$. This limits the small angles, which are $\phi_2 + \phi_6$ and $\phi_2 - \phi_6$, respectively. They are close to the differences of angles in the large triangles $\gamma - \gamma' = \beta' - \beta = \phi_2 - \phi_4$.
  The other two angles are for (7i′) $\sim \beta$ and $\sim \pi - \beta$, and for (7l′) $\sim \gamma$ and $\sim \pi - \gamma$.

- (7g′) and (7j′) have two sides of length λ and one very much shorter of order $A^2\lambda^5$, with a small angle $\phi_4 - \phi_6$ and $\phi_4 + \phi_6$, respectively. Both are of order λ^4.
  The other two angles are for (7j′) $\sim \beta$ and $\sim \pi - \beta$, and for (7g′) $\sim \gamma$ and $\sim \pi - \gamma$.

Tiny differences between the two standard unitarity triangles are $\mathcal{O}(\lambda^2)$ corrections,

$$\begin{array}{l}
A\lambda^3(1 - \rho - i\eta) + \quad\quad - A\lambda^3 \quad\quad + \quad A\lambda^3(\rho + i\eta) = 0 \\
+A\lambda^5(\rho + i\eta - \tfrac{1}{2}) \quad + A\lambda^5(\tfrac{1}{2} - \rho - i\eta) \quad\quad + \mathcal{O}(\lambda^7) \\
\quad + \mathcal{O}(\lambda^7) \quad\quad\quad\quad + \mathcal{O}(\lambda^7)
\end{array} \quad (7\text{h}'')$$

$$\begin{array}{l}
A\lambda^3(\rho + i\eta) + \quad\quad - A\lambda^3 \quad\quad + A\lambda^3(1 - \rho - i\eta) = 0 \\
-\tfrac{1}{2}A\lambda^5(\rho + i\eta) \quad\quad + \mathcal{O}(\lambda^7) \quad\quad + \tfrac{1}{2}A\lambda^5(\rho + i\eta) \\
+ \mathcal{O}(\lambda^7) \quad\quad\quad\quad\quad\quad\quad\quad\quad\quad + \mathcal{O}(\lambda^7)
\end{array} \quad (7\text{k}'')$$

The angles in these two triangles can be estimated from experimental constraints on a 3×3 unitary CKM matrix, but all phase angles are only weakly constrained by these limits, and one of the aims of experiments designed to observe CP violation in B meson decays [19] is a first measurement, and ultimately a precise determination of their values. However, deviations from or extensions to the Standard Model may imply that the two triangles are dissimilar, or even that they are no (closed) triangles at all. Therefore, it is important to distinguish measurements of different parameters, even if they are expected to have identical or close values within the three family Standard Model.

2.1.2 Phases and Observables

The fact that phases of quark fields are unobservable numbers has been used to show that phases in the CKM matrix are not observables either, and there remains some arbitrariness in the parametrization for this matrix. Any valid CKM matrix is obtained from (9) with five independent arbitrary phase angles $\zeta_1 \ldots \zeta_5$ as

$$\mathbf{V} = \begin{pmatrix} |V_{ud}|e^{i\zeta_1} & |V_{us}|e^{i(\zeta_1+\zeta_2)} & |V_{ub}|e^{i(\zeta_1+\zeta_3-\tilde{\gamma})} \\ -|V_{cd}|e^{i(\zeta_4+\phi_4)} & |V_{cs}|e^{i(\zeta_4+\zeta_2-\phi_6)} & |V_{cb}|e^{i(\zeta_4+\zeta_3)} \\ |V_{td}|e^{i(\zeta_5-\tilde{\beta})} & -|V_{ts}|e^{i(\zeta_5+\zeta_2+\phi_2)} & |V_{tb}|e^{i(\zeta_5+\zeta_3)} \end{pmatrix} \tag{10}$$

The freedom to choose quark phases may be extended to antiquarks, with six more phases $\bar{\phi}_u, \bar{\phi}_c, \bar{\phi}_t, \bar{\phi}_d, \bar{\phi}_s, \bar{\phi}_b$. With the new quark states

$$q'_j = e^{i\phi_j} q, \quad \bar{q}'_j = e^{i\bar{\phi}_j} \bar{q}_j, \quad j = u, c, t, d, s, b$$

also the phase induced by the CP operation is changed. The transition

$$\text{CP} |q_j\rangle = e^{i\phi_{\text{CP}\,j}} |\bar{q}_j\rangle \quad \rightarrow \quad \text{CP} |q'_j\rangle = e^{i\phi'_{\text{CP}\,j}} |\bar{q}'_j\rangle$$

requires

$$\phi'_{\text{CP}\,j} = \phi_{\text{CP}\,j} + \phi_j - \bar{\phi}_j$$

This equation leaves $\phi'_{\text{CP}\,j}$ still completely undefined, since all three phases on the right-hand side are not observable, and therefore subject to arbitrary changes. It becomes meaningful, however, if it is applied to observables, like CP eigenvalues. Two CP eigenstates constructed from a meson and antimeson state with eigenvalues ± 1 are related accordingly:

$$|q_j \bar{q}_k\rangle \pm e^{i\phi_{\text{CP}\,jk}} |q_k \bar{q}_j\rangle = e^{-i(\phi_j + \bar{\phi}_k)} \left[|q'_j \bar{q}'_k\rangle \pm e^{i\phi'_{\text{CP}\,jk}} |q'_k \bar{q}'_j\rangle \right]$$

The new states $|q'_j \bar{q}'_k\rangle \pm e^{i\phi'_{\text{CP}\,jk}} |q'_k \bar{q}'_j\rangle$ have the same eigenvalues, and differ by an overall unobservable phase from the old ones.

The CP operation on a meson, e.g. the pseudoscalar B^0 meson $|\bar{b}d\rangle$, is

$$\text{CP}\,|B^0\rangle = e^{i\phi_{\text{CP}B}}|\bar{B}^0\rangle \tag{11}$$

where the phase factor $e^{i\phi_{\text{CP}B}} = \langle \bar{B}^0|\,\text{CP}\,|B^0\rangle$ depends on the parity of the bound-state wave function, and the chosen quark and antiquark phase convention.

Quark phase changes could be compensated by phase changes of the CKM matrix elements according to (2), leaving terms like

$$\langle q_j|V_{jk}|q_k\rangle$$

invariant. However, the phase of this matrix element is **not** an observable. Hence the choice of phases in the CKM matrix parametrization can be made **independent** of the choice of quark phases.

Phase conventions will also enter into relations among decay amplitudes. An amplitude for a weak decay $B^0 \to X$ via a single well defined process can be written as

$$A = \langle X|\mathbf{H}|B^0\rangle = \langle X|\mathbf{O}V|B^0\rangle \tag{12}$$

where V is a product of the appropriate CKM matrix elements and $\mathbf{O}$ is an operator describing the rest of the weak and possibly also subsequent strong interaction processes involved in the transition. Since strong interaction (also weak interaction except for nontrivial phases in V) are CP invariant, the charge conjugate mirror process $\bar{B}^0 \to \bar{X}$ has an amplitude

$$\begin{aligned}
\bar{A} &= \langle \bar{X}|\mathbf{H}|\bar{B}^0\rangle = \langle \bar{X}|\,\text{CP}^+\,\text{CP}\,\mathbf{H}\,\text{CP}^+\,\text{CP}\,|\bar{B}^0\rangle \\
&= e^{i\phi_{\text{CP}X}}\langle X|\,\text{CP}\,\mathbf{O}V\,\text{CP}^+\,e^{-i\phi_{\text{CP}B}}|B^0\rangle \\
&= e^{i(\phi_{\text{CP}X}-\phi_{\text{CP}B})}\langle X|\mathbf{O}V^*|B^0\rangle \\
&= e^{i(\phi_{\text{CP}X}-\phi_{\text{CP}B})}\frac{V^*}{V}A \tag{13}
\end{aligned}$$

where also

$$\frac{V^*}{V} = e^{-2i\arg V}$$

is just a phase. Especially, if X is a CP eigenstate with eigenvalue $\xi_X = \pm 1$,

$$\bar{A} = \xi_X e^{-i(\phi_{\text{CP}B}+2\arg V)}A \tag{14}$$

relates the two amplitudes, and the ratio $\bar{A}/A$ flips sign with the CP eigenvalue.

All physical observables must be independent of the choice of phases. This is the case if only absolute values of amplitudes are involved, but for interference terms the phase convention cancels often in a more subtle way. Some examples will be shown in the following chapters. On the other hand, expressions where the arbitrary phases are still present cannot be observables.

2.1.2.1 Reasonable Phase Conventions

Once the distinction of unobservable and observable phases is clear, it is reasonable to choose phases in a way that simplifies calculations.

So it is sensible to use only one phase in the CKM matrix as in (3), and not six as in (10). If a choice of phases were possible where all CKM matrix elements can be made real, also charged current weak interactions would not violate CP symmetry.

A natural choice for CP phases requires all $J^{PC} = 0^{-+}$ mesons to have $CP|X\rangle = -|\overline{X}\rangle$, fixing $\phi_{CPB} = \pi$. However, it has become fashionable to use the opposite sign convention, i.e. $\phi_{CPB} = 0$.

The appearance of an additional phase factor in $e^{i\phi_{CP} {}^{kj}} \langle \bar{q}_j | V_{jk}^* | \bar{q}_k \rangle$ can be avoided by the restriction $\bar{\phi}_j = -\phi_j$ for quark phase changes, and an appropriate phase convention which makes terms related by a CPT transformation relatively real.

2.1.3 More Parameters

The parametrization (4) can be rewritten as

$$\begin{pmatrix} 1 - \frac{\lambda^2}{2} - \frac{\lambda^4}{8} & \lambda & A\lambda^3(\rho - i\eta) \\ -\lambda - A^2\lambda^5(\rho + i\eta - \frac{1}{2}) & 1 - \frac{\lambda^2}{2} - (\frac{1}{8} + \frac{A}{2})\lambda^4 & A\lambda^2 \\ A\lambda^3[1 - (\bar{\rho} + i\bar{\eta})] & -A\lambda^2 - A\lambda^4(\rho + i\eta - \frac{1}{2}) & 1 - \frac{1}{2}A^2\lambda^4 \end{pmatrix}$$

using the new parameters

$$\bar{\rho} := \rho(1 - \frac{\lambda^2}{2}), \quad \bar{\eta} := \eta(1 - \frac{\lambda^2}{2}) \tag{15}$$

which can be used to write the bd unitarity triangle (7k) as

$$A\lambda^3(\bar{\rho} + i\bar{\eta}) + -A\lambda^3 + A\lambda^3(1 - \bar{\rho} - i\bar{\eta}) = 0 + \mathcal{O}(\lambda^7) \tag{7k'''}$$

'The tip shown in Fig. 1a has the coordinates $(\bar{\rho}, \bar{\eta})$. The sides are given by

$$R_u := \left| \frac{V_{ub}^* V_{ud}}{V_{cb}^* V_{cd}} \right| = \sqrt{\bar{\rho}^2 + \bar{\eta}^2}$$

and

$$R_t := \left| \frac{V_{td} V_{tb}^*}{V_{cb}^* V_{cd}} \right| = \sqrt{(1 - \bar{\rho})^2 + \bar{\eta}^2}.$$

The angles can then be expressed as

$$\sin 2\alpha = \frac{2\bar{\eta}[\bar{\eta}^2 + \bar{\rho}(\bar{\rho} - 1)]}{[\bar{\eta}^2 + (1 - \bar{\rho})^2][\bar{\eta}^2 + \bar{\rho}^2]}, \quad \sin 2\beta = \frac{2\bar{\eta}(1 - \bar{\rho})}{\bar{\eta}^2 + (1 - \bar{\rho})^2}$$

2.1.4 The CKM Matrix and Fermion Mass Generation

The quark mass generating part of the Lagrangian may be written as scalar $\bar{\psi}_L \psi_R + \bar{\psi}_R \psi_L$ couplings

$$\mathcal{L} = \ldots - \sum_{r,g,b} \begin{pmatrix} \bar{u} \\ \bar{c} \\ \bar{t} \end{pmatrix} \mathbf{M}_u \frac{1+\gamma_5}{2} \begin{pmatrix} u \\ c \\ t \end{pmatrix} - \sum_{r,g,b} \begin{pmatrix} \bar{d}' \\ \bar{s}' \\ \bar{b}' \end{pmatrix} \mathbf{M}_d \frac{1+\gamma_5}{2} \begin{pmatrix} d' \\ s' \\ b' \end{pmatrix} + \text{h.c.} \tag{16}$$

where the matrices $\mathbf{M}_u$ and $\mathbf{M}_d$ are applied to the eigenstates of the weak interaction, i.e. with diagonal transitions between (d', s', b') and (u, c, t) through the intermediate W boson. While for one triplet, here the (u, c, t) one, a diagonal mass matrix

$$\mathbf{M}_u = \begin{pmatrix} m_u & 0 & 0 \\ 0 & m_c & 0 \\ 0 & 0 & m_t \end{pmatrix}$$

is sufficient, the other triplet needs a more general one, since the weak partners of (u, c, t) are not the mass eigenstates. This matrix can be diagonalized via

$$\mathbf{M}_d = \mathbf{V} \begin{pmatrix} m_d & 0 & 0 \\ 0 & m_s & 0 \\ 0 & 0 & m_b \end{pmatrix} \mathbf{V}^\dagger$$

where the d-type quarks have to be transformed using the CKM matrix $\mathbf{V}$. In general, both u- and d-type quarks can have their own complex mass matrices, still leaving the W-mediated interaction diagonal. This moves the question of the origin of the CKM matrix elements into the realm of mass generation, which belongs still to the more "mysterious" parts [20] of the Standard Model. Within the Standard Model, the mass of fermions is generated via the couplings to the scalar Higgs field, or, more precisely, to the vacuum expectation value v of the single component of this field that is not eaten up by gauge symmetries:

$$m_i = \frac{c_{ii}}{\sqrt{2}} v$$

where c_{ii} is a coupling constant specific to quark i. The origin of these coupling constants is not understood.

If we employ the Higgs field as the source of particle masses, we must start with the most general couplings of the full four-component Higgs field ϕ to the quarks

$$\mathcal{L} = \ldots - \sum_{r,g,b} (\bar{u}''_R, \bar{c}''_R, \bar{t}''_R) \, \mathbf{C}''_u \, (-\bar{\phi}^0, \phi^-) \begin{pmatrix} \begin{pmatrix} u''_L \\ d''_L \end{pmatrix} \\ \begin{pmatrix} c''_L \\ s''_L \end{pmatrix} \\ \begin{pmatrix} t''_L \\ b''_L \end{pmatrix} \end{pmatrix}$$

$$-\sum_{r,g,b}(\bar{d}''_R, \bar{s}''_R, \bar{b}''_R)\,\mathbf{C}''_d(\phi^+,\phi^0)\begin{pmatrix}\begin{pmatrix}u''_L\\d''_L\end{pmatrix}\\\begin{pmatrix}c''_L\\s''_L\end{pmatrix}\\\begin{pmatrix}t''_L\\b''_L\end{pmatrix}\end{pmatrix} \qquad (17)$$

with arbitrary complex 3×3 matrices $\mathbf{C}''_u$, $\mathbf{C}''_d$ for an arbitrary basis of left- and righthanded quark fields. While the left-handed fields go in doublets and can only be transformed in pairs due to the SU(2) gauge invariance, we can arbitrarily transform the right-handed singlet fields.

There is no *a priori* restriction in the number of parameters in $\mathbf{C}''_u$ and $\mathbf{C}''_d$. However, these matrices can be reduced to a real, diagonal matrix $\mathbf{C}_u$, and a matrix $\mathbf{C}_d$ with a minimum number of parameters that correspond to the mass matrices $\mathbf{M} = \frac{v}{\sqrt{2}}\mathbf{C}$

$$\mathbf{M}_u = \frac{v}{\sqrt{2}}\mathbf{U}_{uR}\mathbf{C}''_u\mathbf{U}_L^\dagger, \quad \mathbf{M}_d = \frac{v}{\sqrt{2}}\mathbf{U}_{dR}\mathbf{C}''_d\mathbf{U}_L^\dagger \qquad (18)$$

via the three unitary transformations

$$\begin{pmatrix}u_R\\c_R\\t_R\end{pmatrix} = \mathbf{U}_{uR}\begin{pmatrix}u''_R\\c''_R\\t''_R\end{pmatrix}, \quad \begin{pmatrix}d'_R\\s'_R\\b'_R\end{pmatrix} = \mathbf{U}_{dR}\begin{pmatrix}d''_R\\s''_R\\b''_R\end{pmatrix}$$

$$\begin{pmatrix}u_L\\c_L\\t_L\end{pmatrix} = \mathbf{U}_L\begin{pmatrix}u''_L\\c''_L\\t''_L\end{pmatrix}, \quad \begin{pmatrix}d'_L\\s'_L\\b'_L\end{pmatrix} = \mathbf{U}_L\begin{pmatrix}d''_L\\s''_L\\b''_L\end{pmatrix}$$

There is only one $\mathbf{U}_L$ since the lefthanded quarks are weak iso-doublets that transform as one entity. Therefore, $\mathbf{C}_u$ and $\mathbf{C}_d$ cannot be diagonalized simultaneously. If $\mathbf{U}_{uR}$ and $\mathbf{U}_L$ are chosen to diagonalize $\mathbf{C}_u$ (which corresponds to a reduction of 18 real parameters to 3, the up-type quark masses, absorbing 9 in $\mathbf{U}_{uR}$ and 6 more in $\mathbf{U}_L$), the remaining $\mathbf{U}_{dR}$ can only be used to reduce the number of parameters of $\mathbf{C}_d$ from 18 real numbers to 7 (absorbing 9 in $\mathbf{U}_{dR}$ and 2 more in $\mathbf{U}_L$, while a common phase of all three $\mathbf{U}$ matrices is the last parameter of $\mathbf{U}_L$). The remaining 7 parameters correspond to the three down-type quark masses and the four non-trivial parameters of the CKM matrix.

These transformations isolate the mass term (16) from the Higgs coupling term (17) using the vacuum expectation value of the Higgs field $(\phi^+, \phi^0) = (0, \frac{v}{\sqrt{2}})$.

The exploration of the Higgs sector is the main motivation for the LHC storage ring, which is built at CERN and will start operation around 2006 [21]. We hope that the experiments will give new insights into the Higgs-quark couplings, which involve the quark masses and the parameters of the CKM matrix as 10 independent numbers that are not related within the theory.

2.2 Oscillation Phenomenology

An unstable meson can be described by the non-relativistic Schrödinger equation $i\partial_t \psi = (m - \frac{i}{2}\Gamma)\psi$, with the solution

$$|\psi\rangle = |\psi_0\rangle e^{-imt} e^{-\frac{1}{2}\Gamma t} \tag{19}$$

which reproduces the exponential law of radioactive decay, since $|\langle\psi_0|\psi\rangle|^2 = e^{-\Gamma t}$. Since it describes the decay of a particle by its "vanishing", the Hamiltonian $H = m - \frac{i}{2}\Gamma$ is not real (i.e. not hermitian).

The four meson pairs $K^0/\overline{K}^0$, $D^0/\overline{D}^0$, $B^0/\overline{B}^0$, and $B_s/\overline{B}_s$ can be described as decaying two-component quantum states obeying the Schrödinger equation

$$i\partial_t \psi = \mathbf{H}\psi$$

with a general, non-hermitian [22] Hamiltonian

$$\mathbf{H} = \mathbf{M} - \tfrac{i}{2}\mathbf{\Gamma} = \begin{pmatrix} m_{11} - \frac{i}{2}\Gamma_{11} & m_{12} - \frac{i}{2}\Gamma_{12} \\ m_{12}^* - \frac{i}{2}\Gamma_{12}^* & m_{22} - \frac{i}{2}\Gamma_{22} \end{pmatrix} \tag{20}$$

written as a sum of the **hermitian** matrices $\mathbf{M}$ and $\mathbf{\Gamma}$. Even when the $\frac{i}{2}\Gamma_{jj}$ of the decay is removed, $\mathbf{H}$ is not hermitian!

If the $B^0/\overline{B}^0$ system is taken as a representative to illustrate the behaviour of oscillating meson pairs, the indices 1 and 2 correspond to base vectors

$$|B^0\rangle = \begin{pmatrix} 1 \\ 0 \end{pmatrix} \quad \text{and} \quad |\overline{B}^0\rangle = \begin{pmatrix} 0 \\ 1 \end{pmatrix}$$

These states are orthogonal, i.e. $\langle B^0|\overline{B}^0\rangle = 0$ and they are assumed to be normalized, i.e. $\langle \overline{B}^0|\overline{B}^0\rangle = \langle B^0|B^0\rangle = 1$.

CPT invariance is assumed for all following calculations unless noted otherwise. This requires $m_{11} = m_{22} := m$ and $\Gamma_{11} = \Gamma_{22} := \Gamma$, reducing the number of real parameters of the Hamiltonian to six.

$$\mathbf{H} = \begin{pmatrix} H & H_{12} \\ H_{21} & H \end{pmatrix} = \begin{pmatrix} m - \frac{i}{2}\Gamma & m_{12} - \frac{i}{2}\Gamma_{12} \\ m_{12}^* - \frac{i}{2}\Gamma_{12}^* & m - \frac{i}{2}\Gamma \end{pmatrix} \tag{21}$$

CPT invariance is one of the indispensable premises of any relativistic field theory within or beyond the Standard Model. This CPT theorem was proven by Lüders [23] in 1954. The generalized phenomenology including CPT violation[5] will therefore not be considered here, but can be found in textbooks [24].

---

[5] Indeed CPT can be violated in modern string or d-brane theories, so it is definitely worthwhile to search for CPT violation in experiment.

The parametrization of the off diagonal elements is convenient for calculation, but it is still the most general case, since 4 real parameters suffice to describe any H_{12} and H_{21}:

$$m_{12} = \tfrac{1}{2}(H_{12} + H_{21}^*)$$
$$\operatorname{Re} m_{12} = \tfrac{1}{2}(\operatorname{Re} H_{12} + \operatorname{Re} H_{21})$$
$$\operatorname{Im} m_{12} = \tfrac{1}{2}(\operatorname{Im} H_{12} - \operatorname{Im} H_{21})$$
$$\Gamma_{12} = i(H_{12} - H_{21}^*)$$
$$\operatorname{Re}\Gamma_{12} = \operatorname{Im} H_{12} + \operatorname{Im} H_{21}$$
$$\operatorname{Im}\Gamma_{12} = \operatorname{Re} H_{12} - \operatorname{Re} H_{21}$$
$$\operatorname{Re} H_{12} = \operatorname{Re} m_{12} + \tfrac{1}{2}\operatorname{Im}\Gamma_{12}$$
$$\operatorname{Re} H_{21} = \operatorname{Re} m_{12} - \tfrac{1}{2}\operatorname{Im}\Gamma_{12}$$
$$\operatorname{Im} H_{12} = \operatorname{Im} m_{12} - \tfrac{i}{2}\operatorname{Re}\Gamma_{12}$$
$$\operatorname{Im} H_{21} = \operatorname{Im} m_{12} - \tfrac{1}{2}\operatorname{Re}\Gamma_{12}$$

The Schrödinger equation for meson pairs is a set of two coupled differential equations. Its solutions can be composed of single particle solutions (19) for the eigenstates of the Hamiltonian (21). Solving the eigenvalue problem $\det(\mathbf{H} - a \cdot \mathbf{1}) = (H-a)^2 - H_{12}H_{21} = 0$, one obtains two eigenstates with eigenvalues

$$a_{L,H} = H \mp \sqrt{H_{12}H_{21}}$$

or explicitly

$$a_L = m_L - \tfrac{i}{2}\Gamma_L = m - \tfrac{i}{2}\Gamma - \sqrt{\left(m_{12} - \tfrac{i}{2}\Gamma_{12}\right)\left(m_{12}^* - \tfrac{i}{2}\Gamma_{12}^*\right)} \quad (22)$$
$$a_H = m_H - \tfrac{i}{2}\Gamma_H = m - \tfrac{i}{2}\Gamma + \sqrt{\left(m_{12} - \tfrac{i}{2}\Gamma_{12}\right)\left(m_{12}^* - \tfrac{i}{2}\Gamma_{12}^*\right)}$$

where L, H stands for "light" and "heavy". It is immediately seen that m and Γ are the average mass $\tfrac{1}{2}(m_H + m_L)$ and width $\tfrac{1}{2}(\Gamma_H + \Gamma_L)$. The differences are

$$\frac{\Delta m}{2} = \frac{m_H - m_L}{2} = \operatorname{Re}\sqrt{H_{12}H_{21}} \quad (23)$$
$$\frac{\Delta\Gamma}{2} = \frac{\Gamma_H - \Gamma_L}{2} = -2\operatorname{Im}\sqrt{H_{12}H_{21}}$$
$$\frac{\Delta a}{2} = \frac{a_H - a_L}{2} = \sqrt{\left(m_{12} - \tfrac{i}{2}\Gamma_{12}\right)\left(m_{12}^* - \tfrac{i}{2}\Gamma_{12}^*\right)}$$
$$= \sqrt{|m_{12}|^2 - \tfrac{1}{4}|\Gamma_{12}|^2 - i\operatorname{Re}(m_{12}\Gamma_{12}^*)}$$

The connection between mass and lifetime (width) differences[6] and the off-diagonal elements in the mass matrix are showing up in these equations, especially $\Delta m = 0$ if $m_{12} = 0$ and $\Delta \Gamma = 0$ if $\Gamma_{12} = 0$. Squaring the last line leads to the useful relation

$$\Delta m \cdot \Delta \Gamma = 4 \operatorname{Re}(m_{12} \Gamma_{12}^*) \qquad (24)$$

which relates the sign of Δm and $\Delta \Gamma$ with the off-diagonal elements m_{12} and Γ_{12}.

It is convenient to define the dimensionless parameters

$$x = \frac{\Delta m}{\Gamma}, \quad y = \frac{\Delta \Gamma}{2\Gamma} = \frac{\Gamma_H - \Gamma_L}{\Gamma_H + \Gamma_L} = \frac{\Gamma - \Gamma_L}{\Gamma} = \frac{\tau_L - \tau_H}{\tau_L + \tau_H} \qquad (25)$$

where x is a non-negative real number, and y may only assume values between -1 and 1. It is an asymmetry parameter in the widths or, equivalently, in the lifetimes τ_L, τ_H.

The widths of the eigenstates are then

$$\Gamma_H = \Gamma \cdot (1 + y), \quad \Gamma_L = \Gamma \cdot (1 - y). \qquad (26)$$

The eigenvectors $|B_{L,H}\rangle = \begin{pmatrix} p \\ \pm q \end{pmatrix}$ are found by inserting (22) into

$$\mathbf{H}|B_{L,H}\rangle = a_{L,H}|B_{L,H}\rangle$$

$$\mathbf{H}\begin{pmatrix} p \\ q \end{pmatrix} = a_L \begin{pmatrix} p \\ q \end{pmatrix}$$

$$(m - \tfrac{i}{2}\Gamma)p + H_{12}\,q = (m - \tfrac{i}{2}\Gamma - \sqrt{H_{12}H_{21}})p$$

$$H_{12}\,q = -\sqrt{H_{12}H_{21}} \cdot p$$

giving the ratio

$$\eta_m := \frac{q}{p} = \frac{1-\epsilon}{1+\epsilon} = \frac{\sqrt{H_{21}H_{12}}}{-H_{12}} = \frac{-H_{21}}{\sqrt{H_{21}H_{12}}} = -2\frac{m_{12}^* - \tfrac{i}{2}\Gamma_{12}^*}{\Delta m - \tfrac{i}{2}\Delta\Gamma} \qquad (27)$$

which includes the phase ambiguity in $H_{12} = \langle B^0|\mathbf{H}|\overline{B}^0\rangle$ and is therefore usually defined to have $\operatorname{Re}\eta_m > 0$. This implies $\operatorname{Re} H_{12} < 0$ and

$$\eta_m = +\sqrt{\frac{H_{21}}{H_{12}}}$$

---

[6] Note that in the literature $\Delta\Gamma \leftrightarrow -\Delta\Gamma$ is often interchanged. This will change the sign of the dimensionless parameter y likewise.

and leads to

$$\epsilon = \frac{1 - \eta_m}{1 + \eta_m} = \frac{p - q}{p + q}$$
$$= \frac{\sqrt{H_{12}} - \sqrt{H_{21}}}{\sqrt{H_{12}} + \sqrt{H_{21}}} = \frac{H_{12} - H_{21}}{H_{12} + H_{21} + 2\sqrt{H_{12}H_{21}}}$$

Normalization requires $|p|^2 + |q|^2 = 1$, i.e.

$$p = \frac{1 + \epsilon}{\sqrt{2(1 + |\epsilon|^2)}} = \frac{1}{\sqrt{1 + |\eta_m|^2}}$$
$$q = \frac{1 - \epsilon}{\sqrt{2(1 + |\epsilon|^2)}} = \frac{\eta_m}{\sqrt{1 + |\eta_m|^2}}$$

and single particle eigenstates are described by one complex parameter η_m. This parameter[7] is defined only up to an arbitrary phase, and only $|\eta_m|$ is a measurable quantity. The value of the phase depends on conventions, one of them is the definition of the phase $\phi_{\text{CP}B} = \arg\langle\overline{B}^0|\,\text{CP}\,|B^0\rangle$. This makes also ϵ (sometimes also denoted $\bar{\epsilon}$ [26], $\tilde{\epsilon}$ [9] or ϵ_m for mixing-epsilon) an arbitrary quantity. The standard choice of the CKM matrix (3) and $\phi_{\text{CP}K} = 0$ make $|\epsilon|$ small in the $K^0/\overline{K}^0$ system, but a consistent convention $\phi_{\text{CP}B} = 0$ leaves it at $\mathcal{O}(0.1\ldots 1)$ in the $B^0/\overline{B}^0$ system.

The dependence on unphysical phases can also be seen if the parameters are expressed in terms of matrix elements as

$$\eta_m = \frac{\langle\overline{B}^0|B_\text{L}\rangle}{\langle B^0|B_\text{L}\rangle} = -\frac{\langle\overline{B}^0|B_\text{H}\rangle}{\langle B^0|B_\text{H}\rangle}$$

and

$$\epsilon = \frac{\langle B^0|B_\text{L}\rangle - \langle\overline{B}^0|B_\text{L}\rangle}{\langle B^0|B_\text{L}\rangle + \langle\overline{B}^0|B_\text{L}\rangle} = \frac{\langle B^0|B_\text{H}\rangle + \langle\overline{B}^0|B_\text{H}\rangle}{\langle B^0|B_\text{H}\rangle - \langle\overline{B}^0|B_\text{H}\rangle}$$

Obviously, relative phases between $|B_\text{L}\rangle$ and $|B_\text{H}\rangle$ are irrelevant, while relative phases between $|B^0\rangle$ and $|\overline{B}^0\rangle$ occur as a phase factor in η_m and in a more involved way in ϵ.

There exist different definitions of ϵ for the kaon system [25] that are independent of arbitrary phases, but not easily generalized to B mesons. However, convention independent parameters can always be defined if specific decays are involved. They can usually be expressed via the unitarity angles (see fig. 1) and will be given for the B and K systems at the appropriate places below.

We can use (27) to rewrite

$$H_{21} = -\eta_m \frac{\Delta m - \frac{i}{2}\Delta\Gamma}{2}, \quad H_{12} = -\frac{1}{\eta_m}\frac{\Delta m - \frac{i}{2}\Delta\Gamma}{2}$$

---

[7] η_m or $-\eta_m$ is sometimes called α in the literature, e.g. in [25,26].

and the original Hamiltonian as

$$\mathbf{H} = \begin{pmatrix} m - \frac{i}{2}\Gamma & -\frac{\Delta m - \frac{i}{2}\Delta\Gamma}{2\eta_m} \\ -\eta_m \frac{\Delta m - \frac{i}{2}\Delta\Gamma}{2} & m - \frac{i}{2}\Gamma \end{pmatrix} = \begin{pmatrix} m - \frac{i}{2}\Gamma & -\frac{\Gamma}{2}\frac{1}{\eta_m}(x - iy) \\ -\frac{\Gamma}{2}\eta_m(x - iy) & m - \frac{i}{2}\Gamma \end{pmatrix} \quad (28)$$

and the mass and flavour eigenstates are related by the equations

$$|B_{\mathrm{L}}\rangle = p|B^0\rangle + q|\overline{B}^0\rangle$$
$$|B_{\mathrm{H}}\rangle = p|B^0\rangle - q|\overline{B}^0\rangle \quad (29)$$
$$|B^0\rangle = \frac{1}{2p}(|B_{\mathrm{L}}\rangle + |B_{\mathrm{H}}\rangle)$$
$$|\overline{B}^0\rangle = \frac{1}{2q}(|B_{\mathrm{L}}\rangle - |B_{\mathrm{H}}\rangle)$$

Due to normalization $|p|^2 + |q|^2 = 1$ we may choose

$$|B^0\rangle = \frac{\sqrt{1 + |\eta_m|^2}}{2}(|B_{\mathrm{L}}\rangle + |B_{\mathrm{H}}\rangle)$$
$$|\overline{B}^0\rangle = \frac{\eta_m^* \sqrt{1 + |\eta_m|^2}}{2|\eta_m|}(|B_{\mathrm{L}}\rangle - |B_{\mathrm{H}}\rangle)$$

The choice of the phase for $|B^0\rangle$ is arbitrary, so there may be a common phase factor in the two latter equations.

The eigenstates for a Hamiltonian with $\Gamma_{12} \neq 0$ are **not orthogonal**:

$$\langle B_{\mathrm{H}}|B_{\mathrm{L}}\rangle = \delta_\epsilon := \begin{pmatrix} p^* \\ -q^* \end{pmatrix} \cdot \begin{pmatrix} p \\ q \end{pmatrix} = |p|^2 - |q|^2 = \frac{1 - |\eta_m|^2}{1 + |\eta_m|^2} = \frac{2\,\mathrm{Re}\,\epsilon}{1 + |\epsilon|^2} \quad (30)$$

In contrast to ϵ the real number δ_ϵ is an observable. The phase of $\langle B_{\mathrm{H}}|B_{\mathrm{L}}\rangle$ is no observable, though. The choice of the same coefficient p in the definitions of $|B_{\mathrm{L}}\rangle$ and $|B_{\mathrm{H}}\rangle$ is a convenient but arbitrary one. If we define instead $|B_{\mathrm{H}}\rangle = e^{i\phi_{HL}}(p|B^0\rangle - q|\overline{B}^0\rangle)$ which is an equally valid solution for B_{H}, we must define

$$\delta_\epsilon = \langle B_{\mathrm{H}}|B_{\mathrm{L}}\rangle \frac{\langle B_{\mathrm{L}}|B^0\rangle}{\langle B_{\mathrm{H}}|B^0\rangle}$$

to be a real number.

Useful exact relations are

$$\frac{|\eta_m|^2}{1 + |\eta_m|^2} = |q|^2 = \frac{1 - \delta_\epsilon}{2}, \quad \frac{1}{1 + |\eta_m|^2} = |p|^2 = \frac{1 + \delta_\epsilon}{2}, \quad |\eta_m|^2 = \frac{1 - \delta_\epsilon}{1 + \delta_\epsilon} \quad (30a)$$

for normalized p, q.

The deviation of $|\eta_m|$ from one (also called d_α [25]) is

$$|\eta_m| - 1 = \sqrt{\frac{1 - \delta_\epsilon}{1 + \delta_\epsilon}} - 1 \approx -\delta_\epsilon$$

2.2.1 Special Cases

We may illustrate the properties of the eigenstates by three special cases. First we assume $\Gamma_{12} = 0$, i.e.

$$\mathbf{H} = \begin{pmatrix} m - \frac{i}{2}\Gamma & m_{12} \\ m_{12}^* & m - \frac{i}{2}\Gamma \end{pmatrix} \tag{31}$$

where the off-diagonal elements are from an Hermitian operator. Then we obtain

$$\Delta m = 2|m_{12}|$$
$$\Delta \Gamma = 0$$
$$\eta_m = -\frac{m_{12}^*}{|m_{12}|}$$
$$|\eta_m| = 1, \quad \delta_\epsilon = 0$$

As a second case we assume $m_{12} = 0$, i.e.

$$\mathbf{H} = \begin{pmatrix} m - \frac{i}{2}\Gamma & H_{12} \\ -H_{12}^* & m - \frac{i}{2}\Gamma \end{pmatrix} \tag{32}$$

Then

$$\Delta m = 0$$
$$\Delta \Gamma = 4|H_{12}|$$
$$\eta_m = -\frac{iH_{12}^*}{|H_{12}|}$$
$$|\eta_m| = 1, \quad \delta_\epsilon = 0$$

In both cases, η_m is a simple phase factor. Since it includes unphysical phases, only its phase relation to other amplitudes is relevant.

Finally, for real m_{12}, Γ_{12} the term $\frac{\Delta a}{2} = \sqrt{(m_{12} - \frac{i}{2}\Gamma_{12})(m_{12}^* - \frac{i}{2}\Gamma_{12}^*)}$ is replaced by $(m_{12} - \frac{i}{2}\Gamma_{12})$ and in the phase convention with $m_{12} < 0$ we have

$$\Delta m = 2|m_{12}| = -2m_{12}$$
$$\Delta \Gamma = 2|\Gamma_{12}|$$
$$\eta_m = +1$$
$$\delta_\epsilon = 0$$
$$|B^0\rangle = \frac{1}{\sqrt{2}}\left(|B_\mathrm{L}\rangle + |B_\mathrm{H}\rangle\right)$$
$$|\overline{B}^0\rangle = \frac{1}{\sqrt{2}}\left(-|B_\mathrm{L}\rangle + |B_\mathrm{H}\rangle\right)$$

2.2.2 Time Evolution

For an arbitrary initial state

$$|\psi(0)\rangle = b_{\rm H}|B_{\rm H}\rangle + b_{\rm L}|B_{\rm L}\rangle = a|B^0\rangle + \bar{a}|\overline{B}^0\rangle$$

where the amplitudes are related via

$$b_{\rm L,H} = \frac{1}{2}\left(\frac{a}{p} \pm \frac{\bar{a}}{q}\right) = \frac{a \pm \bar{a}/\eta_m}{2p}$$
$$a = p(b_{\rm L} + b_{\rm H}), \quad \bar{a} = q(b_{\rm L} - b_{\rm H})$$

its time evolution may be described using a scaled time variable

$$T := \Gamma t \tag{33}$$

where Γ is the average width of the eigenstates $B_{\rm H}$ and $B_{\rm L}$. These states have a simple exponential development with time. Their masses $m_{\rm H,L} = m \pm x\frac{\Gamma}{2}$ and widths $\Gamma_{\rm H,L} = \Gamma(1 \pm y)$ can be expressed with the dimensionless parameters x and y defined in (25).

$$\begin{aligned}|\psi(t)\rangle &= b_{\rm H} e^{-{\rm i}(m_{\rm H} - {\rm i}\Gamma_{\rm H}/2)\,t}|B_{\rm H}\rangle + b_{\rm L} e^{-{\rm i}(m_{\rm L} - {\rm i}\Gamma_{\rm L}/2)\,t}|B_{\rm L}\rangle \\ &= e^{-{\rm i}mt - T/2}\Bigg[\frac{e^{{\rm i}(x-{\rm i}y)T/2} + e^{-{\rm i}(x-{\rm i}y)T/2}}{2}(a|B^0\rangle + \bar{a}|\overline{B}^0\rangle) \\ &\quad + \frac{e^{{\rm i}(x-{\rm i}y)T/2} - e^{-{\rm i}(x-{\rm i}y)T/2}}{2}\left(\frac{\bar{a}}{\eta_m}|B^0\rangle + a\eta_m|\overline{B}^0\rangle\right)\Bigg] \quad (34{\rm a}) \\ &= e^{-{\rm i}mt-T/2}\Bigg[(a|B^0\rangle + \bar{a}|\overline{B}^0\rangle)\cos(x-{\rm i}y)\frac{T}{2} \\ &\quad + {\rm i}\left(\frac{\bar{a}}{\eta_m}|B^0\rangle + a\eta_m|\overline{B}^0\rangle\right)\sin(x-{\rm i}y)\frac{T}{2}\Bigg] \quad (34{\rm b})\end{aligned}$$

Starting with pure B^0 mesons at $T = 0$ corresponds to $\bar{a} = 0$ and

$$|\psi(t)\rangle = a e^{-{\rm i}mt-T/2}\left[\cos(x-{\rm i}y)\frac{T}{2}|B^0\rangle + {\rm i}\eta_m \sin(x-{\rm i}y)\frac{T}{2}|\overline{B}^0\rangle\right] \tag{35}$$

Starting with pure $\overline{B}^0$ mesons at $T = 0$ is described by replacing $\eta_m \longleftrightarrow 1/\eta_m$. This case corresponds to $a = 0$ and

$$|\psi(t)\rangle = \bar{a} e^{-{\rm i}mt-T/2}\left[\cos(x-{\rm i}y)\frac{T}{2}|\overline{B}^0\rangle + \frac{\rm i}{\eta_m}\sin(x-{\rm i}y)\frac{T}{2}|B^0\rangle\right] \tag{36}$$

The numbers of B^0 and $\bar{B}^0$ at time T for N_0 pure B^0 mesons at $T=0$ are[8]

$$\begin{aligned}
N_{B^0}(t) &= N_0 |\langle B^0 | \psi(t, \bar{a}=0, a=1)\rangle|^2 \\
&= N_0 \frac{e^{-T}}{2}(\cosh yT + \cos xT) \\
N_{\bar{B}^0}(t) &= N_0 |\langle \bar{B}^0 | \psi(t, \bar{a}=0, a=1)\rangle|^2 \\
&= N_0 |\eta_m|^2 \frac{e^{-T}}{2}(\cosh yT - \cos xT)
\end{aligned} \quad (37)$$

and the numbers of B^0 and $\bar{B}^0$ at time T for $\bar{N}_0$ pure $\bar{B}^0$ mesons at $T=0$ are

$$\begin{aligned}
\bar{N}_{B^0}(t) &= \bar{N}_0 |\langle B^0 | \psi(t, \bar{a}=1, a=0)\rangle|^2 \\
&= \bar{N}_0 \frac{1}{|\eta_m|^2} \frac{e^{-T}}{2}(\cosh yT - \cos xT) \\
\bar{N}_{\bar{B}^0}(t) &= \bar{N}_0 |\langle \bar{B}^0 | \psi(t, \bar{a}=1, a=0)\rangle|^2 \\
&= \bar{N}_0 \frac{e^{-T}}{2}(\cosh yT + \cos xT)
\end{aligned} \quad (38)$$

These numbers, however, can not be observed. What is accessible by experiment is only the rate of decays to flavour specific final states X and $\bar{X}$ at a given time T. These decay modes are often called **tagging modes**, since they serve as a "tag" to indicate the flavour of the mother particle at decay time. The rates can be obtained from (35) by multiplying with $\langle X|\mathbf{H}$ or $\langle \bar{X}|\mathbf{H}$, respectively, to obtain the amplitudes. They are converted into rates

$$\begin{aligned}
\dot{N}_{B^0 \to X}(t) &= N_0 \int \mathrm{dPS}\, |\langle X|\mathbf{H}|\psi(t,\bar{a}=0)\rangle|^2 \\
&= \tfrac{1}{2} N_0 e^{-T} \Gamma_X (\cosh yT + \cos xT) \\
\dot{N}_{\bar{B}^0 \to \bar{X}}(t) &= N_0 \int \mathrm{dPS}\, |\langle \bar{X}|\mathbf{H}|\psi(t,\bar{a}=0)\rangle|^2 \\
&= \tfrac{1}{2} N_0 |\eta_m|^2 e^{-T} \Gamma_X (\cosh yT - \cos xT)
\end{aligned} \quad (39)$$

where

$$\Gamma_X = \int \mathrm{dPS}\, |\langle X|\mathbf{H}|B^0\rangle|^2 = \int \mathrm{dPS}\, |\langle \bar{X}|\mathbf{H}|\bar{B}^0\rangle|^2$$

is the partial width for a non-oscillating meson. It agrees in value for the two CP conjugate processes if the amplitudes differ only by one phase factor.

---

[8] for example, $\cos u = \tfrac{1}{2}(e^{iu} + e^{-iu})$, $(\cos u)^* = \tfrac{1}{2}(e^{iu^*} + e^{-iu^*})$, therefore
$|\cos u|^2 = \dfrac{1}{4}\left(e^{i(u+u^*)} + e^{i(u-u^*)} + e^{i(u^*-u)} + e^{-i(u+u^*)} \right)$
$= \tfrac{1}{2}(\cos 2\operatorname{Re} u + \cosh 2\operatorname{Im} u)$

Integrating over all times the total number of decays for initial B^0 mesons are

$$N_{B^0 \to X} = \int_0^\infty \dot N_{B^0 \to X}(t)\,dt = N_0 \frac{\Gamma_X}{\Gamma}\left[\frac{1}{2(1-y^2)} + \frac{1}{2(1+x^2)}\right]$$

$$N_{\overline{B}^0 \to \overline{X}} = \int_0^\infty \dot N_{\overline{B}^0 \to \overline{X}}(t)\,dt = N_0 \frac{\Gamma_X}{\Gamma}\left[\frac{|\eta_m|^2}{2(1-y^2)} - \frac{|\eta_m|^2}{2(1+x^2)}\right]$$

The corresponding numbers for initial $\overline{B}^0$ mesons are obtained with the replacement $\eta_m \to 1/\eta_m$. If we ignore CP violating effects in the oscillation, i.e. for $|\eta_m| = 1$, we can define a meaningful branching fraction as

$$\begin{aligned}\mathcal{B}(B^0 \to X) &= \frac{1}{N_0} \int_0^\infty [\dot N_{B^0 \to X}(t) + \dot N_{\overline{B}^0 \to \overline{X}}(t)]\,dt \\ &= \frac{\Gamma_X}{\Gamma(1-y^2)} = \frac{1}{2}\frac{\Gamma_X}{\Gamma_H} + \frac{1}{2}\frac{\Gamma_X}{\Gamma_L}\end{aligned}$$

which agrees with $\mathcal{B}(\overline{B}^0 \to \overline{X})$ defined accordingly for the same number N_0 of $\overline{B}^0$ mesons at $t = 0$.

2.2.3 Mechanical Analogon

Equation (21) characterizes also the mechanical system of two coupled pendula of the same length: Without coupling, they are both described by an oscillation frequency m and a damping constant Γ. They correspond to the meson X^0 and its antiparticle $\overline{X}^0$.

If they are coupled by a spring whose elasticity is proportional to a **non-negative real** number m_{12}, and a **non-negative real** damping constant Γ_{12}, the solutions correspond to a "long-lived" (= low damping), "light" (= low frequency) eigenstate where the pendula oscillate strictly in phase, and a "short-lived" (= high damping), "heavy" (= high frequency) eigenstate where one pendulum oscillates as a mirror image of the other, i.e. with phase difference 180°. The differences in frequency and damping are $\Delta m = 2m_{12}$ and $\Delta\Gamma = 2\Gamma_{12}$, respectively.

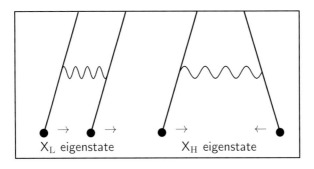

When one pendulum is excited, it will slowly transfer its energy to the other and back. This beating corresponds to the oscillation between a meson X^0 and its antiparticle $\bar{X}^0$. The beat frequency is $2\pi f_{12} = \Delta m$. For the mechanical system, the frequencies m, Δm, and the damping constants Γ, $\Delta\Gamma$ can all be measured.

For mesons, the oscillating part e^{-imt} in (19) is an unobservable phase factor (the mass can, of course, still be measured from kinematics), but in meson antimeson oscillation a mass difference can actually be observed as a frequency! The $B^0\bar{B}^0$ oscillation beat has $2\pi f \approx 0.49/\mathrm{ps}$ or a frequency of $f \approx 78\,\mathrm{GHz}$.

Due to the restriction of m_{12} and Γ_{12} to non-negative real values, this system has always $\Delta m\Delta\Gamma \geq 0$ (in contrast to the oscillating mesons), and can also not simulate CP violation since there are no non-trivial phases.

2.2.4 Standard Model Predictions

The Hamiltonian (21) can be obtained using

$$\mathbf{H} = \mathbf{H}_0 + \mathbf{H}_w$$

where $\mathbf{H}_0$ is the strong and electromagnetic Hamiltonian

$$\mathbf{H}_0 = \begin{pmatrix} E_0 & 0 \\ 0 & E_0 \end{pmatrix}$$

which has the stable flavour eigenstates B^0 and $\bar{B}^0$, and $\mathbf{H}_w$ is the weak interaction perturbation. The Wigner-Weisskopf approximation for small $\mathbf{H}_w$ leads to [27]

$$H_{jk} = H_{0\,jk} + \langle j|\mathbf{H}_w|k\rangle \\ + \sum_X \mathcal{P}\int \mathrm{dPS}\, \langle j|\mathbf{H}_w|X\rangle\langle X|\mathbf{H}_w|k\rangle \left[\frac{1}{E_0 - E_X} - i\pi\delta(E_0 - E_X)\right] \quad (40)$$

where the sum runs over all multiparticle states X which are eigenstates of $\mathbf{H}_0$, and $\mathcal{P}$ denotes the principal value of the integral. The mass (hermitian) and decay (antihermitian) parts defined by (20) are

$$m_{jk} = \tfrac{1}{2}(H_{jk} + H^*_{kj}) = E_0\,\delta_{jk} + \langle j|\mathbf{H}_w|k\rangle + \sum_X \mathcal{P}\int \mathrm{dPS}\, \frac{\langle j|\mathbf{H}_w|X\rangle\langle X|\mathbf{H}_w|k\rangle}{E_0 - E_X}$$

and

$$\Gamma_{jk} = \mathrm{i}(H_{jk} - H^*_{kj}) = 2\pi \sum_X \int \mathrm{dPS}\, \langle j|\mathbf{H}_w|X\rangle\langle X|\mathbf{H}_w|k\rangle\, \delta(E_0 - E_X)$$

The off-diagonal elements $H_{12,21}$ have non-zero contributions in the sum from states X which can be reached in weak decays of both B^0 and $\overline{B}^0$. In contrast to the neutral kaon system, for $B^0/\overline{B}^0$ these are only a small fraction of all B decays, and they contribute with alternating signs. Therefore $H_{12,21}$ are dominated by the leading term $\langle B^0|\mathbf{H}_w|\overline{B}^0\rangle$ which corresponds to the box diagrams

Their evaluation using the Feynman rules is usually simplified in two ways. The inner fermions in the loop may be u-, c- or t-quarks, leading to a total of 9 terms to be summed up (Π = propagator):

$$\begin{aligned}
\mathcal{M} &\sim V_{tb}V_{td}^*\Pi_t \cdot V_{tb}V_{td}^*\Pi_t & tt \\
&+ V_{cb}V_{cd}^*\Pi_c \cdot V_{tb}V_{td}^*\Pi_t & ct \\
&+ V_{ub}V_{ud}^*\Pi_u \cdot V_{tb}V_{td}^*\Pi_t & ut \\
&+ V_{tb}V_{td}^*\Pi_t \cdot V_{cb}V_{cd}^*\Pi_c & tc \\
&+ V_{tb}V_{td}^*\Pi_t \cdot V_{ub}V_{ud}^*\Pi_u & tu \\
&+ V_{cb}V_{cd}^*\Pi_c \cdot V_{cb}V_{cd}^*\Pi_c & cc \\
&+ V_{ub}V_{ud}^*\Pi_c \cdot V_{ub}V_{ud}^*\Pi_u & uu \\
&+ V_{cb}V_{cd}^*\Pi_c \cdot V_{ub}V_{ud}^*\Pi_u & cu \\
&+ V_{ub}V_{ud}^*\Pi_u \cdot V_{cb}V_{cd}^*\Pi_c & uc
\end{aligned}$$

Setting $m_c = m_u = 0$ one can play a trick similar to the GIM mechanism, since this yields $\Pi_c = \Pi_u = \Pi_0$ and

$$\begin{aligned}
\mathcal{M} &\sim V_{tb}V_{td}^*\Pi_t \cdot V_{tb}V_{td}^*\Pi_t \\
&+ (V_{cb}V_{cd}^* + V_{ub}V_{ud}^*)\Pi_0 \cdot V_{tb}V_{td}^*\Pi_t \\
&+ V_{tb}V_{td}^*\Pi_t \cdot (V_{cb}V_{cd}^* + V_{ub}V_{ud}^*)\Pi_0 \\
&+ (V_{cb}V_{cd}^* + V_{ub}V_{ud}^*)\Pi_0 \cdot (V_{cb}V_{cd}^* + V_{ub}V_{ud}^*)\Pi_0
\end{aligned}$$

Using further unitarity (7k) this simplifies to

$$\begin{aligned}
\mathcal{M} &\sim (V_{td}V_{tb}^*)^2[\Pi_t\Pi_t - \Pi_0\Pi_t - \Pi_t\Pi_0 + \Pi_0\Pi_0] \\
&= (V_{td}V_{tb}^*)^2[\Pi_t - \Pi_0][\Pi_t - \Pi_0]
\end{aligned}$$

i.e. the sum corresponds to a single matrix element with inner fermions "$t-u$" where u represents a massless quark.

The second simplification is the vacuum insertion which is corrected for by the *bag parameter* B_B:

$$\langle B^0|J_\mu J^\mu|\overline{B}^0\rangle = \sum_X \langle B^0|J_\mu|X\rangle \langle X|J^\mu|\overline{B}^0\rangle \qquad (41)$$
$$= B_B \cdot \langle B^0|J_\mu|0\rangle \langle 0|J^\mu|\overline{B}^0\rangle = B_B f_B^2 p_\mu p^\mu$$

With these simplifications, the box diagrams give approximately [28]

$$H_{12} = \langle B^0|\mathbf{H}|\overline{B}^0\rangle$$
$$\approx m_{12} = -\frac{G_F^2}{12\pi^2} \mathrm{e}^{-\mathrm{i}\phi_{\mathrm{CPB}}} V_{tb}^2 V_{td}^{*2} m_W^2\, m_B\, [f_B^2\, B_B] \cdot [S(m_t^2/m_W^2) \cdot \eta_{\mathrm{QCD}}] \qquad (42)$$

The CP phase is introduced during the evaluation of the hadronic part of the matrix element. The Inami–Lim function

$$S(x) = x\left[\frac{1}{4} + \frac{9}{4(1-x)} - \frac{3}{2(1-x)^2} - \frac{3x^2 \ln x}{2(1-x)^3}\right] \qquad (43)$$

from the loop [29] is to lowest order a factor m_t^2/m_W^2. An evaluation of the product $S(m_t^2/m_W^2)$ and η_{QCD} within a consistent renormalization scheme yields $S \approx 2.3$, $\eta_{\mathrm{QCD}} \approx 0.55$ [30].

The hadronic part of the matrix element is approximated by

$$\langle B^0|J_\mu J^\mu|\overline{B}^0\rangle = \sum_X \langle B^0|J_\mu|X\rangle \langle X|J^\mu|\overline{B}^0\rangle \qquad (44)$$
$$= B_B \cdot \langle B^0|J_\mu|0\rangle \langle 0|J^\mu|\overline{B}^0\rangle = B_B f_B^2 p_\mu p^\mu$$

where f_B is the B decay constant, and B_B accounts for the corrections to the vacuum insertion approximation. A big uncertainty is the product $f_B^2\, B_B$, where the most reliable calculations now come from lattice gauge theory [31] with values around $f_B\sqrt{B_B} \approx (230 \pm 40)\,\mathrm{MeV}$.

In this approximation, we have for the B system

$$\Delta m = 2|m_{12}|$$

which can be used to determine $|V_{td}|$ (since $V_{tb} = 1$) from experimental results on $B^0/\overline{B}^0$ mixing. The eigenstates are determined by

$$\eta_m = -\frac{m_{12}^*}{|m_{12}|} = \mathrm{e}^{\mathrm{i}\phi_{\mathrm{CPB}}} \frac{V_{tb}^{*2} V_{td}^2}{|V_{tb}^2 V_{td}^2|} = \mathrm{e}^{\mathrm{i}(\phi_{\mathrm{CPB}} - 2\tilde{\beta})} \qquad (45)$$

with $-\tilde{\beta} = \arg V_{tb}^* V_{td}$. This phase depends on the CKM parametrization and is – like the CP phase – not an observable: if we use the more general parametrization (10) we have instead $\arg V_{tb}^* V_{td} = -\tilde{\beta} - \zeta_3$ with an arbitrary

ζ_3 (the arbitrariness cancels only in physical observables, which include decay amplitudes with further CKM elements and a CP phase). The corresponding

$$\epsilon = -\mathrm{i}\frac{\sin\arg\eta_m}{1+\cos\arg\eta_m} = -\mathrm{i}\tan\frac{\arg\eta_m}{2} \qquad (46)$$

is purely imaginary, i.e. $\mathrm{Re}\,\epsilon = 0$ and therefore $\delta_\epsilon = 0$. In the standard parametrization and for $\phi_{CPB} = 0$, it is $\epsilon = \mathrm{i}\tan\tilde{\beta}$.

Within the same framework, for the $B_s/\overline{B}_s$ system

$$\eta_{ms} = \mathrm{e}^{\mathrm{i}(\phi_{CPB_s}+2\phi_2)} \qquad (47)$$

It must be emphasized, however, that there exist common final states for all four meson pairs, and Γ_{12} never vanishes completely, leaving always a small δ_ϵ, and also a small $\Delta\Gamma$. Within the Standard Model Γ_{12} can be approximated by the absorptive part of the box diagram, corresponding to a quark representation of the final states. This is a poor approximation to light hadronic final states which are dominating in the $K/\overline{K}$ system, and may still change the prediction for $B/\overline{B}$ considerably. The box calculation yields [28]

$$\Gamma_{12} \approx -m_{12} \cdot \frac{3\pi}{2S(m_t^2/m_W^2)} \frac{m_b^2}{m_W^2}\left[1 + \frac{8}{3}\frac{m_c^2}{m_b^2}\frac{V_{cb}V_{cd}^*}{V_{tb}V_{td}^*} + \mathcal{O}\left(\frac{m_c^4}{m_b^4}\right)\right] \qquad (48)$$

and $\Delta\Gamma$ and Δm have **opposite signs**[9]. The ratio can be estimated using (24) to be

$$\frac{\Delta\Gamma}{\Delta m} = \frac{2y}{x} \approx -\frac{3\pi}{2}\frac{m_b^2}{m_t^2} \sim -\frac{1}{250} \qquad (49)$$

This ratio applies to both the B^0 and B_s systems.

To leading order in Γ_{12}/m_{12} equation (27) yields

$$|\eta_m| - 1 = -\frac{1}{2}\,\mathrm{Im}\,\frac{\Gamma_{12}}{m_{12}} \approx \frac{2\pi}{S(m_t^2/m_W^2)}\frac{m_c^2}{m_W^2}\,\mathrm{Im}\,\frac{V_{cb}V_{cd}^*}{V_{tb}V_{td}^*} \qquad (50)$$

from (48). This leads to a rough estimate of the convention-independent number

$$\delta_\epsilon \approx 1 - |\eta_m| \approx -2\pi\frac{m_c^2}{m_t^2}\frac{|V_{cb}||V_{cd}|}{|V_{tb}||V_{td}|}\sin\beta \sim -\frac{\sin\beta}{2000} \qquad (51)$$

with $|\delta_\epsilon| \ll 1$ where β is the CKM unitarity angle in Fig. 1a. Since this result is based on a leading order quark diagram, the number should be taken only as an order of magnitude. In particular, at this level of precision it can not be used to measure β.

---

[9] Note that some authors [9] redefine $\Delta\Gamma \mapsto -\Delta\Gamma$ to obtain a positive sign in the Standard Model!

2.2.5 Predictions for x_s, y_s and $\delta_{\epsilon s}$

Since the lifetimes of B^0 and B_s agree within present precision, a first approximation using (42) is

$$\frac{x_s}{x_d} \sim \frac{|V_{ts}|^2}{|V_{td}|^2}$$

which gives an estimate of the expected x_s range between 3 and 100. It suffers from the poor knowledge on V_{td}, which has to be obtained from the measured x_d. With the top mass known and lattice calculations giving more reliable numbers for f_B, f_{B_s}, B_B and B_{B_s}, theoretical predictions for x_s become more precise. The Standard Model now favours numbers between 11 and 40. Recently reported ranges are $13.4\ldots27.8$ [32] and $20.1 \pm \genfrac{}{}{0pt}{}{11.6}{9.7}$ [33], leaving still room above the present experimental limit $x_s > 21$.

The B_s meson eigenstates are expected to have also different widths at an observable level. $\Delta\Gamma$ is caused by final states to which both B and $\bar{B}$ can decay. For the B_s these include with the Cabibbo-allowed final states from the $b \to c\bar{c}s$ decay a substantial fraction, while their number is much smaller in B^0 decays. A value of $2y_s = \Delta\Gamma/\Gamma \approx 0.18 \left(\frac{f_{B_s}}{200\,\text{MeV}}\right)^2$ is predicted from quark level QCD calculations [34], a similar number $2y_s \approx 0.15$ is obtained using exclusive decay channels [35]. In the naive quark model a larger value around 0.20 is expected, and a QCD evaluation gives $2y_s = 0.16 \pm \genfrac{}{}{0pt}{}{0.11}{0.09}$ [36]. It can be related to bag parameters evaluated on the lattice [37] which predicts lower values $2y_s \approx (0.022\ldots0.070)\left(\frac{f_{B_s}}{210\,\text{MeV}}\right)^2$.

The refined ratio (49) is

$$\frac{x_s}{2y_s} \approx -\frac{2}{\pi}\frac{m_t^2}{m_b^2}\frac{0.54\pm0.02}{3-8m_c^2/m_b^2} \approx -(200\ldots250)$$

and corresponds to a lowest order estimate [38] neglecting QCD corrections. The range given reflects only a variation of quark masses.

Finally, replacing d with s, equation (51) can be used to estimate $\delta_{\epsilon s} \sim \sin(\phi_2 + \phi_6)/2000 \sim 10^{-5}$. Again, large corrections to this simple calculation may be expected.

2.2.6 Behaviour of the Four Neutral Meson Antimeson Systems

All four meson pairs $K^0/\bar{K}^0$, $D^0/\bar{D}^0$, $B^0/\bar{B}^0$, and $B_s/\bar{B}_s$ show a different oscillation behaviour, since they have all different relations of Γ, $\Delta\Gamma$, and Δm. The same symbols will be used for all four systems. Only when two specific systems shall be compared, their parameters will be distinguished by the subscripts K, D, d, and s, respectively. The dimensionless parameters

x and y give the ratios of time constants involved: $\tau = 1/\Gamma$ is the harmonic average of the lifetimes, $t_{\rm osc} = 2\pi/\Delta m = 2\pi\tau/x$ is the period of the oscillation, and $t_{\rm rel} = 2/\Delta\Gamma = \tau/y$ is the lifetime of the oscillation amplitude, i.e. the damping time constant of a relaxation process. Numerical values are summarized in Table 1.

Table 1. Parameters of the four neutral oscillating meson pairs [9].

	$K^0/\overline{K}^0$	$D^0/\overline{D}^0$	$B^0/\overline{B}^0$	$B_s/\overline{B}_s$
τ [ps]	89.4 ± 0.1; 51700 ± 400	$0.413 \pm .003$	1.548 ± 0.021	1.49 ± 0.06
Γ [s^{-1}]	$5.61 \cdot 10^9$	$2.4 \cdot 10^{12}$	$(6.41 \pm 0.16) \cdot 10^{11}$	$(6.7 \pm 0.3) \cdot 10^{11}$
$y = \frac{\Delta\Gamma}{2\Gamma}$	-0.9966	$|y| < 0.06$	$|y| \lesssim 0.01^*$	$-(0.01\ldots 0.10)^*$
Δm [s^{-1}]	$(5.300 \pm 0.012) \cdot 10^9$	$< 7 \cdot 10^{10}$	$(4.89 \pm 0.09) \cdot 10^{11}$	$> 15 \cdot 10^{12}$
Δm [eV]	$(3.49 \pm 0.01) \cdot 10^{-6}$	$< 5 \cdot 10^{-6}$	$(3.2 \pm 0.1) \cdot 10^{-4}$	$> 1.0 \cdot 10^{-2}$
$x = \frac{\Delta m}{\Gamma}$	0.945 ± 0.002	< 0.03	0.76 ± 0.02	$21\ldots 40^*$
δ_ϵ	$(3.27 \pm 0.12) \cdot 10^{-3}$		$\sim -10^{-3\,*}$	$|\delta_\epsilon| < 10^{-3\,*}$
$|\eta_m|^2$	0.99348 ± 0.00024	$\approx 1^*$	$1\ldots 1.002^*$	$\approx 1^*$

* Standard Model expectation [39]

While the parameters of the $K^0/\overline{K}^0$ system are well measured [9], theoretical assumptions enter into the B meson columns. Many precise lifetime measurements for neutral B mesons have become available recently, averaging to $\tau_d = (1.548 \pm 0.021)\,{\rm ps}$ [40].

Figs. 3–6 show the number of mesons and antimesons as a function of the scaling lifetime variable $T = t/\tau$ and the asymmetry

$$a(T) = \left.\frac{\dot{N}(X \to X) - \dot{N}(X \to \overline{X})}{\dot{N}(X \to X) + \dot{N}(X \to \overline{X})}\right|_T \qquad (52)$$

$$= \frac{(1 - |\eta_m|^2)\cosh yT + (1 + |\eta_m|^2)\cos xT}{(1 + |\eta_m|^2)\cosh yT + (1 - |\eta_m|^2)\cos xT}$$

for a meson produced at $T = 0$ as a flavour eigenstate X, and decaying to a flavour-specific final state as X or $\overline{X}$ at a later time T. Expressed via the small real parameter δ_ϵ instead of $|\eta_m|$ this reads

$$a(T) = \frac{\cos xT + \delta_\epsilon \cosh yT}{\cosh yT + \delta_\epsilon \cos xT} \qquad (53)$$

For an antimeson produced at $T = 0$ as a flavour eigenstate $\overline{X}$, and decaying to a flavour-specific final state as X or $\overline{X}$ at time T, we obtain a similar expression, where only the $\cos xT$ part changes sign:

$$\bar{a}(T) = \left.\frac{\dot{N}(\overline{X} \to X) - \dot{N}(\overline{X} \to \overline{X})}{\dot{N}(\overline{X} \to X) + \dot{N}(\overline{X} \to \overline{X})}\right|_T = -\frac{\cos xT - \delta_\epsilon \cosh yT}{\cosh yT - \delta_\epsilon \cos xT}$$

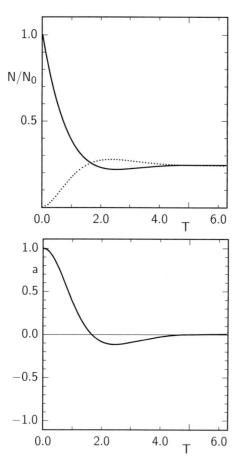

Fig. 3. $K^0/\overline{K}^0$ mixing is determined by the parameters $x = 0.945$, $y = 0.997$, and $|\eta_m|^2 = 0.994$ (see Table 1). $T = t/\bar{\tau}$ is the lifetime in units of $\bar{\tau} \approx 2\tau_S$, the inverse of the average width of K^0_L and K^0_S. The upper diagram shows the number of K^0 (solid) and $\overline{K}^0$ (dotted) as a function of T for a sample starting with 100% K^0 mesons. The lower diagram shows the asymmetry $a = (N_K - N_{\overline{K}})/(N_K + N_{\overline{K}})$. The relaxation process soon dominates, leaving only K^0_L after not much more than one oscillation.

The fraction of "mixed" decays at time T (i.e. decays where the flavour has changed from $T = 0$ to the actual decay time T) is

$$\chi(T) = \frac{\dot{N}(X \to \overline{X})}{\dot{N}(X \to \overline{X}) + \dot{N}(X \to X)}\bigg|_T = \frac{1 - a(T)}{2} = \frac{1 - \delta_\epsilon}{2} \frac{\cosh yT - \cos xT}{\cosh yT + \delta_\epsilon \cos xT}$$

$$\bar{\chi}(T) = \frac{\dot{N}(\overline{X} \to X)}{\dot{N}(\overline{X} \to X) + \dot{N}(\overline{X} \to \overline{X})}\bigg|_T = \frac{1 + \bar{a}(T)}{2} = \frac{1 + \delta_\epsilon}{2} \frac{\cosh yT - \cos xT}{\cosh yT - \delta_\epsilon \cos xT}$$

The approximation $|\eta_m| = 1$ corresponding to $\delta_\epsilon = 0$ leads to simpler expressions

$$a(T) = \frac{\cos xT}{\cosh yT} = -\bar{a}(T) \tag{54}$$

where x is clearly seen as the oscillation parameter, and y as the damping parameter, and

$$\chi(T) = \bar{\chi}(T) = \frac{1}{2} - \frac{\cos xT}{2\cosh yT}$$

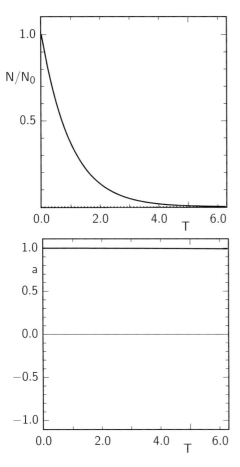

Fig. 4. $D^0/\overline{D}^0$ oscillations have not yet been observed, and are hardly visible even with $x = 0.02$ which is about 10 times the expected value and was used for these plots together with $y = 0$ and $|\eta_m| = 1$.

The kaon has both $x \approx 1$ and $y \approx -1$, i.e. the long-living state K_L^0 is the heavier mass eigenstate K_H. With these parameters one half of a sample of kaons of either flavour decays rapidly, mainly into two pions with CP = +1, and the other half transforms to a sample of the long-living K_L^0 states, which decay (aside from the small CP violation effects) to CP = -1 eigenstates and to flavour-specific states. The ratio of lifetimes of the two states (Table 1) is approximately 580. The time evolution of an initially pure K^0 flavour eigenstate is shown in Fig. 3. The upper diagram shows the number of remaining K^0 and $\overline{K}^0$ after a scaled time $T = \Gamma t$, where $\Gamma \approx \Gamma_S/2 = 1/(2\tau_S)$ is the average width of the short- and long-living state. The decay rate into flavour-specific final states is proportional to these numbers, while the dominant decays to CP eigenstates follow different evolution functions due to CP violation, and will be discussed below.

2.2.6.1 The Neutral D Meson

The D^0 meson decays mainly to flavour specific states with well defined strangeness, with only a few decays to CP $= +1$ eigenstates, as $\pi\pi$, $K\overline{K}$, $K_L^0\pi^0$ and CP $= -1$ states, as $K_S^0\pi^0$ or $K_S^0\omega$. This leads to equal lifetimes for the two eigenstates, i.e. $y \approx 0$. The corresponding box graph has a b quark as the heaviest particle in the loop, which is accompanied by the small CKM elements V_{cb} and V_{ub}. The mass difference induced that way by the Standard Model is very small, corresponding to $x \lesssim 0.002$. Therefore, almost no asymmetry is visible in Fig.4, although the number $x = 0.02$ used for the plot is a factor 10 higher.

Due to the smallness of x and y, equation (54) can be expanded giving to leading order

$$a(T) \approx 1 - \frac{x^2 + y^2}{2} T^2 \tag{55}$$

and

$$\chi(T) = \frac{x^2 + y^2}{4} T^2$$

The value $x = 0.002$ corresponds to a total mixed fraction of initially pure D^0 states given by

$$\chi = \frac{N_{\overline{D}^0 \to X}}{N_{D^0 \to X} + N_{\overline{D}^0 \to \overline{X}}} = \frac{x^2}{2(1+x^2)} \tag{56}$$

as $\chi \approx 2 \cdot 10^{-6}$.

For channels that can also originate from $\overline{D}^0$ decays with a small fraction $r \ll 1$ (e.g. doubly Cabibbo suppressed decays) the observed asymmetry is

$$a(T) \approx 1 - 2r^2 - 2ryT - \frac{x^2 + y^2}{2} T^2$$

where terms of order $r^2 y^2$ have been dropped. The fraction of apparently mixed decays is

$$\chi(T) = r^2 + ryT + \frac{x^2 + y^2}{4} T^2 \tag{57}$$

2.2.6.2 The Neutral B Mesons

The parameters of the $B^0/\overline{B}^0$ system have been introduced above. A good approximation is $y = 0$ and $\delta_\epsilon = 0$, which leads for N_0 pure B^0 at $T = 0$ to

$$N_{B^0}(T) = \tfrac{1}{2} N_0 e^{-T}(1 + \cos xT)$$
$$N_{\overline{B}^0}(T) = \tfrac{1}{2} N_0 e^{-T}(1 - \cos xT) \tag{58}$$

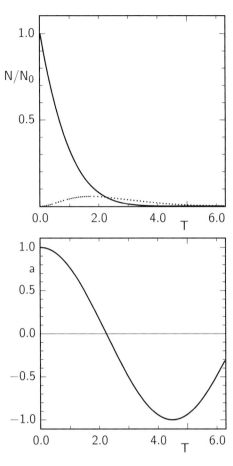

Fig. 5. $B^0/\overline{B}^0$ evolution is dominated by the oscillating part, with the parameters $x = 0.70$, $y = 0$, and $|\eta_m| = 1$. The ratio of the areas under the dotted and solid curve in the upper plot is the mixing probability χ. The zero transition in the asymmetry, which marks the crossover point in the upper plot, is at $T = \frac{\pi}{2x}$.

as shown in Fig. 5. The decay rate for flavour-specific final states (which are the majority of B^0 decays) follows the same time evolution. The asymmetry function is simply

$$a(T) = \cos xT \qquad (59)$$

This asymmetry can be observed using a flavour-tagging decay, like $B^0 \to D^-l^+\nu$. The rate of mesons decaying at time T into the channel X are given by (39) where $y = 1$ makes $\cosh yT = 1$ leading to the same asymmetry function $a(T) = \cos xT$. Integrating over all times, the observed numbers are

$$N_{B^0 \to X} = \int \dot{N}_{B^0 \to X}(T)\,dt = \tfrac{1}{2} N_0 \frac{\Gamma_X}{\Gamma} \frac{2+x^2}{1+x^2}$$

$$N_{\overline{B}^0 \to \overline{X}} = \int \dot{N}_{\overline{B}^0 \to \overline{X}}(T)\,dt = \tfrac{1}{2} N_0 \frac{\Gamma_X}{\Gamma} \frac{x^2}{1+x^2}$$

Their asymmetry becomes

$$a_{\text{int}} = \frac{N_{\overline{B}^0 \to \overline{X}} - N_{B^0 \to X}}{N_{\overline{B}^0 \to \overline{X}} + N_{B^0 \to X}} = \frac{1}{1+x^2}$$

and the mixing probability is as in (56)

$$\chi = \frac{N_{\bar{B}^0 \to \bar{X}}}{N_{B^0 \to X} + N_{\bar{B}^0 \to \bar{X}}} = \frac{x^2}{2(1+x^2)} \tag{60}$$

It was this net effect which gave the first proof for a sizeable mixing parameter $x \approx 0.7$ in the B^0 meson system in 1987 [41]. The time-dependent particle antiparticle oscillations of the neutral B meson have been first seen six years later by experiments at LEP [42]. With $x \approx 0.7$, about one period is visible before most of the mesons are decayed.

If we assume the Standard Model predictions to be true, the B_s meson is a very interesting case. There will be a small y yet significantly different from zero and a very large x. Fig. 6 is plotted with $x_s = 15$, which is close to the lower limit of the theoretical range and already too low to fit present data. The time-integrated mixing probability is for $|\eta_m| = 1$

$$\chi = \frac{x^2 + y^2}{2(1+x^2)}$$

For $x_s \gg 1$, this approaches its maximum value of 0.5, where a measurement of this quantity has no sensitivity on x any more. To observe the rapid oscillations, a very good lifetime resolution will be required. Experimentally, a lower limit $x_s > 21$ has been found at LEP (see below).

In the general case $|\eta_m| \neq 1$, the integrated mixing probability depends on the initial flavour. It is

$$\chi = \frac{|\eta_m|^2(x^2+y^2)}{2 + x^2(1+|\eta_m|^2) - y^2(1-|\eta_m|^2)} = \frac{(1-\delta_\epsilon)(x^2+y^2)}{2[1+x^2+\delta_\epsilon(1-y^2)]} \tag{61a}$$

for an initial B and

$$\bar{\chi} = \frac{(x^2+y^2)}{2|\eta_m|^2 + x^2(1+|\eta_m|^2) + y^2(1-|\eta_m|^2)} = \frac{(1+\delta_\epsilon)(x^2+y^2)}{2[1+x^2-\delta_\epsilon(1-y^2)]} \tag{61b}$$

for an initial $\bar{B}$ (which is χ with $|\eta_m|$ replaced by $1/|\eta_m|$ or, correspondingly, δ_ϵ by $-\delta_\epsilon$). This exhibits already CP violation, since the probabilities $P(X \to \bar{X})$ and $P(\bar{X} \to X)$ are different. It is also T violation, since the transition $X \to \bar{X}$ is the time reversed process $\bar{X} \to X$.

2.2.7 Oscillation at the $\Upsilon(4S)$

The $B\bar{B}$ system from strong interaction $\Upsilon(4S)$ decay is in an odd C and P eigenstate with angular momentum $L = 1$, retaining the quantum numbers $J^{PC} = 1^{--}$ of the mother particle. This system has to be treated as a coherent quantum state. The time evolution of a state with odd symmetry is

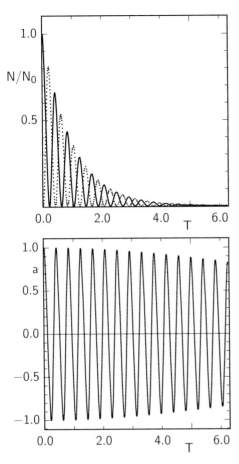

Fig. 6. $B_s/\overline{B}_s$ is expected to be the most rapidly oscillating system, with a longer relaxation time. This plot assumes $x_s = 15$, $y_s = 0.10$, and $|\eta_m| = 1$.

different from that of one with even symmetry. This is due to the fact, that only one antisymmetric $X\overline{X}$ state,

$$|\overline{X}(1)X(2)\rangle - |X(1)\overline{X}(2)\rangle$$

is possible, so it has to stay constant. There are, however, three symmetric states,

$$|\overline{X}(1)X(2)\rangle + |X(1)\overline{X}(2)\rangle$$
$$|X(1)X(2)\rangle$$
$$|\overline{X}(1)\overline{X}(2)\rangle$$

and their relative amplitudes may change with time. The quantum numbers characterizing the two different mesons, which are represented by (1) and (2) here, can be thought of as the spatial wave functions $\psi(\boldsymbol{x})$ and $\psi(-\boldsymbol{x})$ or alternatively the states in momentum space $|\boldsymbol{p}\rangle$ and $|-\boldsymbol{p}\rangle$.

Explicitly, for initial $B\bar{B}$ states of well defined symmetry,

$$\psi(0) = |B^0(1)\bar{B}^0(2)\rangle \pm |B^0(2)\bar{B}^0(1)\rangle$$

the time evolution from (34b) translates into

$$\psi(t) = e^{-2imt}e^{-T} \cdot$$
$$\left[\left(c^2|B^0(1)\bar{B}^0(2)\rangle + i\eta_m sc|\bar{B}^0(1)\bar{B}^0(2)\rangle \right. \right.$$
$$\left. + \frac{i}{\eta_m} sc|B^0(1)B^0(2)\rangle - s^2|\bar{B}^0(1)B^0(2)\rangle \right)$$
$$\pm \left(c^2|\bar{B}^0(1)B^0(2)\rangle + i\eta_m sc|\bar{B}^0(1)\bar{B}^0(2)\rangle \right.$$
$$\left. \left. + \frac{i}{\eta_m} sc|B^0(1)B^0(2)\rangle - s^2|B^0(1)\bar{B}^0(2)\rangle \right) \right] \quad (62)$$

for $-$:
$$\psi_-(t) = e^{-2imt}e^{-T}\left[|B^0(1)\bar{B}^0(2)\rangle - |\bar{B}^0(1)B^0(2)\rangle\right] \quad (62\text{a})$$

for $+$:
$$\psi_+(t) = e^{-2imt}e^{-T}\left[\cos(x-iy)T\left(|B^0(1)\bar{B}^0(2)\rangle + |\bar{B}^0(1)B^0(2)\rangle\right)\right.$$
$$\left. + i\sin(x-iy)T\left(\frac{1}{\eta_m}|B^0(1)B^0(2)\rangle + \eta_m|\bar{B}^0(1)\bar{B}^0(2)\rangle\right)\right] \quad (62\text{b})$$

where the shorthand notation $s = \sin(x-iy)T/2$, $c = \cos(x-iy)T/2$ has been used in (62). This means, the antisymmetric state stays always a 100% correlated $B\bar{B}$, as long as none of them has decayed. It is a typical example of an entangled quantum state, where both mesons always have exactly opposite flavour, although none of the single mesons is in a flavour eigenstate. Only when one decays into a state revealing its flavour (not necessarily the first one that decays) the state "collapses" into either $|B^0(1)\bar{B}^0(2)\rangle$ or $|\bar{B}^0(1)B^0(2)\rangle$ and the second meson continues as a one-particle state evolving in time according to (34).

The second case (62b) of an even wave function leads to a probability oscillating with twice the single-B frequency between a like-sign (BB or $\bar{B}\bar{B}$) and opposite-sign ($B\bar{B}$) flavour state.

For different times T_1 and T_2 of B meson (1) and (2) we have for the antisymmetric state

$$|\psi_-(T_1, T_2)\rangle = e^{-(im/\Gamma + \frac{1}{2})(T_1+T_2)} \cdot$$
$$\left[\cos(x-iy)\frac{T_1-T_2}{2}\left(|B^0(1)\bar{B}^0(2)\rangle - |B^0(2)\bar{B}^0(1)\rangle\right)\right.$$

$$-\mathrm{i}\sin(x-\mathrm{i}y)\frac{T_1-T_2}{2}\left(\frac{1}{\eta_m}|B^0(1)B^0(2)\rangle-\eta_m|\overline{B}^0(1)\overline{B}^0(2)\rangle\right)\Bigg]$$
(63 a)

$$=\frac{\mathrm{e}^{-(\mathrm{i}m/\Gamma+\frac{1}{2})(T_1+T_2)}}{2pq}\cdot$$

$$\left[\cos(x-\mathrm{i}y)\frac{T_1-T_2}{2}\left(-|B_\mathrm{L}(1)B_\mathrm{H}(2)\rangle+|B_\mathrm{L}(2)B_\mathrm{H}(1)\rangle\right)\right.$$
$$\left.-\mathrm{i}\sin(x-\mathrm{i}y)\frac{T_1-T_2}{2}\left(|B_\mathrm{L}(1)B_\mathrm{H}(2)\rangle+|B_\mathrm{L}(2)B_\mathrm{H}(1)\rangle\right)\right]$$
(63b)

This is observable if we associate the times T_1 and T_2 with the times of decay of the two B mesons. Again it is seen that for $T_1=T_2$ only the antisymmetric state is present, and mixed states, i.e. two final states indicating the same beauty flavour, will show up only at $T_1 \neq T_2$. The mixing probability (B and $\overline{B}$ denote the flavour at decay time)

$$\frac{N(BB+\overline{B}\overline{B})}{N(BB+B\overline{B}+\overline{B}B+\overline{B}\overline{B})}=\chi \qquad (64)$$

is identical to that for a single B meson. This can be understood from the fact that the second B meson is in a flavour eigenstate exactly when the first one decays into a tagging mode, and then evolves in time as a single oscillating B meson until it decays itself. The probability can also be obtained from equation (63a) using $N(BB) = \mathcal{N}\int|\langle B^0(1)B^0(2)|\psi_-(T_1,T_2)\rangle|^2\,\mathrm{d}T_1\,\mathrm{d}T_2$, and $N(\overline{B}\overline{B})$, $N(B\overline{B})$, $N(\overline{B}B)$ accordingly. The normalization factor $\mathcal{N}$ depends on the branching fractions into tagging modes and in general on the parameters y, δ_ϵ and x, but cancels anyway in the ratio.

For incoherent $B^0\overline{B}^0$ pair production, e.g. in $b\bar{b}$ jet fragmentation, the integrated mixed-rate is determined by two independent mixing probabilities

$$\frac{N(BB+\overline{B}\overline{B})}{N(BB+B\overline{B}+\overline{B}B+\overline{B}\overline{B})}=2\chi(1-\chi)$$

The actual distribution of observables is obtained from (63) multiplying with an appropriate bra vector, and computing the absolute square, $|\langle \mathit{final} \leftarrow BB|\psi_-\rangle|^2$. One is left with a function of T_1, T_2, or rather the more useful variables $T_+ = T_1+T_2$ and $T_- = T_1-T_2$. Integrating over all T_+ in the allowed range $|T_1-T_2|$ to infinity yields

$$\int_{|T_-|}^\infty \mathrm{e}^{-T_+} f(T_-)\,\mathrm{d}T_+ = \mathrm{e}^{-|T_-|} f(T_-)$$

which adds a factor $\mathrm{e}^{-|T_-|}$ to the function of the decay time difference. Thus, assuming $T=T_-$ and replacing the factor e^{-T} with $\mathrm{e}^{-|T|}$ will transform most distributions for single B^0 time dependence into distributions in time difference at the $\Upsilon(4S)$. This feature is used below to write universal distributions valid for two definitions of T: $T = t_s/\tau$ defined by the signal B lifetime t_s for incoherent $b\bar{b}$ production, and $T = (t_s - t_t)/\tau$ for coherent $B\bar{B}$ production on the $\Upsilon(4S)$, where t_t denotes the second B meson in the $\Upsilon(4S)$ decay in a flavour tagging decay mode.

Equation (63b) is an expansion in the two mass eigenstates. The antisymmetric wave function is always composed of two different states, there will be never $B_H B_H$ or $B_L B_L$, even at different decay times.

There is no wave function of CP eigenstates, because the CP properties are undefined before both B mesons have decayed. They involve the phases in decay amplitudes, and include all effects of CP violation which will be discussed in detail below.

For the symmetric state, the wave function is

$$|\psi_+(T_1,T_2)\rangle = e^{-(im/\Gamma + \frac{1}{2})(T_1+T_2)} \cdot$$
$$\left[\cos(x-iy)\frac{T_1+T_2}{2}\left(|B^0(1)\bar{B}^0(2)\rangle + |B^0(2)\bar{B}^0(1)\rangle\right)\right.$$
$$\left.+ i\sin(x-iy)\frac{T_1+T_2}{2}\left(\frac{1}{\eta_m}|B^0(1)B^0(2)\rangle + \eta_m|\bar{B}^0(1)\bar{B}^0(2)\rangle\right)\right] \tag{65a}$$

$$= \frac{e^{-(im/\Gamma+\frac{1}{2})(T_1+T_2)}}{2pq} \cdot$$
$$\left[\cos(x-iy)\frac{T_1+T_2}{2}\left(|B_L(1)B_L(2)\rangle - |B_H(1)B_H(2)\rangle\right)\right.$$
$$\left.+ i\sin(x-iy)\frac{T_1+T_2}{2}\left(|B_L(1)B_L(2)\rangle + |B_H(1)B_H(2)\rangle\right)\right] \tag{65b}$$

This is very similar to the function of an antisymmetric state, but the oscillation is in the **sum** of the two lifetimes instead of the lifetime difference.

In the approximation $|\eta_m| = 1$ and $y = 0$, the integrated mixed-rate is

$$\frac{N(BB+\bar{B}\bar{B})}{N(BB+B\bar{B}+\bar{B}B+\bar{B}\bar{B})} = \frac{x^2(3+x^2)}{2(1+x^2)^2} = \chi(3-4\chi)$$

In the general case, it is

$$\frac{N(BB+\bar{B}\bar{B})}{N(BB+B\bar{B}+\bar{B}B+\bar{B}\bar{B})} = \frac{(1+\delta_\epsilon^2)(x^2+y^2)[3-y^2+x^2(1+y^2)]}{2[(1+x^2)^2(1+y^2)-\delta_\epsilon^2(1-x^2)(1-y^2)^2]}$$

but cannot be related to the mixing probabilities of single mesons χ and $\bar{\chi}$.

The expansion in mass eigenstates shows, that the symmetric wave functions consists always of two eigenstates with the same mass, i.e. $B_H B_H$ or $B_L B_L$.

2.3 Experimental Determination of the Mixing Parameters of B Mesons

The asymmetry (52) in its simplified form (59) can be detected, if the flavour of B^0 mesons is known at production and at decay time. The flavour at decay time can be inferred from the reconstruction of its decay products. The flavour at production time can not directly be observed, but can be deduced from the flavour of the second b-hadron in $b\bar{b}$ pair production. At the $\Upsilon(4S)$, this is another neutral B meson and their coherent oscillation can be observed as a function of their decay time difference.

While only a small fraction of B meson decays can be fully reconstructed, the flavour of a B meson can be identified by various "tags". These are reconstructed particles from B meson decays that determine the beauty (or bottomness) flavour of the B meson by a measureable property, predominantly by their electric charge.

The first observation of a then unexpected large $B^0\bar{B}^0$ mixing by ARGUS in 1987 [41] used the best flavour tags available: In multihadron events on the $\Upsilon(4S)$ resonance, like-sign lepton pairs were observed which could not be attributed to other sources but semileptonic B decays. The charge of the lepton from $\bar{b} \to l^+ \nu \bar{c}$ in these decays is identical to the beauty quantum number of the meson, and these events had to be attributed to $B^0 B^0$ and $\bar{B}^0 \bar{B}^0$ final states from $\Upsilon(4S)$ decays.

2.3.1 Flavour Tagging

Flavour tagging exploits always a correlation between the beauty flavour of the parent b-hadron and a charge – in most cases the electric charge of the tagging particle, but e.g. for Λ hyperons the baryon number or strangeness and for D^0 mesons the charm.

This correlation is perfect for all fully reconstructed beauty baryons and charged B mesons, and almost all fully reconstructed neutral B mesons. However, a complete reconstruction of B decays will only be possible in a very limited number of events. A more universal approach is to collect this information via certain characteristics of the particles which are able to identify the flavour. In the first measurement of $B\bar{B}$ mixing only leptons have been used, with a charge correlated to the beauty flavour via semileptonic decays $\bar{b} \to l^+ \nu \bar{c}$. This is not a perfect correlation since a substantial fraction of leptons from other sources have the "wrong" charge and thereby dilute the oscillation amplitude. If the fraction of oppositely charged leptons is w, the observed numbers of pairs with one unambiguously identified B meson and one lepton are

$$N_1 \equiv N(Bl^- + \bar{B}l^+) = (1-w) \cdot N(B\bar{B}) + w \cdot N(BB + \bar{B}\bar{B})$$
$$N_2 \equiv N(Bl^+ + \bar{B}l^-) = (1-w) \cdot N(BB + \bar{B}\bar{B}) + w \cdot N(B\bar{B})$$

and the observed asymmetry is

$$a_\text{obs} = \frac{N_1 - N_2}{N_1 + N_2} = (1 - 2w) \cdot a \qquad (66)$$

i.e. the true asymmetry a is **diluted** by a factor $D_t = 1 - 2w$. Likewise, the observed mixing probability

$$\chi_\text{obs} = \frac{N_2}{N_1 + N_2} = (1 - 2w) \cdot \chi + w$$

is a function of w and the true mixing probability χ. In the case of lepton pairs, we have

$$N_1 \equiv N(l^+l^-) = [(1-w)^2 + w^2] \cdot N(B\bar{B}) + 2w(1-w) \cdot N(BB + \bar{B}\bar{B})$$
$$N_2 \equiv N(l^+l^+ + l^-l^-) = [(1-w)^2 + w^2] \cdot N(BB + \bar{B}\bar{B}) + 2w(1-w) \cdot N(B\bar{B})$$

and

$$\chi_\text{obs} = \frac{N_2}{N_1 + N_2} = (1 - 2w)^2 \cdot \chi + 2w(1-w)$$

is reduced by the square of the lepton-tag dilution factor and also shifted by an offset $2w(1-w)$.

For the determination of a mixing probability, a fit of the lepton momentum distribution is used to determine the fraction of primary leptons.

For a lifetime-dependent measurement of mixing via the oscillation frequency, the time-dependence of the asymmetry is observed. Although this asymmetry will experience a constant dilution from each tag, this dilution is irrelevant for the measurement of the frequency.

2.3.1.1 Lepton Tags

Electrons and muons occur in b hadron decays from several sources:

$$
\begin{aligned}
&b \to c(e^-/\mu^-)\bar{\nu} &&\text{(a)}\\
&b \to c\tau^-\bar{\nu},\, \tau^- \to e^-/\mu^- &&\text{(b)}\\
&b \to cX,\, c \to e^+/\mu^+ &&\text{(c)}\\
&b \to cD_s^-,\, D_s^- \to e^-/\mu^- &&\text{(d)}\\
&b \to (c\bar{c})X,\, (c\bar{c}) \to e^+e^-/\mu^+\mu^- &&\text{(e)}
\end{aligned}
$$

Using the electric charge as a priori flavour assignment, processes a,b,d yield "right sign" tags with a total branching fraction of about 24%, while the cascade process c yields "wrong-sign" tags with about 20% branching fraction, and process e (mainly J/ψ decays) yields leptons of both signs with no discriminating power, but their total contribution per b is less than

1%. The same holds for most other sources of leptons, like Dalitz pairs or converted photons.

In cases a–d, if the decay path is known, the electric charge Q determines uniquely the flavour of the mother B. Otherwise, since the number of positive and negative leptons in B final states is approximately the same, there is no information on the B flavour.

Fortunately, even without complete reconstruction, different decay paths lead to different distributions in many particle or event variables. The most prominent discriminating variable at the $\Upsilon(4S)$ is the momentum p^* in the tag-B cms, and at $b\bar{b}$ jet production the transverse momentum $p_\perp$ of the lepton. Cutting on this variable yields rather pure tags with dilution factors $D_t > 0.8$.

Misidentification of hadrons as leptons is an additional source of dilution, and calls for an efficient particle identification – specifically for leptons – in the oscillation experiments.

2.3.1.2 Kaon Tags

Sources for charged kaons are

$$
\begin{aligned}
&b \to cW^-, c \to s \to K^- &\text{(a)}\\
&b \to XW^-, W^- \to (\bar{c}/\bar{u})s \to K^- &\text{(b)}\\
&b \to XW^-, W^- \to \bar{c}(s/d), \bar{c} \to \bar{s} \to K^+ &\text{(c)}\\
&b \to Xs\bar{s}, s \to K^- &\text{(d)}\\
&b \to Xs\bar{s}, \bar{s} \to K^+ &\text{(e)}\\
&b \to s, s \to K^-\ (\text{penguin}) &\text{(f)}
\end{aligned}
$$

The strangeness is the obvious quantum number to tag the beauty flavour. It is identical to the electric charge Q of the kaon, hence the a priori assignment for the beauty estimator being Q, processes a,b,d,f yield "right-sign" kaons, and c,e yield "wrong-sign" kaons.

In addition to the charged kaon sources mentioned there are two more in high energy $b\bar{b}$ jet experiments: A kaon from the spectator s quark in B_s decays will have the wrong-sign charge. A kaon can also have the wrong charge, if it originates from the fragmentation chain, and not from b decay. This involves not only a dilution, since the s flavour is not correlated to the tag-b, but in hadron collisions also an intrinsic asymmetry due to leading particle effects. An impact parameter with respect to the main vertex is a useful discriminating variable that can reduce the mistag fraction of kaons in these experiments.

2.3.1.3 Other Particle Tags

Further improvements on kaon tagging are possible via partial or full reconstruction of $D \to K(n\pi)$ and $D_s \to KK(n\pi)$, where also a common vertex of the charged particles is very precious information. Sources for D and D_s mesons are

$$b \to cW^-, c \to D \quad \text{(a)}$$
$$b \to cs\bar{s}W^-, c\bar{s} \to D_s^+ \quad \text{(b)}$$
$$b \to XW^-, W^- \to \bar{c}s \to D_s^- \quad \text{(c)}$$
$$b \to XW^-, W^- \to \bar{c}sX \to \bar{D}KX \quad \text{(d)}$$
$$b \to XW^-, W^- \to \bar{c}d \to \bar{D} \quad \text{(e)}$$
$$\bar{B}_s \to D_s^+ W^- \quad \text{(f)}$$

Processes a and c are the dominating ones, favouring D^+, D^0 and D_s^- as tags for a b quark ($\bar{B}$ meson). However, the other sign can occur via b, d or e, making the dilution factor smaller than one. CLEO found from Dl correlations [43] that $(9.1 \pm 2.6)\%$ of all D mesons have the "wrong" charge. If this is the same for charged and neutral B mesons, it corresponds to a dilution factor $D_t = 0.82 \pm 0.05$.

Tagging with D_s mesons is diluted from process b by an unknown amount assumed to be small. Process f contributes to "wrong" sign D_s mesons at b jet experiments.

There is still considerable improvement to kaon tags alone. Especially kaons from $D_s \to KK(n\pi)$ give no flavour information, while a fully reconstructed D_s is a very powerful flavour tag. The suppression of $s\bar{s}$ production and the inclusive branching fraction of $\sim 10\%$ make it very likely, that its origin is $b \to cW^-, W^- \to D_s^-$, with a strong correlation between its charge and the beauty flavour. However, there exist no measurements to quantify this statement.

The D meson charge flavour correlation is exploited in the charge dipole tagging technique by the SLD collaboration [44]: Typically, the B and charm vertex cannot be separated, but the point of closest approach of a track to the common vertex carries still information on the cascade, since charm decay products tend to be further downstream than direct B decay products. The charge dipole moment

$$d_Q = \sum_{\text{tracks}} w_i l_i Q_i$$

where l_i is the path of flight from the primary vertex (interaction point) to the point of closest approach of track i to the secondary vertex, Q_i is its charge, and $w_i \propto \sin^2 \theta_i / \sigma(d_i)$ are weights normalized to $\sum w_i = 1$ separately for positive and negative tracks. θ_i is the angle between track i and the axis from the main vertex to the secondary vertex, and $\sigma(d_i)$ the error on the impact parameter with respect to the common vertex. The distribution of these dipole moments shows a statistical correlation with the B flavour, which increases with the absolute value $|d_Q|$. The dilution factor has been estimated at SLD to be

$$D(|d_Q|) \approx 0.68 \exp\left(-\left|\frac{|d_Q|}{4.2\,\text{mm}} - 0.8\right|\right).$$

A typical B event consists of many different tags, which may be combined to a global optimum. All particles which can be cleanly associated with a b-hadron either via a common vertex or, at the $\Upsilon(4S)$, all particles not coming from a fully reconstructed neutral B meson, are potential tags. They can be combined into a flavour estimator using statistical discrimination methods, e.g. via neural network techniques.

In addition to leptons and kaons, tagging information comes from charged pions with a significant flavour correlation at the high end of the momentum spectrum due to two-body decays with a charged pion or ρ meson, and at the low end due to pions from $D^{*+} \to \pi^+ D^0$.

Some less important tags are protons and Λ hyperons from baryonic B decays and, in b jets, from beauty baryon decays. It should be emphasized, that a simple diquark antidiquark pair creation model employed in B decay simulation programs may be wrong in reproducing baryon-flavour correlations, so Monte Carlo "predictions" should be taken with reservation.

At b jet experiments, tagging with pions and protons is expected to give very little information, since these particles are abundant in fragmentation, and in hadron collisions they show intrinsic asymmetries which reduce further their usefulness as flavour tags.

2.3.1.4 Tag Jet Charge

Another technique with b jets is the idea to determine the charge of a quark by statistical methods [45], defining a jet charge for the jet with the tag b-hadron. This is a weighted sum of charges of all particles in a jet,

$$Q_J = \frac{\sum w_i Q_i}{\sum w_i}$$

where the weights are typically chosen proportional to the rapidity or momentum of a particle. An example [46] is $w_i = p_{\|i}^{0.6}$, where $p_\|$ is the momentum component parallel to the sphericity axis. Using a cut $|Q_J| > 0.1$, a dilution factor $D_t = 0.376 \pm 0.004$ at 67.5% efficiency was obtained.

2.3.1.5 Same Jet Tags

If the primary meson from fragmentation is an excited state "B^{**}", which decays hadronically as

$$B^{**+} \to \pi^+ B^{(*)0}, \qquad B^{**-} \to \pi^- \overline{B}^{(*)0}$$

into the signal B, the charge of the accompanying pion uniquely tags the flavour of the $B^0/\overline{B}^0$ at production time. The pion follows in direction and velocity closely its partner B meson, which allows to isolate it from other pions in the event. Its identification can be completed by a cut in the invariant mass of the $B\pi$ system, which shows a structure of several resonant states around 5.7 GeV [47–49].

Even the nonresonant $B\pi$ systems in this phase space region show a correlation between the pion charge and the B flavour, because the fragmentation process produces $B^0(\bar{b}d)$ next to a $\bar{d}$ quark, which may hadronize into $\pi^+(\bar{d}u)$, while a $\overline{B}^0(b\bar{d})$ has close to it a d quark, which may hadronize into $\pi^-(d\bar{u})$. At the Tevatron, CDF [50] has used this technique, and determined a dilution factor of $0.18 \pm 0.03 \pm 0.02$.

A natural extension of the idea of the jet charge is the effective charge of the remnant jet where the B^0 meson was split off as leading particle. This is on average the charge of the residual d or $\bar{d}$ quark. The method is similar to the tag jet charge determination described above, but is applied to the jet accompanying the signal B meson excluding its decay products from the calculation.

A very special tag is the polarization tag used at SLC. Here, polarized Z^0 bosons are produced by longitudinally polarized e^+e^- pairs. Due to weak parity violation, the probability that the $\bar{b}$ (beauty +1) of the $b\bar{b}$ pair is found at polar angle θ with respect to the positron direction is

$$P(B = +1, \cos\theta) = \frac{1}{2} + \frac{A_b(A_e - P_e)\cos\theta}{(1 - A_e P_e)(1 + \cos^2\theta)}$$

where P_e is the longitudinal polarization of the beams ($P_e > 0$ for right-handed electrons), and $A_e = 0.152 \pm 0.004$ and $A_b = 0.91 \pm 0.05$ [9] are the parity-violating asymmetry parameters of the Z-fermion-fermion coupling to electrons and b quarks, respectively. Thereby, the SLD experiment achieved flavour tagging by the polar angle measurement of the b jet. The dilution at angle θ is

$$D(\cos\theta) = 2P - 1 = 2\frac{A_b(A_e - P_e)\cos\theta}{(1 - A_e P_e)(1 + \cos^2\theta)}$$

with an average $\langle D_t \rangle \approx 0.52$ for the highest polarization of 77%.

2.3.2 Mixing of B and $\bar{B}$

The first observation of $B^0\bar{B}^0$ mixing by ARGUS in 1987 [41] was the observation of 25 like-sign lepton pairs at the $\Upsilon(4S)$ energy. These could not be attributed to other sources but to B^0B^0 and $\bar{B}^0\bar{B}^0$ final states with semileptonic decay of both B mesons.

In addition, this observation was supported by four events with one fully reconstructed B meson plus a lepton of "wrong" sign, and one exclusive event $\Upsilon(4S) \to B^0B^0$ with both B^0 mesons reconstructed.

This measurement was the first in a series of "mixing" measurements, where the fraction χ of oscillated events is determined without lifetime information. The mixing probability χ can be calculated from the number of like- and opposite-sign dilepton events as

$$\chi = \frac{[N(l^+l^+) + N(l^-l^-)] \cdot (1+f)}{N(l^+l^+) + N(l^-l^-) + N(l^+l^-)}$$

where N are numbers corrected for mistags from secondary leptons, and the ratio of semileptonic branching fractions of neutral and charged B mesons and of their production rates enters as

$$f = \frac{[\mathcal{B}(B^+ \to l^+\nu X)]^2 \cdot \mathcal{B}(\Upsilon(4S) \to B^+B^-)}{[\mathcal{B}(B^0 \to l^+\nu X)]^2 \cdot \mathcal{B}(\Upsilon(4S) \to B^0\bar{B}^0)}$$

in a background dilution factor $D_\pm = 1/(1+f)$. In all analyses, the factor f is taken to be 1 consistent with measured numbers and theoretical expectation.

Results on mixing obtained on the $\Upsilon(4S)$ are summarized in Table 2. In addition to leptons and fully reconstructed B mesons as flavour tags also fully reconstructed $D^{*\pm}$ mesons, partially reconstructed $D^{*\pm} \to (D^0/\bar{D}^0)\pi^\pm$, partially reconstructed $B^0 \to D^{*-}l^+\nu$, and charged kaons have been used. The most precise single measurement [51] uses $B^0 \to D^{*-}\pi^+(\pi^0)$ where the $D^{*-} \to \bar{D}^0\pi^-$ is partially reconstructed using the soft π^-. The flavour of the second B is determined by a lepton with $p > 1.4\,\text{GeV}$.

2.3.3 Oscillations in Time-Dependent Measurements

While only the integrated effect can be observed on the $\Upsilon(4S)$ at symmetric colliders, an observation of the oscillating behaviour was first possible at the Z^0, where the lifetime can be measured. This yields directly the frequency ν as $2\pi\nu = \Delta m$ from the asymmetry

$$a(t) = \left.\frac{\dot{N}(B) - \dot{N}(\bar{B})}{\dot{N}(B) + \dot{N}(\bar{B})}\right|_t = \cos\Delta m\, t$$

Results are summarized in Table 3. They have been recently augmented by results from asymmetric colliders on the $\Upsilon(4S)$ by the B meson factories PEP II/BABAR and KEKB/BELLE.

Table 2. B^0 mixing parameter χ, using $f = 1$. Tagging particles are given, where B^0 means a fully reconstructed B meson, and (D^*) means partial reconstruction of the $D^{*+} \to D^0\pi^+$ decay via the soft π^+.

0.10 0.20 0.30	χ	tags	experiment
	0.16 ± 0.05	ll	ARGUS 87 [41]
	$0.144 \pm 0.036 \pm 0.036$	ll	CLEO 89 [52]
	0.180 ± 0.050	ll, D^*l, B^0l	ARGUS 92 [53]*
	$0.157 \pm 0.016 \pm 0.018 \pm ^{0.028}_{0.021}$	ll	CLEO 93 [54] *
	$0.149 \pm 0.023 \pm 0.021$	B^0l	CLEO 93 [54] *
	$0.162 \pm 0.044 \pm 0.039$	$(D^*)ll$	ARGUS 93 [55]*
	$0.20 \pm 0.13 \pm 0.12$	D^*K	ARGUS 96 [56]
	$0.19 \pm 0.07 \pm 0.09$	B^0K	ARGUS 96 [56]*
	$0.198 \pm 0.013 \pm 0.014$	B^0l	CLEO 01 [51] *
	0.179 ± 0.014		all (*) averaged

The measurements differ mostly in the tagging method used. Reconstructing hadronic final states of $B^0/\overline{B}^0$ decays is denoted "B^0" in Table 3. This technique requires a large number of B mesons and has therefore only been used at the asymmetric B factories at SLAC and KEKB. The same holds for the channel $B^0 \to D^{*-}l^+\nu$.

The experiments at LEP use charged leptons (l), fully (D^*) or partially (π) reconstructed decays of $D^{*+} \to D^0\pi^+$, alone and associated with a lepton indicating a $B^0 \to D^{*-}(X)l^+\nu$ final state, and various definitions of a jet charge. The C and P asymmetry of polarized Z^0 decays is exploited at the Stanford Linear Collider by SLD.

As in the time integrated measurement, also in measurements of the lifetime dependent oscillation there is a background fraction $f_\pm$ from other b hadrons, predominantly charged $B^\pm$ which do not oscillate. This modifies (58) for B mesons at $T = 0$ to

$$N_B(T) = \tfrac{1}{2}N_0 e^{-T}(1 + \cos xT + f_\pm)$$
$$N_{\overline{B}}(T) = \tfrac{1}{2}N_0 e^{-T}(1 - \cos xT) \qquad (67)$$

which corresponds to an asymmetry

$$a(T) = \frac{\cos xT}{1 + f_\pm/2} + \frac{f_\pm}{2 + f_\pm} = D_\pm \cos xT + I_\pm \qquad (68)$$

corresponding to an effective dilution factor $D_\pm = \frac{1}{1+f_\pm/2}$ and an offset $I_\pm = \frac{f_\pm}{2+f_\pm}$. The value $a(0) = D_\pm + I_\pm = 1$ is independent of $f_\pm$.

If the flavour of one B meson is assigned wrong with probability w, the asymmetry is given by

$$N_B(T) = \tfrac{1}{2}N_0 e^{-T}[(1-w)(1 + \cos xT + f_\pm) + w(1 - \cos xT)]$$

Table 3. $B^0/\bar{B}^0$ eigenstate mass difference from the oscillation frequency. Tagging particles are given, where B^0 means a fully reconstructed B meson, Q_J is a jet charge technique, π^+ is a same-jet tag, and NN is a neural network exploiting many particles in the event.

0.40 0.50 0.60	Δm [ps^{-1}]	tags	experiment
	0.446 ± 0.032	$D^*, l/Q_J; l/l$	ALEPH 96 [57]
	$0.441 \pm 0.026 \pm 0.029$	Q_J	ALEPH 97 prel. [40]
	$0.496 \pm 0.026 \pm 0.023$	$D^*, l, \pi l/Q_J; l/l$	DELPHI 97 [46]
	$0.548 \pm 0.050 \pm ^{0.023}_{0.019}$	$D^*/l; D^*l/Q_J$	OPAL 96 [58]
	$0.444 \pm 0.029 \pm ^{0.020}_{0.017}$	$l/Q_J, l$	OPAL 97 [59]
	$0.430 \pm 0.043 \pm ^{0.028}_{0.030}$	l/l	OPAL 97 [60]
	$0.497 \pm 0.024 \pm 0.025$	$D^*l/Q_J, l$	OPAL 00 [61]
	$0.444 \pm 0.028 \pm 0.028$	$l/l, Q_J$	L3 98 [62]
	$0.58 \pm 0.07 \pm 0.08$	K ^{a}	SLD 96 prel. [44]
	$0.56 \pm 0.08 \pm 0.04$	Q_J^a	SLD 96 prel. [44]
	$0.520 \pm 0.072 \pm 0.035$	l ^{a}	SLD 96 prel. [63]
	$0.452 \pm 0.074 \pm 0.049$	Q_J, l ^{a}	SLD 96 prel. [64]
	$0.471 \pm ^{0.078}_{0.068} \pm 0.034$	$\pi^+, D/l$	CDF 97 [50]
	$0.450 \pm 0.045 \pm 0.051$	e/μ	CDF 98 prel. [65]
	$0.500 \pm 0.052 \pm 0.043$	l/Q_J	CDF 99 [66]
	$0.503 \pm 0.064 \pm 0.071$	μ/μ	CDF 99 [67]
	$0.516 \pm 0.099 \pm ^{0.029}_{0.035}$	D^*l/l	CDF 99 [68]
	$0.562 \pm 0.068 \pm ^{0.041}_{0.050}$	$D^{(*)}/l$	CDF 99 prel. [40]
	$0.463 \pm 0.008 \pm 0.016$	l/l	BELLE 01 [69]
	0.522 ± 0.026	$D^*l\nu/l, K, \pi$	BELLE 01 prel. [70]
	0.527 ± 0.032	$B^0/l, K, \pi$	BELLE 01 prel. [70]
	$0.493 \pm 0.012 \pm 0.009$	l/l	BABAR 01 [71]
	$0.508 \pm 0.020 \pm 0.022$	$D^*l\nu/l, K, \mathrm{NN}$	BABAR 00 prel. [72]
	$0.516 \pm 0.016 \pm 0.010$	$B^0/l, K, \mathrm{NN}$	BABAR 01 [73]
	0.487 ± 0.009		average^b 01 [40]
	0.489 ± 0.009		average

^{a} using the Z^0 polarization asymmetry at SLC

$$N_{\bar{B}}(T) = \tfrac{1}{2} N_0 e^{-T} \left[(1-w)(1-\cos xT) + w(1 + \cos xT + f_\pm) \right]$$

$$a(T) = (1-2w) \left[\frac{\cos xT}{1 + f_\pm/2} + \frac{f_\pm}{2 + f_\pm} \right]$$

$$= D_t D_\pm \cos xT + D_t I_\pm$$

with a tagging dilution factor $D_t = a(0) = 1 - 2w$.

Two examples are shown in Fig. 7.

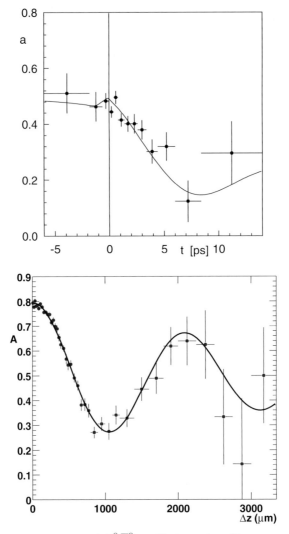

Fig. 7. Two measurements of $B^0\overline{B}^0$ oscillation using dilepton events at LEP (ALEPH [57], top) and the PEP II B factory (BABAR [71], bottom). Both show the asymmetry defined in (59), the left figure has been derived from one showing the like-sign fraction $\chi_{\text{obs}}(t)$ using $a(t) = 1 - 2\chi_{\text{obs}}(t)$.

2.3.3.1 Lifetime Resolution

A limited error in vertex reconstruction leads to a smearing of the T distribution. This implies an amplitude reduction factor D_r. Assuming a

Gaussian resolution in Δt of σ_t, we have[10]

$$D_r = e^{-\frac{1}{2}(x\sigma_t/\tau)^2} \tag{69}$$

which is multiplied to the tagging dilution. A resolution function described by two Gaussians as

$$r(t) = \frac{f}{\sqrt{2\pi}\sigma_n} e^{-\frac{(t-t')^2}{2\sigma_n^2}} + \frac{1-f}{\sqrt{2\pi}\sigma_w} e^{-\frac{(t-t')^2}{2\sigma_w^2}}$$

yields

$$D_r = f e^{-\frac{1}{2}(x\sigma_n/\tau)^2} + (1-f) e^{-\frac{1}{2}(x\sigma_w/\tau)^2} \tag{70}$$

This approach to time smearing ignores the asymmetry entered by the exponential decay distribution. If the true distributions

$$f(T) = e^{-|T|}(1 \pm \cos xT)$$

are convoluted with a Gaussian, and the asymmetry is calculated from the smeared distributions, there is in addition to the reduction in amplitude also a distortion.

At the b jet experiments, the distance between the primary vertex of $b\bar{b}$ jet production and the B decay vertex is measured with high precision silicon detectors. The primary vertex is determined from the interaction region, which has an RMS in the transverse plane of $150\,\mu\text{m} \times 10\,\mu\text{m}$ at LEP, $1\,\mu\text{m} \times 1\,\mu\text{m}$ at SLC and $25\,\mu\text{m} \times 25\,\mu\text{m}$ at the Tevatron. The wide x region at LEP is compensated by the use of fragmentation tracks to fit an event by event vertex. The time resolution is dominated by the B^0 decay vertex which is reconstructed with accuracies varying between 50 and $200\,\mu\text{m}$. These values are small compared to a typical distance of flight of $\sim 3\,\text{mm}$ (both at the Z^0 and the Tevatron). A detailed description of the experimental analyses is found in a recent review [74].

Since

$$t = \frac{\Delta s \cdot m_B}{p}$$

---

[10] the measured asymmetry function is

$$f(\Delta t) = \int_{-\infty}^{\infty} \cos x \frac{\Delta t'}{\tau} \cdot e^{-\frac{(\Delta t - \Delta t')^2}{2\sigma_t^2}} \, d(\Delta t')$$

$$= e^{-\frac{1}{2}(x\sigma_t/\tau)^2} \cos x \frac{\Delta t}{\tau}$$

where Δs is the distance between the secondary and primary vertices, and p the momentum of the B meson, the error is

$$\frac{\delta t}{t} = \frac{\delta \Delta s}{\Delta s} \oplus \frac{\delta p}{p}, \qquad \delta t = \delta \Delta s \frac{m_B}{p} \oplus \frac{\delta p}{p} t$$

with a constant term from vertex resolution and a linear term from momentum resolution. The measurements with B^0 mesons in b jets have been performed without full reconstruction of the B^0 meson. In this case, an important effect on the lifetime resolution is the error on the B^0 momentum, and since $\delta p/p \approx$ const ($\gtrsim 10\%$ at LEP) or $\delta t \propto t$, the lifetime becomes increasingly smeared. This leads to a progressive reduction in the amplitude of the observable asymmetry.

In contrast to this behaviour, oscillation measurements at asymmetric colliders at the $\Upsilon(4S)$ energy have a dominantly constant resolution

$$\delta \Delta t \approx \frac{\delta \Delta z}{\beta \gamma} \approx \text{const}$$

due to the short distance between the two B decay vertices of $\sim 260\,\mu\text{m}$ (BABAR) and $\sim 200\,\mu\text{m}$ (BELLE). The velocity of the centre of mass system (the $\Upsilon(4S)$) fluctuates by less than one per mille, the momentum uncertainty due to the proper motion of the B in the $\Upsilon(4S)$ cms causes a small contribution to the lifetime error that is visible at large lifetimes.

The constant and time-dependent resolution effects differ from experiment to experiment, and depend on the method of B identification. They are convoluted with the full time dependence, including the exponential which makes the effective "smearing" of the true lifetime asymmetric. These together with a possible bias from cascade charm decays, resolution tails and background effects are modelled for each experiment and are taken into account in the fit functions, leading to shapes, of which two examples are shown in Fig. 7. The proper modelling of the distortions is crucial for a bias-free extraction of the oscillation frequency.

2.3.4 Summary of Experimental Results on the B^0 Meson

The average of all measurements in Table 3 is

$$\Delta m(B^0) = (0.489 \pm 0.009)/\text{ps}$$

corresponding to an oscillation frequency $\nu = \Delta m/2\pi = (77.8 \pm 1.6)\,\text{GHz}$ and a mass difference $\Delta m = (0.322 \pm 0.007)\,\text{meV}$. This is a fraction of $6 \cdot 10^{-14} \cdot m(B^0)$.

The dimensionless mixing parameter x can be calculated from the mixing probability using (60) and from the oscillation frequency as $x = \Delta m\, \tau$ which requires also precise knowledge on the average lifetime $\tau_d = (1.548 \pm 0.021)\,\text{ps}$ [9] of the B^0 meson. It is

$$x = \Delta m\,\tau = 0.757 \pm 0.017$$
$$x = \sqrt{\frac{\chi}{\frac{1}{2} - \chi}} = 0.747 \pm 0.045$$
$$x = 0.756 \pm 0.016 \qquad \text{common average}$$

The two independent methods agree very well. The common value of the scaled mass difference is dominated by the direct measurements of the oscillation frequency and has reached a precision of 2%.

There is no observation of $\Delta\Gamma$ of the B^0 mass eigenstates. In the Standard Model, a value $y \sim 10^{-3}$ is expected. CLEO finds $|y| < 0.41$ from their mixing analysis [51]. BELLE [69] gives an upper limit of $|y| < 0.08$ at 90%CL.

2.3.5 Experimental Results on the B_s Meson

The first hint on large B_s mixing was obtained by UA1 [75] even before $B^0\overline{B}^0$ oscillation was established. They observed an average mixing probability $\chi = 0.12 \pm 0.05$ in b jets from $\bar{p}p$ annihilation. From this quantity at e^+e^- annihilation and Z^0 decay, a value $\chi_s \approx 0.5$ can be inferred with large errors due to the small B_s fraction in b jets. Direct (non-) observations of the oscillation leads to more stringent limits on the frequency, summarized in Table 4. To determine a limit, the LEP Working group on B oscillations [40] combined the fit results of the B_s asymmetry amplitude A (which is 1 for true oscillations) for a series of assumed values of Δm_s from 13 measurements by the five experiments ALEPH, CDF, DELPHI, OPAL, and SLD. All measurements have been adjusted to a common set of inputs before averaging. Systematic correlations are taken into account. The average amplitudes are shown in Fig. 8.

A 95%CL lower limit is calculated using the point where the hypothesis $A = 1$ is excluded at 5% significance level in a one-sided Gaussian test. This means all values of Δm_s for which the combined amplitude A plus $1.645 \cdot \sigma(A)$ is smaller than 1 are excluded at 95%CL, where $\sigma(A)$ is the total error on A. The procedure excludes all values with $\Delta m_s < 15.0\,\text{ps}^{-1}$ leading to a lower limit

$$\Delta m_s > 15.0\,\text{ps}^{-1} \quad (95\%\text{CL}).$$

The combined sensitivity for 95%CL is given by the range where $1.645 \cdot \sigma(A) \leq 1$, i.e. the limit if an average observed amplitude of $A = 0$ would be obtained, and extents to $18.1\,\text{ps}^{-1}$.

For the $B_s/\overline{B}_s$ system, using this lower limit on the oscillation frequency and the B_s lifetime value $\tau_s = (1.47 \pm 0.06)\,\text{ps}$ [9], the present lower limit is

$$x_s > 21 \quad (95\%\text{CL})$$

Table 4. Experimental limits on the $B_s/\overline{B}_s$ eigenstate mass difference from the oscillation frequency.

Δm_s [ps^{-1}]	experiment
> 1.8 (95%CL)	ALEPH 94 [76]
> 6.1 (95%CL)	ALEPH 95 [77]
> 6.6 (95%CL)	ALEPH 96 [78]
$> 3.1, \notin [5.0, 7.6]$ (95%CL)	OPAL 97 [59]
$> 6.5, \notin [8.2, 9.4]$ (95%CL)	DELPHI 97 [79]
> 15.0 (95%CL)	combined 00 [40], incl. preliminary results
> 14.9 (95%CL)	combined 02 [80], incl. preliminary results

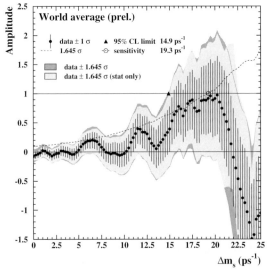

Fig. 8. Average amplitude for $B_s\overline{B}_s$ oscillation fits as a function of the oscillation frequency Δm_s from 13 measurements by the five experiments ALEPH, CDF, DELPHI, OPAL and SLD [80]. The shaded areas are the 90%CL intervals (with and without systematics), corresponding to a significance level of 5% for a one-sided test.

which is already above the lowest expected values. It corresponds to a mixing probability $\chi_s > 0.498$.

The available results on the B_s lifetime are average lifetimes which can all have different weights of τ_H and τ_L due to the different mixture of final states used. The experimental information on the lifetime difference of the mass eigenstates is still weak and no significant lifetime difference has been observed. All flavour specific decays – like the semileptonic decays – have a time distribution (39)

$$\dot{N}(T) \propto e^{-T} \cosh yT = e^{-t/\tau_\mathrm{L}} + e^{-t/\tau_\mathrm{H}}$$

in the approximation $\delta_\epsilon = 0$. The kink in this distribution cannot be observed with the presently available data samples. However, assuming equal lifetime $\tau = 1/\Gamma$ of B_s and B_0, the deviation of the fit result for semileptonic decays from this average can be used to derive an upper limit on y_s. Upper limits reported from this method are $2|y_s| < 0.47$ (95%CL) [81] and $2|y_s| < 0.31$ (95%CL) [82].

Table 5. $B_s/\overline{B}_s$ eigenstate lifetime difference. Limits are on $|\Delta\Gamma/\Gamma|$, while values with errors are signed results assuming the Standard Model description ($\tau_{0,s}$ = lifetime of B^0, B_s). Note that $y = \frac{1}{2}\Delta\Gamma/\Gamma$.

0 0.5 1.0	$-\Delta\Gamma/\Gamma$	method	experiment
⊢—	< 0.67 (95%CL)	incl, $\tau_s = \tau_0$	L3 98 [83]
⊢—	< 0.47 (95%CL)	$D_s l, \tau_s = \tau_0$	DELPHI 00 [81]
⊢—	< 0.70 (95%CL)	$D_s, \tau_s = \tau_0$	DELPHI^a 00 [84]
⊢	< 0.30 (95%CL)	$D_s l, \tau_s = \tau_0$	combined 01 [78,82,85,86]
⊢+⊣	$0.26 \pm ^{0.30}_{0.15}$	$\phi\phi$	ALEPH^a 00 [87]
⊢———+———⊣	$0.43 \pm ^{0.81}_{0.48}$	$\phi\phi, \tau_s = \tau_0$	ALEPH 00 [87]
⊢——+——⊣	$0.33 \pm ^{0.45}_{0.42}$	$J/\psi\phi$	CDF 99 [88]
+⊣	$0.24 \pm ^{0.16}_{0.13}$		combined [82]
+	$0.16 \pm ^{0.08}_{0.09}$	$\tau_s = \tau_0$	combined [82]

^{a} assuming dominance of $B_{sL} \to D_s^{(*)+} D_s^{(*)-}$, CP = +1.

Other methods rely even more on Standard Model assumptions:
- the CP = +1 states dominate in the $D_s^{(*)+} D_s^{(*)-}$ and $J/\psi\phi$ final states,
- the light eigenstate B_{sL} is the short lived one as suggested by the sign in (49), and
- this state decays dominantly into the CP = +1 final states.

The latter one is discussed in more detail in the next chapter on CP violation. A summary of all experimental information is given in Table 5. The combination of all results has been calculated by the LEP $\Delta\Gamma_s$ Working Group [82]. They have also re-evaluated the constraint $\tau_s = \tau_0$ for individual experiments.

3 CP Violation

Standard Model weak interactions are long known to violate parity and charge conjugation symmetries, in most cases even maximally. However, the combined symmetry operation CP leads generally to transitions identical to the original ones, i.e. CP symmetry is conserved. A typical example is the weak decay of a τ lepton into $\pi\nu_\tau$, as shown in Fig. 9. While a τ^- lepton can decay into a left-handed neutrino and a pion, the charge-conjugate decay of a τ^+ into a left-handed antineutrino is forbidden. However, if one looks at the mirror-image, i.e. one applies a parity transformation at the same time, the decay is allowed, and even more the amplitudes for both decays are equal. If we extend our definition of "antiparticle" to mean not only sign-flip of all charge-like quantities, but also of the spin, we have the CP operation and a perfect symmetry even for most weak-interaction processes. CP violation, on the contrary, is a true violation of particle antiparticle symmetry, which can not be restored by a mirror.

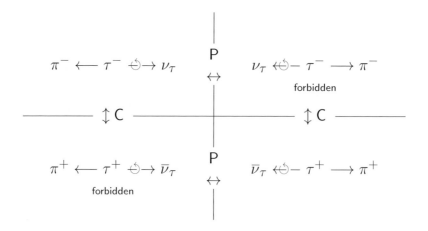

Fig. 9. Parity (P) and charge conjugation (C) operations on $\tau^- \to \pi^- \nu_\tau$. The upper right and lower left processes are forbidden.

How is violation of symmetry detected? As a simple example, let us investigate the symmetry operation P corresponding to space coordinate inversion

$$\mathrm{P}\,\psi(\boldsymbol{x}) = \psi(-\boldsymbol{x})$$

which has – as do the C and CP operations – have $\mathrm{P}^2 = 1$, i.e. two successive applications of the parity operator restores the original state. There exist eigenstates with eigenvalues $+1$ (even parity) and -1 (odd parity).

Parity conservation in an interaction means

$$[H_{IA}, \mathrm{P}] = 0$$

where the interaction is described by the operator H_{IA}. Equivalently, $\mathrm{P}^{-1} H_{IA} \mathrm{P} = H_{IA}$, and thereby $\mathrm{P}(H_{IA}|\pm\rangle) = \pm(H_{IA}|\pm\rangle)$, i.e. the product of the interactions has the same eigenvalue as the initial state (if it was an eigenstate).

This implies that parity violation can be established as different effects
1) $\mathrm{P}(H_{IA}|\pm\rangle) = \mp(H_{IA}|\pm\rangle)$, i.e. the eigenvalue has changed during the interaction[11]
2) $\mathrm{P}(H_{IA}|\psi\rangle) \neq H_{IA}|\mathrm{P}\,\psi\rangle$, e.g. the interaction in a left-handed systems differs from the mirror-image of the interaction in a right-handed system.
3) Starting from a parity eigenstate of any eigenvalue, a parity-odd observable $\langle H_{IA}\rangle$ with $\mathrm{P}\,H_{IA}\,\mathrm{P}^{-1} = -H_{IA}$ has a non-zero expectation value. An example is any pseudoscalar variable like the helicity.

The violation of more complex symmetries, as CP, can be observed accordingly as either a change of the CP eigenvalue during an interaction or a difference between the interaction probability of CP-transformed states with respect to the original process.

If all interactions were CP symmetric, we had no way to distinguish left-/right-handedness, positive/negative charge etc. Parity and charge conjugation violation in weak interaction connects handedness with charge, but still does not allow a distinction between the two members of a pair. CP violating K^0 decays, however, provide a different decay rate function of time for K^0 and $\overline{K}^0$, which could be used to explicitly distinguish them by a dip or bump in this function.

Although we are presently not able to observe the difference of matter and antimatter at far regions of the universe, the absence of regions of matter antimatter annihilation boundaries suggests that the whole universe is made of matter, violating CP asymmetry to a large extent. Small asymmetries of the order 10^{-10} at the early universe are sufficient to explain this present situation, however, it is difficult to create these from the CP violation in the Standard Model which has particle antiparticle asymmetry only in mesons, whereas baryon number violation is observed in the universe. If one assumes baryon number violating processes at phase boundaries of the early universe, e.g. at the symmetry breaking phase transition to the electroweak interaction in the Standard Model, still the CP asymmetry via the CKM phase is many orders of magnitude smaller than the observed number of baryons per background photon [89]. Thus, CP violating mechanisms beyond the Standard Model are likely to exist, and we will possibly observe them as small deviations from the Standard Model predictions.

The origin of CP violation in K and B mesons may be only within the

---

[11] Nota bene: Applying this case to CP, one thinks of the decay of K_L^0 to CP eigenstates with $\xi = +1$. This in itself is only an indirect manifestation of CP violation, since it is no change in eigenvalue, for the K_L^0 is not a physical CP eigenstate. It indicates, however, the existence of a CP-violating transition from the CP-odd three-pion state to a CP-even state.

Standard Model, but other possibilities are not ruled out. The observation of CP violation in B mesons can either yield a consistent picture with one set of Standard Model parameters, or produce contradictory results, making extensions or an alternative theory unavoidable. Almost any extension of the Standard Model introduces new phases which result in CP violation. This can only be avoided by some fine tuning of parameters which seems unnatural.

Complementary searches for CP violation will give additional constraints: Only small CP violating effects are predicted by the Standard Model in weak decays of other particles, like D^0 mesons or strange baryons [90]. CP asymmetries in neutrino oscillations [91] or CP violation in lepton decay [92] are potential windows to alternative models. All these are examples of C violating processes, while P does not compensate for that as in Fig. 9.

A different way to violate the CP symmetry is by C conserving parity violation. The search for magnetic monopoles or electric dipole moments [93] in pointlike or spherically symmetric particles, e. g. in leptons, quarks or the neutron, is a way to find non-standard CP violation of this kind. At present, only upper limits on these effects exist, and no glimpse beyond the Standard Model has been obtained.

3.1 CP Eigenstates Versus Mass Eigenstates

The following discussion will again use $B^0/\overline{B}^0$ as an example, but is applicable to each of the four systems accordingly. Independent of any convention two orthogonal CP eigenstates

$$|B^0_+\rangle = \frac{1}{\sqrt{2}}\left(|B^0\rangle + \text{CP}\,|B^0\rangle\right), \quad |B^0_-\rangle = \frac{1}{\sqrt{2}}\left(|B^0\rangle - \text{CP}\,|B^0\rangle\right) \qquad (71)$$

with $\text{CP}\,|B^0_+\rangle = |B^0_+\rangle$ and $\text{CP}\,|B^0_-\rangle = -|B^0_-\rangle$ can be defined. If a state agrees with one of these except for a phase factor, it will be a CP eigenstate.

The mass eigenstates of the $B^0/\overline{B}^0$ system are not CP eigenstates (this statement is valid for any $X^0/\overline{X}^0$ system including $K^0/\overline{K}^0$). Using $\text{CP}\,|B^0\rangle = e^{i\phi_{\text{CPB}}}|\overline{B}^0\rangle$, they are transformed by a CP operation as

$$\text{CP}\,|B_{\text{L}}\rangle = \left(\frac{e^{i\phi_{\text{CPB}}}}{2\,\eta_m} + \frac{e^{-i\phi_{\text{CPB}}}\eta_m}{2}\right)|B_{\text{L}}\rangle - \left(\frac{e^{i\phi_{\text{CPB}}}}{2\,\eta_m} - \frac{e^{-i\phi_{\text{CPB}}}\eta_m}{2}\right)|B_{\text{H}}\rangle$$
$$\approx \cos 2\tilde{\beta}\,|B_{\text{L}}\rangle - \sin 2\tilde{\beta}\,|B_{\text{H}}\rangle$$
$$\text{CP}\,|B_{\text{H}}\rangle = -\left(\frac{e^{i\phi_{\text{CPB}}}}{2\,\eta_m} + \frac{e^{-i\phi_{\text{CPB}}}\eta_m}{2}\right)|B_{\text{H}}\rangle + \left(\frac{e^{i\phi_{\text{CPB}}}}{2\,\eta_m} - \frac{e^{-i\phi_{\text{CPB}}}\eta_m}{2}\right)|B_{\text{L}}\rangle$$
$$\approx -\cos 2\tilde{\beta}\,|B_{\text{H}}\rangle + \sin 2\tilde{\beta}\,|B_{\text{L}}\rangle$$

where the approximation for the $B^0/\overline{B}^0$ system depends on the phase convention for the CKM matrix which determines the angle $\tilde{\beta}$. If $\tilde{\beta} = 0$

is chosen by an appropriate phase redefinition (that may be accompanied by a phase redefinition for the b and/or d field), these states would be eigenstates with CP $= \pm 1$, respectively. Still, there would be CP violation in their decay, and the CP eigenvalue of the final state would be different. Therefore, the question of which of the mass eigenstates is closest to which CP eigenstate has no convention independent answer. Only the CP eigenvalue of a decay product of one of these states is an observable. Since the weak interaction does not conserve CP, it is not legitimate to deduce from the final state's eigenvalue an eigenvalue of the decaying state.

A meaningful question is which of B_H or B_L **decays** more often into CP $= \pm 1$ eigenstates. In contrast to the neutral kaon system, most final states from B decays are flavour-specific, and both mass eigenstates decay into them via either their B^0 or their $\overline{B}^0$ component. The small fraction of states that can be reached both by B^0 and $\overline{B}^0$ includes the contribution from CP eigenstates which appear mainly through three processes. On the tree level, there are two main decay channels that can produce CP eigenstates: $b \to c\bar{c}d$ with the $c\bar{c}d\bar{d}$ final state, and $b \to u\bar{u}d$ with the quark content $u\bar{u}d\bar{d}$. A state of the first kind with CP eigenvalue ξ_X will have decay amplitudes

$$A = \langle X_{c\bar{c}}|\mathbf{H}|B^0\rangle = V_{cb}^* V_{cd} A_0, \quad \overline{A} = \langle X_{c\bar{c}}|\mathbf{H}|\overline{B}^0\rangle = \xi_X e^{-i\phi_{CPB}} V_{cb} V_{cd}^* A_0$$

where $\xi_X = \pm 1$ is the CP eigenvalue of the state. The corresponding decay amplitudes of B_H and B_L are

$$A_{L,H} = \langle X_{c\bar{c}}|\mathbf{H}|B_{L,H}\rangle = pA \pm q\overline{A}$$
$$= pA\left(1 \pm \eta_m \xi_X e^{-i\phi_{CPB}} \frac{V_{cb} V_{cd}^*}{V_{cb}^* V_{cd}}\right)$$
$$= pA\left(1 \pm |\eta_m|\xi_X e^{-2i\beta}\right)$$

The decay ratio is then (in the approximation $|\eta_m| = 1$)

$$\frac{|A_H|^2}{|A_L|^2} = \frac{|1 - \xi_X e^{-2i\beta}|^2}{|1 + \xi_X e^{-2i\beta}|^2} = \frac{1 - \xi_X \cos 2\beta}{1 + \xi_X \cos 2\beta} = (\tan^2 \beta)^{\xi_X} \qquad (72)$$

which is for $\beta < \frac{\pi}{4}$ less than 1 for $\xi_X = +1$ and *vice versa*. In this case, the heavier state B_H will decay more often into states with negative CP eigenvalue, $\xi_X = -1$.

The value $\tan \beta$ is also a parameter of CP violation in the interference of mixed and unmixed decay amplitudes: One convention-independent parameter that describes CP violation for the given final state is $\eta_X = i\xi_X \tan \beta$ which corresponds to the small "ϵ" paramter for the K meson.

Accordingly, for the $u\bar{u}d\bar{d}$ states

$$A_{L,H} = \langle X_{u\bar{u}}|\mathbf{H}|B_{L,H}\rangle = pA\left(1 \pm |\eta_m|\xi_X e^{2i\alpha}\right)$$

and the ratio

$$\frac{|A_\mathrm{H}|^2}{|A_\mathrm{L}|^2} = \frac{|1-\xi_X e^{2i\alpha}|^2}{|1+\xi_X e^{2i\alpha}|^2} = \frac{1-\xi_X \cos 2\alpha}{1+\xi_X \cos 2\alpha} = (\tan^2\alpha)^{\xi_X} \qquad (73)$$

depends on the angle α, which is likely to be larger than $\frac{\pi}{4}$. This would give the opposite answer, i.e. the heavier state B_H will decay more often into states with positive CP eigenvalue, $\xi_X = +1$.

Some decays with an intermediate state $c\bar{c}d\bar{s}$ or $c\bar{c}\bar{d}s$ proceed into K^0 or $\bar{K}^0$, which finally result in $c\bar{c}d\bar{d}$ via a K_L^0 or K_S^0 sequential decay. Among those is the favourite decay $B^0 \to J/\psi K_\mathrm{S}^0$. The total decay chain involves almost the same CKM element phase factors as the direct $b \to c\bar{c}d$ decay, leading to the same answers as for this decay mode (a more detailed discussion follows below).

For decays via W exchange, like $b\bar{d} \to c\bar{c}$ or $b\bar{d} \to u\bar{u}$, the same CKM elements are involved, and the same arguments lead to the same answers as above. Also, the favoured penguin-type transition $b \to s$ with subsequent hadronization into a K_L^0 or K_S^0 has a net phase close to β' leading to the ratio (72).

CP eigenstates with quark content $d\bar{d}$ can be reached via CKM-suppressed penguin-type loops. Due to the top quark dominance the amplitudes are

$$A = \langle X_{d\bar{d}}|\mathbf{H}|B^0\rangle \approx V_{tb}^* V_{td} A_0, \quad \bar{A} = \langle X_{d\bar{d}}|\mathbf{H}|\bar{B}^0\rangle \approx \xi_X e^{-i\phi_{CPB}} V_{tb} V_{td}^* A_0$$

and the CKM element phases cancel, which gives

$$\frac{|A_\mathrm{H}|^2}{|A_\mathrm{L}|^2} = \frac{1-\xi_X}{1+\xi_X}$$

i.e. B_H decays exclusively into states with negative CP eigenvalue, $\xi_X = -1$, and B_L into states with $\xi_X = +1$.

All these results receive corrections from non-leading terms, like c quark loops in the last case, or $b \to d$ penguin corrections to the $b \to u$ transition final states. The general case for an arbitrary ratio

$$r := \frac{\eta_m \bar{A}}{A}$$

leads to

$$A_\mathrm{L,H} = pA \pm q\bar{A} = pA(1 \pm r)$$

and a ratio of rates

$$\frac{|A_\mathrm{H}|^2}{|A_\mathrm{L}|^2} = \frac{1-\Omega_0}{1+\Omega_0} \quad \text{with} \quad \Omega_0 := \frac{2\operatorname{Re} r}{1+|r|^2} \qquad (74)$$

Since systems with a $c\bar{c}$ pair probably constitute the major part for both CP eigenvalues, the heavy mass eigenstate can be said to be the one which decays more often into final states with CP $= -1$, and the light one into those with CP $= +1$, but both have also substantial branching fractions into final states with the opposite CP value. This is a consequence of the CP violating phase in the CKM matrix, but is not a CP violating decay, since none of the two B mass eigenstates was a CP eigenstate before it decayed.

The dominant decays of the B_s to CP eigenstates is to the quark state $c\bar{c}s\bar{s}$, which yields

$$\frac{|A_{\rm H}|^2}{|A_{\rm L}|^2} \approx (\tan^2 \phi_2)^{\xi_X} \tag{75}$$

Since $\tan^2 \phi_2 \ll 1$, the heavy state, which is supposed to have the longer lifetime, will decay dominantly to $c\bar{c}s\bar{s}$ states with CP $= -1$, and the light, short-lived into CP $= +1$. This property is used to estimate the lifetime difference in the second group of experiments in Table 5, which assume to measure τ_{sL} from CP $= +1$ final states.

3.1.1 The Formalism for Conserved CP

The situation would be different if a purely real CKM matrix could be achieved by choice of appropriate unphysical (quark) phases. In this case, all unitarity triangles would be degenerate to lines, and their angles would be 0 or π. Therefore, $\cos 2\alpha = \cos 2\beta = 1$, and the heavier state would be the **only** to decay to CP $= -1$, while CP $= +1$ final states would be reached exclusively via decays of $B_{\rm L}$. For decay products which are CP eigenstates this situation would correspond to a perfectly predictable CP eigenvalue corresponding to the mass eigenstate $B_{\rm L}$ or $B_{\rm H}$. A natural choice of phases in this case would force all terms of the weak interaction Hamiltonian to be real, corresponding to $\eta_m = e^{i\phi_{\rm CPB}}$. Then CP$|B_{\rm L}\rangle = +|B_{\rm L}\rangle = |B_+\rangle$ and CP$|B_{\rm H}\rangle = -|B_{\rm H}\rangle = -|B_-\rangle$, and CP is conserved in decays where this quantum number is meaningful.

In general, if CP is conserved η_m is a phase factor

$$\eta_m = e^{-i\phi_{\eta_m}}$$

as in (46). The Hamiltonian is then

$$\mathbf{H} = \begin{pmatrix} m - \frac{i}{2}\Gamma & -\frac{e^{i\phi_{\eta_m}}}{2}\left[\Delta m - \frac{i}{2}\Delta\Gamma\right] \\ -\frac{e^{-i\phi_{\eta_m}}}{2}\left[\Delta m - \frac{i}{2}\Delta\Gamma\right] & m - \frac{i}{2}\Gamma \end{pmatrix}$$
$$= \begin{pmatrix} m - \frac{i}{2}\Gamma & -e^{i\phi_{\eta_m}}\frac{\Gamma}{2}(x - iy) \\ -e^{-i\phi_{\eta_m}}\frac{\Gamma}{2}(x - iy) & m - \frac{i}{2}\Gamma \end{pmatrix} \tag{76}$$

and

$$m_{12} = \tfrac{1}{2}(H_{12} + H_{21}^*) = -e^{i\phi_{\eta m}}\frac{\Delta m}{2}$$
$$\Gamma_{12} = i(H_{12} - H_{21}^*) = -e^{i\phi_{\eta m}}\frac{\Delta \Gamma}{2}$$

i.e. m_{12} and Γ_{12} have the same (unobservable) phase and the ratio Γ_{12}/m_{12} is real.

Starting from this situation, CP violation can be introduced as a perturbation via a small phase difference between m_{12} and Γ_{12}.

3.1.2 The Kaon as Approximate CP Invariant

Exactly this situation is **almost** true for the $K^0/\overline{K}^0$ system. The light K_S^0 decays to about 99.9% into the CP = +1 eigenstates $\pi^+\pi^-$ and $\pi^0\pi^0$, while the K_L^0 decays to one third into a CP = −1 eigenstate with 3 pions, the rest being mainly flavour specific semileptonic decays, and only 0.3% are to the CP = +1 two pion state [9]. Therefore, a parametrization is chosen where $K_S^0 \approx K_+$ and $K_L^0 \approx K_-$. If we have a K_S^0 as decay product of the B, we are used to assign it a CP = +1 eigenvalue contribution to the whole final state. To be precise, this is only correct if the K_S^0 decays into a CP = +1 final state. In this case also a $K_L^0 \to \pi\pi$ will be assigned the same CP = +1 eigenvalue, i.e. the "K_S^0" denotes its final state rather than the undecayed particle, and a $K_S^0 \to \pi l \nu$ as a flavour specific state is not included in this use of the label K_S^0.

3.2 CP Violating Interference Effects in B Decays

The $B^0/\overline{B}^0$ meson system has a simple description in the Standard Model: One parameter x is sufficient to parametrize the oscillation, since $y = 0$ and $|\eta_m| = 1$ are good approximations.

CP violation in B decays (as in K decays) occurs always via interference of (at least) two amplitudes with different CP even and CP odd phases, in three different ways:
1) Direct CP violation $\Gamma(B \to X) \neq \Gamma(\overline{B} \to \overline{X})$ can be observed by final state counting experiments. It occurs from the interference of two decay amplitudes with different phases that transform as CP $\phi = -\phi$ (CP odd phases from the CKM matrix), and with different phases that transform as CP $\delta = +\delta$ (CP even phases from the strong interaction). Direct CP violation is not restricted to neutral mesons, but may also be observed in charged meson or baryon decays.
2) CP violation induces a small asymmetry in the oscillation probability $P(B^0 \to \overline{B}^0) \neq P(\overline{B}^0 \to B^0)$ due to $|\eta_m| \neq 1$. This is due to the interference of other amplitudes with the leading box diagram of $B/\overline{B}$ mixing.

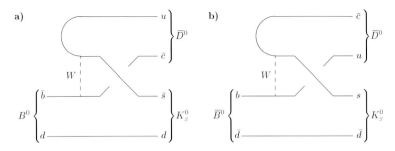

Fig. 10. Diagrams for $B^0 \to \overline{D}^0 K_S^0$ (a) and $\overline{B}^0 \to D^0 K_S^0$ (b).

3) The interference of mixed and unmixed amplitudes leads to lifetime dependent differences $\Gamma(B^0|_{t=0} \to X|t) \neq \Gamma(\overline{B}^0|_{t=0} \to X|t)$ for a common final state of B and $\overline{B}$ with asymmetry amplitude modulation $\propto \sin\Delta m\, t$. This is also called "CP violation from interference of oscillation and decay", or "mixing-induced CP violation". Here, the two interfering phases are phases from the CKM matrix that transform as CP $\phi = -\phi$ and the phase $\frac{\pi}{2}$ between the coefficients $\cos x\frac{T}{2}$ and $i\sin x\frac{T}{2}$ from (35).

The final state X can be a CP eigenstate, like $J/\psi K_S^0$ (CP $= -1$) or $\pi^+\pi^-$ (CP $= +1$), or a state that can be reached from both B^0 mesons via different processes, like $B^0 \to \overline{D}^0 K_S^0$ and $\overline{B}^0 \to D^0 K_S^0$ (Fig. 10).

In the Standard Model, CP violating interference can lead to almost maximum asymmetries. In many cases, large values are expected, and the time-dependence is a further handle to avoid misinterpretation of data. Therefore, all present and proposed future experiments focus on these effects, which will be described below.

A unique case of this interference can be observed in coherent antisymmetric $B\overline{B}$ states, i.e. in $\Upsilon(4S) \to B^0\overline{B}^0$, as a single event

$$\Upsilon(4S) \to B\overline{B} \to XY$$

with $\mathrm{CP}(XY) = \mathrm{CP}(X)\,\mathrm{CP}(Y)\,(-1)^L = -1 \neq \mathrm{CP}(\Upsilon(4S)) = +1$. Here X and Y are CP eigenstates. The expected rate for such events varies with the lifetime difference of the two B mesons. A special case is $X = Y$ and $\mathrm{CP}(XX) = -1$.

3.3 Direct CP Violation

Decays with direct CP asymmetries require in the Standard Model at least two interfering channels with different CKM phase $\phi_{1,2}$ and different strong phases $\delta_{1,2}$. This defines the amplitudes

$$\begin{aligned} A(B^0 \to X) &= |A_1|e^{i\phi_1 + i\delta_1} + |A_2|e^{i\phi_2 + i\delta_2} \\ \overline{A}(\overline{B}^0 \to \overline{X}) &= |A_1|e^{-i\phi_1 + i\delta_1} + |A_2|e^{-i\phi_2 + i\delta_2} \end{aligned} \quad (77)$$

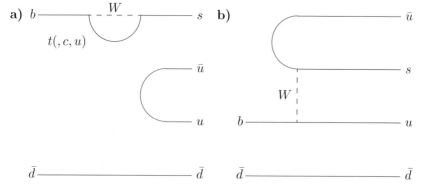

Fig. 11. Diagrams for $B^0 \to K^+\pi^-$: penguin (a) and tree (b). The CKM elements for both amplitudes give a suppression of the tree diagram by a factor $\sim \lambda^2$.

where $|A_1|e^{i\delta_1}$ and $|A_2|e^{i\delta_2}$ is unchanged due to CP invariance of the strong interaction. They contribute to the rates as

$$|A|^2 = |A_1|^2 + |A_2|^2 + 2|A_1||A_2|\cos(\phi_1 - \phi_2 + \delta_1 - \delta_2)$$
$$|\bar{A}|^2 = |A_1|^2 + |A_2|^2 + 2|A_1||A_2|\cos(\phi_2 - \phi_1 + \delta_1 - \delta_2)$$

which are different if $\cos(\phi_2 - \phi_1 + \delta_1 - \delta_2) \neq \cos(\phi_1 - \phi_2 + \delta_1 - \delta_2)$. This is not the case if $\delta_1 = \delta_2$ or if $\phi_1 = \phi_2$. The difference is

$$|\bar{A}|^2 - |A|^2 = 2|A_1||A_2|[\cos(\phi_2 - \phi_1 + \delta_1 - \delta_2) - \cos(\phi_1 - \phi_2 + \delta_1 - \delta_2)]$$
$$= 4|A_1||A_2|\sin(\phi_1 - \phi_2)\sin(\delta_1 - \delta_2)$$

The ratio of the amplitudes can be expressed as

$$\left|\frac{\bar{A}}{A}\right| = \frac{1 - \epsilon'_X}{1 + \epsilon'_X} \tag{78}$$

with a real parameter ϵ'_X. Due to the arbitrary phase factor $e^{-i\phi_{CPB}}$ and the free CKM phases in A and $\bar{A}$ the complex ratio $\bar{A}/A$ itself is not an observable.

The asymmetry is

$$a = \frac{N(\bar{B} \to \bar{X}) - N(B \to X)}{N(\bar{B} \to \bar{X}) + N(B \to X)}$$
$$= \frac{|\bar{A}|^2 - |A|^2}{|\bar{A}|^2 + |A|^2}$$
$$= \frac{2|A_1||A_2|\sin(\phi_1 - \phi_2)\sin(\delta_1 - \delta_2)}{|A_1|^2 + |A_2|^2 + 2|A_1||A_2|\cos(\phi_1 - \phi_2)\cos(\delta_1 - \delta_2)}$$

An example is the B decay to $K\pi$, where a rate asymmetry

$$a = \frac{N(\bar{B}^0 \to K^-\pi^+) - N(B^0 \to K^+\pi^-)}{N(\bar{B}^0 \to K^-\pi^+) + N(B^0 \to K^+\pi^-)}$$

with $|a| \lesssim 0.1$ is expected. In this example, the first amplitude is from a $b \to s$ penguin diagram (Fig. 11a) which has a dominant contribution from the t quark in the loop, with a CKM phase $\arg(V_{tb}^*V_{ts})$ and a $K\pi$ state with isospin $\frac{1}{2}$. The second amplitude occurs via a tree diagram $b \to u + \bar{u}s$ transition (Fig. 11b), with $\arg(V_{ub}^*V_{us})$ and isospin $\frac{1}{2}$ and $\frac{3}{2}$ amplitudes. The interference terms are proportional to $\sin\gamma' \sin(\delta_1 - \delta_2)$. Asymmetries of this type can also be observed in charged B decays, e.g. $B^\pm \to K^\pm \pi^0$.

The asymmetry is limited, however, by the ratio of amplitudes. This is approximately $P : \lambda^2 T$, where $P : T < 1$ reflects the suppression of the loop penguin diagram (P) with respect to a tree diagram (T) of same order. This ratio has been believed to be small, $P : T \sim \lambda^2$, which would allow indeed for large asymmetries, but is now known to be $\sim \lambda$ from the measurement of the $K\pi$ and $\pi\pi$ branching fractions, as can be seen in Table 6.

Table 6. Tree (T) and leading penguin (P) contributions to two-body final states. The experimental branching fractions are averages over charged and neutral B decays (which implies that corrections from the colour suppressed contribution to the $B^+ \to \pi^+\pi^0$ amplitude are ignored).

channel	amplitude	$\mathcal{B}$
$B \to K\pi$	$A\lambda^2 P + A\lambda^4(\rho - i\eta)T$	$(14.9 \pm 1.0) \cdot 10^{-6}$ [94,95,96]
$B \to \pi\pi$	$A\lambda^3(1 - \rho - i\eta)P + A\lambda^3(\rho - i\eta)T$	$(4.4 \pm 0.8) \cdot 10^{-6}$ [94,95,96]

CKM unitarity angles can only be extracted from those asymmetries when the strong phase difference is known. This can be obtained, however, from flavour SU(3) and isospin relations on a set of results from related channels [97]. Ratios of branching fractions for individual $K\pi$ and $\pi\pi$ channels can be used to extract information on the angle γ in the unitarity triangle [98] and provide an additional constraint.

3.3.1 Experimental Results

First experimental measurements of asymmetries of this type have been reported by CLEO, BABAR and BELLE and are summarized in Table 7.

There is presently no evidence for CP violation. The experimental precision is not yet sensitive to the range of asymmetries predicted by the Standard Model, but large effects from new CP violating interactions could have shown up already and have not been observed.

Table 7. Results on direct CP violation in B meson decays: the negative asymmetry between the decay rates for the given channel and its CP (C) conjugate is measured.

channel	asymmetry a	experiment
$B^+ \to J/\psi K^+$	$+0.018 \pm 0.043 \pm 0.004$	CLEO 00 [99]
	$-0.009 \pm 0.027 \pm 0.005$	BABAR 01 prel.
$B^+ \to \psi(2S) K^+$	$+0.020 \pm 0.091 \pm 0.010$	CLEO 00 [99]
$B^+ \to D^0_{CP+} K^+$	$+0.04 \pm ^{0.40}_{0.35} \pm 0.15$	BELLE 01 [100]
$B^0 \to K^+\pi^-$	-0.04 ± 0.16	CLEO 99 [101]
	$(-0.07 \pm 0.08 \pm 0.02)$	BABAR 01 [102]
	$-0.05 \pm 0.06 \pm 0.01$	BABAR 02 prel. [103]
	$0.04 \pm ^{0.19}_{0.17} \pm 0.02$	BELLE 01 [104]
$B^+ \to K^+\pi^0$	-0.29 ± 0.23	CLEO 99 [101]
	$0.00 \pm 0.18 \pm 0.04$	BABAR 01 [95]
	$-0.06 \pm ^{0.22}_{0.20} \pm ^{0.06}_{0.02}$	BELLE 01 [104]
$B^+ \to K^0_S \pi^+$	$+0.18 \pm 0.24$	CLEO 99 [101]
	$-0.21 \pm 0.18 \pm 0.03$	BABAR 01 [95]
	$0.10 \pm ^{0.43}_{0.34} \pm ^{0.02}_{0.06}$	BELLE 01 [104]
$B^+ \to K^+\eta$	$+0.03 \pm 0.12$	CLEO 99 [101]
$B^+ \to K^+\eta'$	$-0.11 \pm 0.11 \pm 0.02$	BABAR 01 [105]
$B^+ \to K^+\phi$	$-0.05 \pm 0.20 \pm 0.03$	BABAR 01 [105]
$B^+ \to K^{*+}\phi$	$-0.43 \pm ^{0.36}_{0.30} \pm 0.06$	BABAR 01 [105]
$B^0 \to K^{*0}\phi$	$0.00 \pm 0.27 \pm 0.03$	BABAR 01 [105]
$B^+ \to \omega\pi^+$	-0.34 ± 0.25	CLEO 99 [101]
	$-0.01 \pm ^{0.29}_{0.31} \pm 0.03$	BABAR 01 [105]
$B^0, B^+ \to K^*\gamma$	$+0.08 \pm 0.13 \pm 0.03$	CLEO 99 [106]
	$-0.035 \pm 0.076 \pm 0.012$	BABAR 01 prel.
$B^0/B^+ \to s\gamma$	$-0.079 \pm 0.108 \pm 0.022$	CLEO 01 [107]

3.4 CP Violation in the Oscillation

CP violation induces a small asymmetry in the oscillation probability $P(B^0 \to \overline{B}^0) \neq P(\overline{B}^0 \to B^0)$ due to $|\eta_m| \neq 1$. This is due to the interference of other amplitudes with the leading box diagram of $B/\overline{B}$ mixing, i.e. replacing one t quark in the loop with a c quark.

The oscillation asymmetry (53) starting with an initial B^0 meson can be expanded in δ_ϵ as

$$a(T) = \left.\frac{\dot{N}(B) - \dot{N}(\overline{B})}{\dot{N}(B) + \dot{N}(\overline{B})}\right|_T = \frac{\cos xT}{\cosh yT} + \delta_\epsilon \left(1 - \frac{\cos^2 xT}{\cosh^2 yT}\right) + \mathcal{O}(\delta_\epsilon^2) \quad (79)$$

which is for $y = 0$

$$a(T) \approx \cos xT + \delta_\epsilon \sin^2 xT = \cos xT + \frac{\delta_\epsilon}{2}(1 - \cos 2xT)$$

Starting with a $\overline{B}^0$ at $T = 0$ gives for the same asymmetry

$$\bar{a}(T) = \left.\frac{\dot{N}(B) - \dot{N}(\overline{B})}{\dot{N}(B) + \dot{N}(\overline{B})}\right|_T \approx -\cos xT + \frac{\delta_\epsilon}{2}(1 - \cos 2xT)$$

i.e. if $\delta_\epsilon < 0$ as indicated by (51) there are always more $B \to \overline{B}$ than $\overline{B} \to B$ oscillations. Using leptons as flavour tag the net asymmetry can be observed at the $\Upsilon(4S)$ by counting like-sign lepton pairs which originate directly from semileptonic neutral B meson decays. They occur in final states with mixing and show an asymmetry

$$a = \frac{N(B\overline{B} \to l^+l^+) - N(B\overline{B} \to l^-l^-)}{N(B\overline{B} \to l^+l^+) + N(B\overline{B} \to l^-l^-)} = \frac{1 - |\eta_m|^4}{1 + |\eta_m|^4} = \frac{2\delta_\epsilon}{1 + \delta_\epsilon^2} \quad (80)$$

constant in time. This asymmetry should be very small in the Standard Model, where (50) predicts $|2\delta_\epsilon| \lesssim 10^{-3}$.

The quantity δ_ϵ is typically quoted as $2\operatorname{Re}\epsilon_B$ which is incorrect, since the mixing parameter ϵ is convention-dependent. Neither $\operatorname{Re}\epsilon$ nor $\operatorname{Im}\epsilon$ or $|\epsilon|$ are observables, only one function of this complex parameter

$$\delta_\epsilon = \frac{2\operatorname{Re}\epsilon}{1 + |\epsilon|^2}$$

is a measurable quantity and can indeed be observed as an CP violating asymmetry in $B/\overline{B}$ oscillation.

3.4.1 The Total Decay Rate

This same parameter can also been observed as an asymmetry in the time-dependent total decay rate of B and $\overline{B}$. This rate for an initially pure B^0 sample can be calculated as

$$\frac{\mathrm{d}N}{\mathrm{d}t} = N_0 \sum_X \int \mathrm{dPS}\,|\mathcal{M}|^2\,\delta(m_B - E_X)$$

The amplitude for $B|_{t=0} \to X|_{t=t}$ is derived from (35) as

$$\mathcal{M}(B^0 \to X) = e^{-imt - T/2} A \left\{ \cos(x - iy)\frac{T}{2} - ir\sin(x - iy)\frac{T}{2} \right\} \quad (81)$$

using the ratio
$$r := \frac{\eta_m \overline{A}}{A} \tag{82}$$

where $A = A(B^0 \to X)$ and $\overline{A} = A(\overline{B}^0 \to X)$ are the decay amplitudes of the flavour eigenstates, and the latter occurs in combination with the oscillation $B^0 \to \overline{B}^0$. The amplitude for $\overline{B}|_{t=0} \to X|_{t=t}$ is correspondingly

$$\overline{\mathcal{M}}(\overline{B}^0 \to X) = e^{-imt-T/2} \overline{A} \left\{ \cos(x-iy)\frac{T}{2} - \frac{i}{r}\sin(x-iy)\frac{T}{2} \right\} \tag{83}$$

The decay rates are proportional to

$$|\mathcal{M}|^2 = e^{-T}\left\{ |A|^2 \left|\cos(x-iy)\frac{T}{2}\right|^2 + |A|^2|r|^2 \left|\sin(x-iy)\frac{T}{2}\right|^2 \right.$$
$$+ i\sin x\frac{T}{2}\cos x\frac{T}{2}(A^*\overline{A}\eta_m - A\overline{A}^*\eta_m^*)$$
$$\left. + \sinh y\frac{T}{2}\cosh y\frac{T}{2}(A^*\overline{A}\eta_m + A\overline{A}^*\eta_m^*) \right\} \tag{84}$$

$$= e^{-T}|A|^2 \left\{ \frac{1+|r|^2}{2}\cosh yT + \frac{1-|r|^2}{2}\cos xT \right.$$
$$\left. + |r|\cos(\arg r)\sinh yT - |r|\sin(\arg r)\sin xT \right\}$$

$$= e^{-T}|A|^2 \left\{ \frac{1+|r|^2}{2}\cosh yT + \frac{1-|r|^2}{2}\cos xT \right.$$
$$\left. + \operatorname{Re} r \sinh yT - \operatorname{Im} r \sin xT \right\} \tag{84a}$$

$$|\overline{\mathcal{M}}|^2 = e^{-T}|\overline{A}|^2 \left\{ \frac{1+|r|^2}{2|r|^2}\cosh yT - \frac{1-|r|^2}{2|r|^2}\cos xT \right.$$
$$\left. + \frac{\cos(\arg r)}{|r|}\sinh yT + \frac{\sin(\arg r)}{|r|}\sin xT \right\}$$

$$= e^{-T}\frac{|A|^2}{|\eta_m|^2} \left\{ \frac{1+|r|^2}{2}\cosh yT - \frac{1-|r|^2}{2}\cos xT \right.$$
$$\left. + \operatorname{Re} r \sinh yT + \operatorname{Im} r \sin xT \right\} \tag{84b}$$

For the sum over all decays, we replace $\sum_X |A|^2 \to \Gamma$, $\sum_X |\overline{A}|^2 \to \Gamma$, $|r|^2 \to |\eta_m|^2$ and $\sum_X A^*\overline{A} \to \sum_X \langle B^0|\mathbf{H}_w|X\rangle\langle X|\mathbf{H}_w|\overline{B}^0\rangle = \Gamma_{12}$. From (28) one can write

$$\sum_X \eta_m A^*\overline{A} = \eta_m \Gamma_{12} = i\eta_m(H_{12} - H_{21}^*)$$
$$= \frac{\Gamma}{2}\left[-ix(1-|\eta_m|^2) - y(1+|\eta_m|^2)\right]$$
$$\approx \Gamma(1-\delta_\epsilon)\left[-ix\delta_\epsilon - y\right]$$

where the last line is an approximation for $\delta_\epsilon \ll 1$, which is good for all four meson pairs. This yields a total rate

$$\frac{dN}{dt} = N_0 \Gamma \frac{1}{1+\delta_\epsilon} e^{-T} \left\{ \cosh yT + \delta_\epsilon \cos xT - y \sinh yT + x\delta_\epsilon \sin xT \right\} \quad (85)$$

using $(1 + |\eta_m|^2) \cdot (1 + \delta_\epsilon) \equiv 2$. The total number of decays is the initial number of mesons N_0, as can be verified from

$$\int_0^\infty \frac{dN}{dt} dt = \frac{1}{\Gamma} \int_0^\infty \frac{dN}{dt} dT$$
$$= N_0 \frac{1+|\eta_m|^2}{2} \left\{ \frac{1}{1-y^2} + \delta_\epsilon \frac{1}{1+x^2} - y \frac{y}{1-y^2} + x\delta_\epsilon \frac{x}{1+x^2} \right\}$$
$$= N_0 \quad (86)$$

In the approximation $y = 0$ expected to be good for the B^0 the rate is

$$\frac{dN}{dt} = N_0 \Gamma e^{-T} \left[1 + \delta_\epsilon \cdot (-1 + \cos xT + x \sin xT) \right] + \mathcal{O}(\delta_\epsilon^2) \quad (87)$$

An initially pure $\overline{B}^0$ sample gives the same rates with the replacement $\eta_m \longleftrightarrow 1/\eta_m$ and $\delta_\epsilon \longleftrightarrow -\delta_\epsilon$. The asymmetry (where the B flavour is understood as the one at $T = 0$) is therefore

$$a(T) = \left. \frac{\dot{N}(\overline{B}^0 \to \text{anything}) - \dot{N}(B^0 \to \text{anything})}{\dot{N}(\overline{B}^0 \to \text{anything}) + \dot{N}(B^0 \to \text{anything})} \right|_T$$
$$= \delta_\epsilon \left(1 - \cos xT - x \sin xT \right)$$
$$= 2\delta_\epsilon \left(-\frac{x}{2} \sin xT + \sin^2 \frac{xT}{2} \right) \quad (88)$$

Another approximation, $\delta_\epsilon = 0$, gives the total decay rate of the B_s from (85) as

$$\frac{dN}{dt} = N_0 \Gamma e^{-T} \left\{ \cosh yT - y \sinh yT \right\} \propto \Gamma_H e^{-\Gamma_H t} + \Gamma_L e^{-\Gamma_L t}$$

that has been used in [83] to obtain an upper limit on y_s.

3.4.2 Experimental Results

CLEO [54] has obtained an asymmetry (80) of $a = 0.031 \pm 0.096 \pm 0.032$ from lepton pairs, corresponding to $\delta_\epsilon = 0.016 \pm 0.048 \pm 0.016$. This first upper limit of $\delta_\epsilon < 0.09$ (90%CL) was still far above expectation. A recent update [108] results in a value of $a = 0.013 \pm 0.050 \pm 0.005$. An independent method [51] using reconstructed $B^0 \to D^{*-} \pi^+ (\pi^0)$ on one side and leptons

Table 8. Results on $\delta_\epsilon(B^0)$.

	$\delta_\epsilon\ [10^{-3}]$	method	experiment
	$-11 \pm 15 \pm 5$	incl.	DELPHI prel. 97 [110]
	$4 \pm 14 \pm 6$	ll	OPAL 97 [59]
	$3 \pm 28 \pm 6$	incl.	OPAL 99 [111]
	$-18 \pm 16 \pm 3$	ll	ALEPH 00 [112]
	$8 \pm 17 \pm 4$	incl.	ALEPH 00 [112]
	$7 \pm 21 \pm 3$	ll, Bl	CLEO 01 [54,108]
	$2.4 \pm 5.8 \pm 7.2$	ll	BABAR 02 [109]
	-0.7 ± 5.7		average

from the other B gives $a = 0.017 \pm 0.070 \pm 0.014$. The average of both analyses corresponds to $\delta_\epsilon = 0.007 \pm 0.021 \pm 0.003$. The B factories allow for much smaller statistical errors. A dilepton analysis by BABAR [109] gives $a = 0.005 \pm 0.012 \pm 0.014$ and dominates the average in Table 8.

With B^0 mesons from jets, the interpretation of any charge asymmetry in lepton production is more difficult. OPAL at LEP gives a value $\delta_\epsilon = 0.004 \pm 0.014 \pm 0.006$ [59]. A measurement by ALEPH using semileptonic B^0 decays [112] gave the result $a = -0.037 \pm 0.032 \pm 0.007$ corresponding to $\delta_\epsilon = -0.018 \pm 0.016 \pm 0.003$.

A first measurement of the total asymmetry at LEP gave the preliminary result $\delta_\epsilon = -0.011 \pm 0.015 \pm 0.005$ [110]. OPAL obtains $a = 0.005 \pm 0.055 \pm 0.013$ corresponding to $\delta_\epsilon = 0.003 \pm 0.028 \pm 0.006$ [111]. ALEPH has published an asymmetry $a = 0.016 \pm 0.034 \pm 0.009$ corresponding to $\delta_\epsilon = 0.008 \pm 0.017 \pm 0.004$ [112]. This is an alternative way to measure CP violation in $B^0/\overline{B}^0$ oscillation, but it implies the danger of a possible bias due to the B^0 event selection.

CPT conservation is assumed in these results. Averaging gives

$$\delta_\epsilon = -0.0007 \pm 0.0057$$

or $-0.012 < \delta_\epsilon < 0.010$ at 95%CL.

3.5 CP Violation in Common Final States of B^0 and $\overline{B}^0$

The most pronounced manifestation of CP violation in the $B^0/\overline{B}^0$ system is expected in interference of oscillation and decay to final states common to B^0 and $\overline{B}^0$ [113,114]. The effect is largest for CP eigenstates, but may occur at any final state where the amplitudes of the mixed and unmixed decay can interfere:

$$B^0 \underset{\nearrow}{\longrightarrow} \overset{\overline{B}^0}{\searrow} X \tag{89}$$

The simplest situation is the evolution of an isolated B^0 meson produced (e.g. incoherently in $b\bar{b}$ fragmentation) at $t = 0$ as a flavour eigenstate. An unambiguous flavour tag for the state at production time may be a charged state from the second b, which cannot mix. The amplitude for $B|_{t=0} \to X|_{t=t}$ is (81) where the ratio of the upper and lower path's amplitudes in (89) is

$$r := \eta_m \frac{\bar{A}}{A} = \frac{\langle \bar{B}^0 | B_L \rangle}{\langle B^0 | B_L \rangle} \frac{\langle X | \mathbf{H} | \bar{B}^0 \rangle}{\langle X | \mathbf{H} | B^0 \rangle} \tag{90}$$

The cos-term in (81) describes the lower path with pure $B/\bar{B}$ oscillation, while the sin-term is a true interference term that vanishes if $r = 0$. The amplitude for $\bar{B}|_{t=0} \to X|_{t=t}$ is (83). If X is a CP eigenstate, the ratio $\bar{A}/A$ is often just a phase, which includes the sign of the CP eigenvalue of X. The phase of the product r is independent of conventions, and is in fact an observable, as will be shown below. More general, A and $\bar{A}$ can have also different magnitudes, which in the absence of oscillation would still imply an asymmetry and corresponds to the direct CP violation of non-oscillating particles. This direct CP violation is responsible for the cos-term in (81) for final states that can be reached by both B^0 and $\bar{B}^0$. Fig. 10 shows an example for this case, where the diagrams for $B^0 \to \bar{D}^0 K_S^0$ and $\bar{B}^0 \to \bar{D}^0 K_S^0$ are different. Another example is a mixture of CP eigenstates, as in the final state $D^{*+}D^{*-}$ which is CP = -1 for $L = 1$ and CP = $+1$ for $L = 0$ or 2.

The corresponding decay rates are proportional to $|\mathcal{M}|^2$ of (84a) and $|\bar{\mathcal{M}}|^2$ of (84b) where the oscillating interference term is proportional to

$$\tfrac{1}{2}(A^* \bar{A} \eta_m - A \bar{A}^* \eta_m^*) = \text{Im}(A^* \bar{A} \eta_m) = |A|^2 |r| \sin \arg r = |A|^2 \cdot \text{Im}\, r$$

and the relaxation part of the interference is proportional to $|r| \cos \arg r = \text{Re}\, r$. For decays to CP eigenstates, $|r| = |\eta_m| \approx 1$ and the amplitude of the asymmetry oscillation is the sine of the phase angle $\arg r$. For decays like $B^0/\bar{B}^0 \to D^0 K_S^0$ where $|A| \neq |\bar{A}|$ and consequently $|r| \neq 1$, the asymmetry oscillation gets a physical dilution factor

$$D_P = \frac{2|r|}{1 + |r|^2} = \frac{2|A||\bar{A}|}{|A|^2 + |\bar{A}|^2}$$

to the $\sin xT$ term, and a mixing contribution as an additional $\cos xT$ term. Using the parameters[12]

$$\Omega_0 := \frac{2 \text{Re}\, r}{1 + |r|^2} = D_P \cos \arg r, \quad \Lambda_0 := \frac{2 \text{Im}\, r}{1 + |r|^2} = D_P \sin \arg r,$$

$$\Theta_0 := \frac{|r|^2 - 1}{|r|^2 + 1} \tag{91}$$

---

[12] Another set of parameters in use is

$$S = \Lambda_0, \quad C = \frac{1 - |r|^2}{1 + |r|^2} = -\Theta_0$$

which are given by the two real numbers $\mathrm{Re}\, r$ and $\mathrm{Im}\, r$ and related via

$$\Omega_0^2 + \Lambda_0^2 + \Theta_0^2 = 1 \tag{92}$$

so that always two define the third one up to a sign. We can rewrite

$$|\mathcal{M}|^2 = e^{-T} |A|^2 \frac{1+|r|^2}{2} \Big\{ \cosh yT - \Theta_0 \cos xT$$
$$+ \Omega_0 \sinh yT - \Lambda_0 \sin xT \Big\} \tag{84a'}$$

$$|\overline{\mathcal{M}}|^2 = e^{-T} \frac{|A|^2}{|\eta_m|^2} \frac{1+|r|^2}{2} \Big\{ \cosh yT + \Theta_0 \cos xT$$
$$+ \Omega_0 \sinh yT + \Lambda_0 \sin xT \Big\} \tag{84b'}$$

This corresponds to a general asymmetry function[13]

$$a(T) = \frac{\dot{N}(\overline{B} \to X) - \dot{N}(B \to X)}{\dot{N}(\overline{B} \to X) + \dot{N}(B \to X)}\bigg|_T = \frac{|\overline{\mathcal{M}}|^2 - |\mathcal{M}|^2}{|\overline{\mathcal{M}}|^2 + |\mathcal{M}|^2}$$
$$= \frac{\Theta_0 \cos xT + \Lambda_0 \sin xT + \delta_\epsilon (\cosh yT + \Omega_0 \sinh yT)}{\cosh yT + \Omega_0 \sinh yT + \delta_\epsilon (\Theta_0 \cos xT + \Lambda_0 \sin xT)} \tag{93}$$

The slightly simplified form as expected for the $B_s/\overline{B}_s$ system is

$$a(T) = \frac{\dot{N}(\overline{B}_s \to X) - \dot{N}(B_s \to X)}{\dot{N}(\overline{B}_s \to X) + \dot{N}(B_s \to X)}\bigg|_T$$
$$= \frac{\Theta_0 \cos xT + \Lambda_0 \sin xT}{\cosh yT + \Omega_0 \sinh yT} = \frac{a_0 \sin(xT + \phi_0)}{\cosh yT + \Omega_0 \sinh yT} \tag{94}$$

where $\delta_{\epsilon s} = 0$ (corresponding to $|\eta_{ms}| = 1$) is used, which is believed to be a very good approximation. Under this condition, Θ_0 describes direct CP violation, while Λ_0 parametrizes the interference of oscillation and decay.

The parameter-triplet $(\Omega_0, \Lambda_0, \Theta_0)$ is forced to be on the unit sphere through (92), and can be given by the polar angle and azimuth (see Fig. 12)

$$\cos \theta_{\Omega\Lambda\Theta} = \frac{|r|^2 - 1}{|r|^2 + 1} = \sqrt{1 - D_P^2}, \quad \phi_{\Omega\Lambda\Theta} = \arg r$$

---

[13] This definition of the asymmetry is consistent with the definition of r, however, the opposite sign $a \to -a$ is also used in the literature, leading to flipped signs in the coefficients. To add to the confusion, when checking signs one has also to distinguish $N(\overline{B} \to X) = N(B + X)$, where the right hand side denotes the flavour of the tag, and the left hand side the flavour of the signal at $T = 0$.

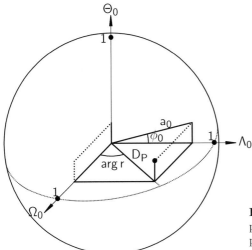

Fig. 12. The $(\Omega_0, \Lambda_0, \Theta_0)$ parameter space. All valid parameter sets are on the unit sphere.

The physical dilution factor is $D_P = \sqrt{\Omega_0^2 + \Lambda_0^2}$. The asymmetry amplitude at $T = 0$ is

$$a_0 = \sqrt{\Lambda_0^2 + \Theta_0^2} = \sqrt{1 - \Omega_0^2}$$

and the phase is given by

$$\tan \phi_0 = \frac{\Theta_0}{\Lambda_0}$$

In the approximation $|\eta_m| = 1$ and $y = 0$ the rates are given by

$$|\mathcal{M}|^2 = e^{-T}\left\{\frac{|A|^2 + |\overline{A}|^2}{2} + \frac{|A|^2 - |\overline{A}|^2}{2}\cos xT - |A||\overline{A}|\sin(\arg r)\sin xT\right\}$$

$$|\overline{\mathcal{M}}|^2 = e^{-T}\left\{\frac{|A|^2 + |\overline{A}|^2}{2} - \frac{|A|^2 - |\overline{A}|^2}{2}\cos xT + |A||\overline{A}|\sin(\arg r)\sin xT\right\}$$

(95)

These lead to an oscillating asymmetry as a function of the proper lifetime of the signal-B

$$a(T) = \left.\frac{\dot{N}(\overline{B}^0 \to X) - \dot{N}(B^0 \to X)}{\dot{N}(\overline{B}^0 \to X) + \dot{N}(B^0 \to X)}\right|_T = \Theta_0 \cos xT + \Lambda_0 \sin xT \quad (96)$$

where the B flavours are taken at $T = 0$, and the amplitudes are given by (91). If in addition $|A| = |\overline{A}|$, especially if X is a CP eigenstate, this simplifies further to $|r| = 1$:

$$|\mathcal{M}|^2 = e^{-T}|A|^2\left\{1 - \sin(\arg r)\sin xT\right\}$$
$$|\overline{\mathcal{M}}|^2 = e^{-T}|A|^2\left\{1 + \sin(\arg r)\sin xT\right\}$$

(97)

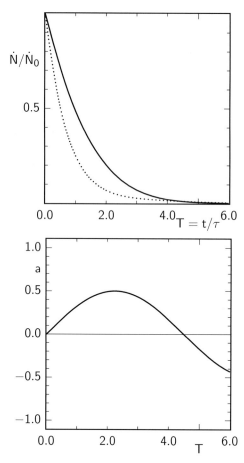

Fig. 13. Time dependent rate of $\overline{B}^0 \to J/\psi K^0_S$ (———) and $B^0 \to J/\psi K^0_S$ (········) for $\Lambda_0 = \sin 2\beta = 0.5$ and $x = 0.70$. The lower plot shows the asymmetry $a(T)$.

The corresponding rates $\dot{N}(B^0|_{T=0} \to X) \propto |\mathcal{M}|^2$ and $\dot{N}(\overline{B}^0|_{T=0} \to X) \propto |\overline{\mathcal{M}}|^2$ are illustrated in Fig. 13. They show a time-dependent asymmetry

$$a(T) = \left.\frac{\dot{N}(\overline{B}^0 \to X) - \dot{N}(B^0 \to X)}{\dot{N}(\overline{B}^0 \to X) + \dot{N}(B^0 \to X)}\right|_T = \Lambda_0 \sin xT \qquad (98)$$

with $\Lambda_0 = \sin\arg r$.

The dimensionless time variable is $T = t_s/\tau$ defined by the signal B lifetime t_s for incoherent $b\bar{b}$ production. For coherent $B\overline{B}$ production on the $\Upsilon(4S)$, it has to be replaced by $T = (t_s - t_t)/\tau$ where t_t denotes the second B meson in the $\Upsilon(4S)$ decay in a flavour tagging decay mode. Also, the factor e^{-T} has to be replaced by $e^{-|T|}$, which originates from the integration over the unobserved $t_s + t_t$. This is in full analogy to the mixing situation described in section 2.2.7, and will be discussed in more detail below.

3.5.1 CP Violation Described by r

The ratio r in (90) may include in fact all three types of CP violation:
- direct CP violation if $|\bar{A}/A| \neq 1$,
- CP violation in the oscillation if $|\eta_m| \neq 1$,
- and CP violation from interference of mixed and unmixed amplitudes if $\operatorname{Im} r \neq 0$.

If the small effect of CP violation in the oscillation is ignored, one can split the complex number r into two pieces, $|r| - 1$ describing direct CP violation and $\arg r$ describing the CP violating interference of mixed and unmixed amplitudes. The first can be derived from Θ_0 as

$$|r| = \sqrt{\frac{1 + \Theta_0}{1 - \Theta_0}} \tag{99}$$

Since

$$D_P = \sqrt{1 - \Theta_0^2}$$

the phase is obtained as

$$\arg r = \arcsin \frac{\Lambda_0}{\sqrt{1 - \Theta_0^2}} \tag{100}$$

and is dominated by Λ_0 for small values of Θ_0.

3.5.2 The Eigenstate Parametrization

Instead of the ratio (90) one may use in analogy to the terminology of kaon physics

$$\eta := \frac{A_{\mathrm{H}}}{A_{\mathrm{L}}} \tag{101}$$

where $A_{\mathrm{H,L}} = A(B_{\mathrm{H,L}} \to X)$ are the decay amplitudes of the mass (and width) eigenstates. If one leaves a free relative phase between B_{H} and B_{L}, one has to define

$$\eta := \frac{A_{\mathrm{H}}}{A_{\mathrm{L}}} \frac{\langle B^0 | B_{\mathrm{L}} \rangle}{\langle B^0 | B_{\mathrm{H}} \rangle} = -\frac{A_{\mathrm{H}}}{A_{\mathrm{L}}} \frac{\langle \bar{B}^0 | B_{\mathrm{L}} \rangle}{\langle \bar{B}^0 | B_{\mathrm{H}} \rangle} \tag{102}$$

which is determined unambiguously. It is related to the ratio r via

$$\eta = \frac{1 - r}{1 + r} = \mathrm{i} \tan \mathrm{i} \frac{\ln r}{2}, \quad r = \frac{1 - \eta}{1 + \eta} = \mathrm{i} \tan \mathrm{i} \frac{\ln \eta}{2}$$

For small δ_ϵ and $|\bar{A}| = |A|$ (no direct CP violation) we have $r = |\eta_m| e^{\mathrm{i}\phi_r}$ and

$$\eta = -\mathrm{i} \tan \frac{\phi_r}{2} \left(1 - \frac{\delta_\epsilon^2}{4}\right) + \frac{1}{2 \cos^2 \frac{\phi_r}{2}} \delta_\epsilon + \mathcal{O}(\delta^3) \tag{103}$$

The asymmetry parameters can be calculated as

$$\Omega_0 := \frac{1-|\eta|^2}{1+|\eta|^2}, \quad \Lambda_0 := -\frac{2\operatorname{Im}\eta}{1+|\eta|^2}, \quad \Theta_0 := -\frac{2\operatorname{Re}\eta}{1+|\eta|^2} \tag{104}$$

which implies

$$a_0 = \frac{2|\eta|}{1+|\eta|^2}, \quad \phi_0 = -\frac{\pi}{2} - \arg\eta \Leftrightarrow \arg\eta = -\frac{\pi}{2} - \phi_0$$

or $\eta = -\mathrm{i}|\eta|e^{-\mathrm{i}\phi_0}$

For a final state f that is a CP eigenstate, this parameter is related to ϵ as can be seen from (102)

$$\begin{aligned}
\eta &:= \frac{\langle f|\mathbf{H}|B_\mathrm{H}\rangle}{\langle f|\mathbf{H}|B_\mathrm{L}\rangle} \frac{\langle B^0|B_\mathrm{L}\rangle}{\langle B^0|B_\mathrm{H}\rangle} \\
&= \frac{\langle f|\mathbf{H}|B^0\rangle\langle B^0|B_\mathrm{H}\rangle + \langle f|\mathbf{H}|\overline{B}^0\rangle\langle \overline{B}^0|B_\mathrm{H}\rangle}{\langle f|\mathbf{H}|B^0\rangle\langle B^0|B_\mathrm{L}\rangle + \langle f|\mathbf{H}|\overline{B}^0\rangle\langle \overline{B}^0|B_\mathrm{L}\rangle} \frac{\langle B^0|B_\mathrm{L}\rangle}{\langle B^0|B_\mathrm{H}\rangle} \\
&= \frac{\langle f|\mathbf{H}|B^0\rangle\langle B^0|B_\mathrm{L}\rangle + \langle f|\mathbf{H}|\overline{B}^0\rangle\langle \overline{B}^0|B_\mathrm{H}\rangle\frac{\langle B^0|B_\mathrm{L}\rangle}{\langle B^0|B_\mathrm{H}\rangle}}{\langle f|\mathbf{H}|B^0\rangle\langle B^0|B_\mathrm{L}\rangle + \langle f|\mathbf{H}|\overline{B}^0\rangle\langle \overline{B}^0|B_\mathrm{L}\rangle} \\
&= \frac{\langle f|\mathbf{H}|B^0\rangle\langle B^0|B_\mathrm{L}\rangle - \langle f|\mathbf{H}|\overline{B}^0\rangle\langle \overline{B}^0|B_\mathrm{H}\rangle\frac{\langle \overline{B}^0|B_\mathrm{L}\rangle}{\langle \overline{B}^0|B_\mathrm{H}\rangle}}{\langle f|\mathbf{H}|B^0\rangle\langle B^0|B_\mathrm{L}\rangle + \langle f|\mathbf{H}|\overline{B}^0\rangle\langle \overline{B}^0|B_\mathrm{L}\rangle} \\
&= \frac{\langle f|\mathbf{H}|B^0\rangle\langle B^0|B_\mathrm{L}\rangle - \langle f|\mathbf{H}|\overline{B}^0\rangle\langle \overline{B}^0|B_\mathrm{L}\rangle}{\langle f|\mathbf{H}|B^0\rangle\langle B^0|B_\mathrm{L}\rangle + \langle f|\mathbf{H}|\overline{B}^0\rangle\langle \overline{B}^0|B_\mathrm{L}\rangle}
\end{aligned}$$

Since $|\langle f|\mathbf{H}|B^0\rangle| = |\langle f|\mathbf{H}|\overline{B}^0\rangle|$ and one can choose a phase ϕ_{CPB} so that $\langle f|\mathbf{H}|B^0\rangle = \langle f|\mathbf{H}|\overline{B}^0\rangle$ and

$$\eta_f = \frac{\langle B^0|B_\mathrm{L}\rangle - \langle \overline{B}^0|B_\mathrm{L}\rangle}{\langle B^0|B_\mathrm{L}\rangle + \langle \overline{B}^0|B_\mathrm{L}\rangle} = \epsilon$$

This is usually exploited in the neutral kaon system, where there is one dominant final state, the two pion state, and thus a convenient phase convention yields

$$\epsilon_m = \eta_{\pi\pi,I=0} \approx \eta_{+-} \approx \eta_{00}$$

3.5.3 Parameters for Conserved CP

The parameters $\Omega_0, \Lambda_0, \Theta_0$ describe also oscillation phenomena if CP is conserved.

The special cases $r = 0$ and $r = \infty$ correspond to a final state that occurs only from one flavour. The corresponding parameter set

$$\Omega_0 = 0, \quad \Lambda_0 = 0, \quad \Theta_0 = \mp 1$$

describes the pure $B\bar{B}$ oscillation

$$a(T) = \mp \cos xT$$

(for $\delta_\epsilon = 0$).

If there is a small, suppressed amplitude from the second flavour, but no CP violating phase difference, we have a real ratio $r \ll 1$. In this case, the parameters are

$$\Omega_0 = \frac{2r}{1+r^2} \approx 2r, \quad \Lambda_0 = 0, \quad \Theta_0 = \frac{r^2-1}{r^2+1} \approx -1$$

corresponding to an asymmetry

$$a(T) = \frac{-\cos xT}{\cosh yT + 2r \sinh yT}$$

In the case of final CP eigenstates and conserved CP, the values are $r = \pm 1$ and

$$\Omega_0 = \pm 1, \quad \Lambda_0 = 0, \quad \Theta_0 = 0$$

corresponding to

$$a(T) = 0$$

i.e. equal differential decay rates for B and $\bar{B}$.

3.5.4 The $B_s/\bar{B}_s$ Case

For the B_s meson, y is not negligible, and the asymmetry oscillation (96) is modulated as given by (94). For the simpler case $|r| = 1$ the asymmetry is

$$a(T) = \left.\frac{\dot{N}(\bar{B}_s \to X) - \dot{N}(B_s \to X)}{\dot{N}(\bar{B}_s \to X) + \dot{N}(B_s \to X)}\right|_T = \frac{\Lambda_0 \sin xT}{\cosh yT + \Omega_0 \sinh yT} \tag{105}$$

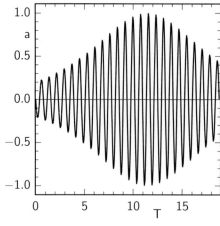

Fig. 14. Time dependent asymmetry for $B_s/\overline{B}_s \to$ CP eigenstate, assuming for $\Lambda_0 = -0.2$, $\Omega_0 > 0$, $x = 8$ and $y = -0.2$ (these unrealistic values are chosen for illustration of the effects).

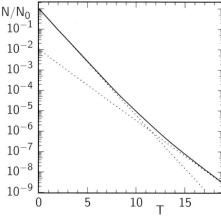

Fig. 15. The two components B_{sH} and B_{sL} of the $B_s + \overline{B}_s$ decay to a CP eigenstate for the parameter set of Fig. 14. The maximum CP asymmetry occurs at the point where the B_{sH} and B_{sL} components of this decay are equal.

with $\Lambda_0 = \sin \arg r$ and $\Omega_0 = \cos \arg r$.

The envelope function modulating the asymmetry amplitude

$$f(T) = \frac{1}{\cosh yT + \Omega_0 \sinh yT} \tag{106}$$

has a maximum

$$f(T_0) = \frac{1}{\sqrt{1 - \Omega_0^2}} \quad \text{at} \quad T_0 = \frac{1}{2y} \ln \frac{1 - \Omega_0}{1 + \Omega_0} = -\frac{\operatorname{Artanh} \Omega_0}{y}$$

and a width

$$\text{FWHM} = \frac{2 \ln(2 + \sqrt{3})}{|y|} \approx \frac{2.634}{|y|}$$

The maximum asymmetry amplitude at $T = T_0$ is

$$\frac{a_0}{\sqrt{1 - \Omega_0^2}} = 1$$

due to (92). In the case $|r| = 1$ (where $\Theta_0 = 0$) the asymmetry amplitude is

$$\frac{\Lambda_0}{\sqrt{1 - \Omega_0^2}} = 1$$

This effect is illustrated in Fig. 14, using $x = 8$, $y = -0.2$, $\Theta_0 = 0$, and $\Lambda_0 = 0.2$, corresponding to $\Omega_0 = 0.9798$. The asymmetry amplitude is amplified to its maximum possible value of 1 at $T_0 \approx 11.5$, i.e. more than 11 mean lifetimes from production. At that time, only a fraction $\mathrm{e}^{-T_0} \approx 10^{-5}$ of all B_s mesons is left. For more realistic smaller $|y|$ values, the maximum is even further out.

Note that both the sign of Λ_0 and Ω_0 change with the CP eigenvalue of an eigenstate. Therefore one of a pair of two always has the maximum on the negative T axis, which is not accessible in a jet production experiment.

The modulation function (106) can be rewritten as

$$f(T) = \frac{2}{(1 + \Omega_0)\mathrm{e}^{yT} + (1 - \Omega_0)\mathrm{e}^{-yT}}$$

The maximum corresponds to the minimum of the denominator, which is the position where

$$(1 + \Omega_0)\mathrm{e}^{yT} = (1 - \Omega_0)\mathrm{e}^{-yT}$$

or, multiplying with e^{-T} and using (26), where

$$(1 + \Omega_0)\mathrm{e}^{-t/\tau_\mathrm{L}} = (1 - \Omega_0)\mathrm{e}^{-t/\tau_\mathrm{H}}$$

This is exactly the point, where $B_{s\mathrm{L}}$ and $B_{s\mathrm{H}}$ contribute equally to the decay probability

$$|\mathcal{M}|^2 + |\overline{\mathcal{M}}|^2 = 2\mathrm{e}^{-|T|}|A|^2 \left\{\cosh yT + \cos\arg r \cdot \sinh yT\right\}$$
$$= |A|^2 \left\{(1 + \Omega_0)\mathrm{e}^{-t/\tau_\mathrm{L}} + (1 - \Omega_0)\mathrm{e}^{-t/\tau_\mathrm{H}}\right\} \quad (107)$$

as illustrated in Fig. 15. A completely equivalent point of view to the interference of $B_s \to X$ and $B_s \to \overline{B}_s \to X$ amplitudes is the picture of interfering $B_{s\mathrm{L}}$ and $B_{s\mathrm{H}}$ states. In this picture, it becomes obvious that the interference is maximum at the point where both states contribute with equal magnitude.

The time evolution of an equal, untagged mixture of B_s and $\overline{B}_s$ mesons is given by

$$|\mathcal{M}|^2 + |\overline{\mathcal{M}}|^2 = 2\mathrm{e}^{-|T|}|A|^2 \frac{1 + |r|^2}{2} \left\{\cosh yT + \Omega_0 \sinh yT\right\} \quad (108)$$

For decays into CP eigenstates ($|r| = 1$) the time dependence of the decay probability is simply (107), i.e. it is a sum of two exponential distributions

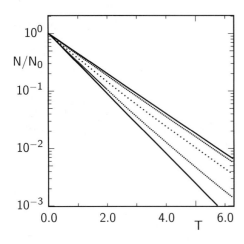

Fig. 16. Time dependent rate of $B_s + \bar{B}_s \to X$ for $\arg r = 0$ or π (———), for $\arg r = \pi/4$ or $3\pi/4$ (- - - -), and for $\arg r = \pi/2$ (······), using $y = -0.2$.

with weights $(1 \pm \Omega_0) = (1 \pm \cos \arg r)$. This opens an alternative way to measure the CP violation parameter $\arg r$, as illustrated in Fig. 16 on a logarithmic scale. The upper and lower solid curve correspond to the CP conserving case $r = \pm 1$, where e.g. $B_{s\mathrm{L}}$ decays into either CP $= +1$ or CP $= -1$ eigenstates exclusively, and $B_{s\mathrm{H}}$ into the opposite one. The central curve corresponds to maximum CP violation, i.e. $\mathrm{Re}\, r = 0$, where both $B_{s\mathrm{L}}$ and $B_{s\mathrm{H}}$ decay into CP $= +1$ and CP $= -1$ eigenstates with the same probability, The other two curves correspond to $\cos \arg r = \pm 0.7$. Fig. 15 shows an example with $\Omega_0 \approx 0.98$.

If $\arg r = 2\phi_{\mathrm{CKM}}$ is a large angle, a measurement of $\cos 2\phi_{\mathrm{CKM}}$ via the mixture of short and long lived states is complementary to a measurement of $\sin 2\phi_{\mathrm{CKM}}$ via an oscillating asymmetry function (105). Due to the large value of x_s, the latter requires a very precise measurement of the individual lifetimes and flavour tagging, while the decomposition of the short and long lived fractions can be done with untagged events and a modest resolution, but requires a large data sample.

3.5.5 CP Violation at the $\Upsilon(4S)$

B meson pairs from $\Upsilon(4S)$ decay are initially in a CP $= +1$, P $= -1$, C $= -1$ eigenstate

$$|B^0(1)\bar{B}^0(2)\rangle - |B^0(2)\bar{B}^0(1)\rangle$$

with angular momentum $L = 1$. Their time evolution is described by (63a), where the two scaled times T_1 and T_2 may be taken as the decay times of the two mesons. By multiplying this state function with $\langle X(1)X(2)|\mathbf{H}(1)\mathbf{H}(2)$ where

$$\langle X(1)X(2)|\mathbf{H}(1)\mathbf{H}(2)|B^0(1)B^0(2)\rangle = \langle X(1)|\mathbf{H}|B^0(1)\rangle \langle X(2)|\mathbf{H}|B^0(2)\rangle$$
$$= A_1 A_2$$

are the decay amplitudes of two B^0 mesons and amplitudes of other mixtures of $B(i)$ and $\overline{B}(j)$ yield products of A_i and $\overline{A}_j$ accordingly, one obtains an amplitude

$$\mathcal{M}_- := e^{i\phi_0} e^{-\frac{1}{2}(T_1+T_2)} \left[C_- \cos(x-iy) \frac{T_1-T_2}{2} + iS_- \sin(x-iy) \frac{T_1-T_2}{2} \right] \tag{109}$$

where ϕ_0 is a common, unobservable phase including the imt phases of two free B mesons, and the coefficients are

$$C_- = A_1 \overline{A}_2 - \overline{A}_1 A_2, \quad S_- = \eta_m \overline{A}_1 \overline{A}_2 - \frac{A_1 A_2}{\eta_m}$$

There can be always two non-zero amplitude factors separated, leaving coefficients like

$$\frac{C_-}{A_1 \overline{A}_2} = 1 - \frac{\overline{A}_1}{A_1} \frac{A_2}{\overline{A}_2}, \quad \frac{S_-}{A_1 \overline{A}_2} = \frac{\eta_m \overline{A}_1}{A_1} - \frac{A_2}{\eta_m \overline{A}_2}$$

These coefficients are **convention independent** factors similar to r: CP phases common to $A/\overline{A}$ and η_m cancel, and the exchange of quarks with antiquarks ensure that the product of CKM elements has each quark index in as many V as V^* (or, with the same phase, $1/V$) factors. From the general amplitude, we can derive various special cases listed in Table 9.

The square $|\mathcal{M}_-|^2$ leads to a general formula [115], which reads on the $\Upsilon(4S)$ with the final states X_1, X_2 from the two B^0 mesons

$$\dot{N}(B\overline{B} \to X_1 X_2) \propto e^{-2\Gamma t_1} e^{-T} (g_1 \cosh yT + g_2 \sinh yT + h_1 \cos xT + h_2 \sin xT) \tag{110}$$

with $T = T_2 - T_1 = \Gamma(t_2 - t_1)$ and

$$g_1 = |C_-|^2 + |S_-|^2$$
$$g_2 = 2\operatorname{Re}(S_-^* C_-)$$
$$h_1 = |C_-|^2 - |S_-|^2$$
$$h_2 = 2\operatorname{Im}(S_-^* C_-)$$

For B^0 mesons, we can assume $\delta_\epsilon = 0$ and $y = 0$, which corresponds to the simpler equation

$$\dot{N}(B\overline{B} \to X_1 X_2) \propto e^{-2\Gamma t_1} e^{-T} (g_1 + h_1 \cos xT + h_2 \sin xT) \tag{111}$$

This includes the case of $B\overline{B}$ oscillation, if we set e.g. $\overline{A}_1 = A_2 = 0$. Then $C_- = A_1 \overline{A}_2$, $S_- = 0$, $g_1 = h_1 = |A_1 \overline{A}_2|^2$, and $g_2 = h_2 = 0$. For a mixed mode we use e.g. $\overline{A}_1 = \overline{A}_2 = 0$. Then $C_- = 0$, $S_- = -A_1 A_2/\eta_m$, $g_1 = -h_1 = |A_1 A_2|^2$, and $g_2 = h_2 = 0$. The corresponding asymmetry is

$$a(T) = \frac{\dot{N}(X_1 \overline{X}_2) - \dot{N}(X_1 X_2)}{\dot{N}(X_1 \overline{X}_2) + \dot{N}(X_1 X_2)} = \frac{\dot{N}(\overline{X}_1 X_2) - \dot{N}(\overline{X}_1 \overline{X}_2)}{\dot{N}(\overline{X}_1 X_2) + \dot{N}(\overline{X}_1 \overline{X}_2)} = \cos xT \tag{112}$$

corresponding to Fig. 5 for positive T as well as for negative T with $T \to -T$.

Table 9. Coefficients for $B\bar{B} \to X_1 X_2$ decays from $\Upsilon(4S)$. $\xi_{1,2}$ are the CP eigenvalues of final states 1, 2, respectively, and $\xi_{12} = \xi_1 \xi_2$. $d\bar{d}$ denotes a penguin mode with a t quark in the loop, "com." denotes a final state that can be reached from both B^0 and $\bar{B}^0$ and needs not to be a CP eigenstate. In this case, also strong phases are involved, which can be resolved using the charged conjugate final state, too. The coefficients of the time-dependent rate according to (111) are also given for $\delta_\epsilon = 0$.

X_1	X_2	AA	C_-/AA	S_-/AA	$g_1/|AA|^2$	$h_1/|AA|^2$	$h_2/|AA|^2$
B-tag	B-tag	$A_1 A_2$	0	$-1/\eta_m$	1	-1	0
$\bar{B}$-tag	$\bar{B}$-tag	$\bar{A}_1 \bar{A}_2$	0	η_m	1	-1	0
B-tag	$\bar{B}$-tag	$A_1 \bar{A}_2$	1	0	1	1	0
$\bar{B}$-tag	B-tag	$\bar{A}_1 A_2$	-1	0	1	1	0
com.	B-tag	$\bar{A}_1 A_2$	-1	$-\dfrac{1}{r} = -\dfrac{A'_1}{\eta_m A_1}$	$\dfrac{1+|r|^2}{|r|^2}$	$\dfrac{1-|r|^2}{|r|^2}$	$\dfrac{2\,\text{Im}\,r}{|r|^2}$
com.	$\bar{B}$-tag	$A'_1 \bar{A}_2$	1	$r = \dfrac{\eta_m A_1}{A'_1}$	$1+|r|^2$	$1-|r|^2$	$-2\,\text{Im}\,r$
$c\bar{c}$	B-tag	$\bar{A}_1 A_2$	-1	$-\xi_1 e^{2i\beta}$	2	0	$-2\xi_1 \sin 2\beta$
$c\bar{c}$	$\bar{B}$-tag	$A_1 \bar{A}_2$	1	$\xi_1 e^{-2i\beta}$	2	0	$2\xi_1 \sin 2\beta$
$u\bar{u}$	B-tag	$\bar{A}_1 A_2$	-1	$-\xi_1 e^{-2i\alpha}$	2	0	$2\xi_1 \sin 2\alpha$
$u\bar{u}$	$\bar{B}$-tag	$A_1 \bar{A}_2$	1	$\xi_1 e^{2i\alpha}$	2	0	$-2\xi_1 \sin 2\alpha$
$d\bar{d}$	B-tag	$\bar{A}_1 A_2$	-1	$-\xi_1$	2	0	0
$d\bar{d}$	$\bar{B}$-tag	$A_1 \bar{A}_2$	1	ξ_1	2	0	0
$c\bar{c}$	$c\bar{c}$	$2 A_1 \bar{A}_2$	$\dfrac{1-\xi_{12}}{2}$	$\dfrac{\xi_1}{2} e^{2i\beta} + \dfrac{\xi_2}{2} e^{-2i\beta}$	$1 - \xi_{12} \cos^2 2\beta$	$-\xi_{12} \sin^2 2\beta$	0
$c\bar{c}$	$u\bar{u}$	$2 A_1 \bar{A}_2$	$\dfrac{1}{2} - \dfrac{\xi_{12}}{2} e^{2i(\beta+\alpha)}$	$\dfrac{\xi_1}{2} e^{2i\beta} + \dfrac{\xi_2}{2} e^{2i\alpha}$	$1 - \xi_{12} \cos 2\beta \cos 2\alpha$	$\xi_{12} \sin 2\beta \sin 2\alpha$	0
$c\bar{c}$	$d\bar{d}$	$2 A_1 \bar{A}_2$	$\dfrac{1}{2} - \dfrac{\xi_{12}}{2} e^{2i\beta}$	$\dfrac{\xi_1}{2} e^{2i\beta} + \dfrac{\xi_2}{2}$	$1 - \xi_{12} \cos 2\beta$	0	0

For a CP eigenstate with eigenvalue $\xi_1 = \pm 1$ as meson (1), $\bar{A}_1/A_1 = \xi_1 e^{-2i\tilde{\phi}}$, where $\tilde{\phi}$ is the phase of the CKM matrix elements involved. If the state (2) is a tagging mode, e.g. $A_2 \neq 0$, $\bar{A}_2 = 0$ we have the situation of a tagged decay with

$$C_- = \bar{A}_1 A_2 \cdot (-1)$$
$$S_- = \bar{A}_1 A_2 \cdot (-\xi_1 e^{2i(\tilde{\phi}+\tilde{\beta})})$$
$$g_1 = 2|\bar{A}_1|^2|A_2|^2$$
$$h_1 = 0$$
$$h_2 = -2|\bar{A}_1|^2|A_2|^2 \xi_1 \sin 2(\tilde{\phi}+\tilde{\beta})$$

which is the typical situation of CP violation in the oscillation/decay interference, with a T dependence as shown in Fig. 17. The dotted curve corresponds to a $\bar{B}$ tag $A_2 = 0$, $\bar{A}_2 \neq 0$ with

$$C_- = A_1 \bar{A}_2$$
$$S_- = A_1 \bar{A}_2 \cdot (\xi_1 e^{-2i(\tilde{\phi}+\tilde{\beta})})$$
$$g_1 = 2|A_1|^2|\bar{A}_2|^2$$
$$h_1 = 0$$
$$h_2 = 2|A_1|^2|\bar{A}_2|^2 \xi_1 \sin 2(\tilde{\phi}+\tilde{\beta})$$

and the asymmetry of both is described by equation (98) with $\Lambda_0 = -\xi_1 \sin 2(\tilde{\phi}+\tilde{\beta})$. The flavour of the signal-B meson (1) is uniquely defined at the time of decay t_2 of the second B meson, since the flavour of the latter can be identified by its final state.

To derive this immediately from (109) for the special case of a decay to a CP eigenstate where r is just a phase factor, together with a $\bar{B}$ tag, one inserts $\frac{C_-}{A_1 \bar{A}_2} = 1$, $\frac{S_-}{A_1 \bar{A}_2} = r$, and for $y = 0$ finds

$$\mathcal{M}_- = A_1 \bar{A}_2 e^{i\phi_0} e^{-\frac{1}{2}(T_1+T_2)} \left[\cos x \frac{T_1-T_2}{2} - ir \sin x \frac{T_1-T_2}{2} \right]$$

$$|\mathcal{M}_-|^2 = |A_1 \bar{A}_2|^2 e^{-(T_1+T_2)} \left[1 + 2 \operatorname{Im} r \cos x \frac{T_1-T_2}{2} \sin x \frac{T_1-T_2}{2} \right]$$
$$= |A_1 \bar{A}_2|^2 e^{-(T_1+T_2)} \left[1 + \operatorname{Im} r \sin x (T_1-T_2) \right]$$

If both final states are CP eigenstates with eigenvalues $\xi_{1,2} = \pm 1$, $\bar{A}_{1,2}/A_{1,2} = \xi_{1,2} e^{-2i\tilde{\phi}_{1,2}}$, where $\tilde{\phi}_{1,2}$ are the phases of the CKM matrix elements involved, the coefficients are

$$C_- = A_1 \bar{A}_2 (1 - \xi_1 \xi_2 e^{2i(\tilde{\phi}_1 - \tilde{\phi}_2)})$$
$$S_- = A_1 \bar{A}_2 (\xi_1 e^{-2i(\tilde{\phi}_1+\tilde{\beta})} - \xi_2 e^{2i(\tilde{\phi}_2+\tilde{\beta})})$$
$$g_1 = 4|A_1|^2|A_2|^2 [1 - \xi_1 \xi_2 \cos 2(\tilde{\phi}_1+\tilde{\beta}) \cos 2(\tilde{\phi}_2+\tilde{\beta})]$$
$$h_1 = -4|A_1|^2|A_2|^2 \xi_1 \xi_2 \sin 2(\tilde{\phi}_1+\tilde{\beta}) \sin 2(\tilde{\phi}_2+\tilde{\beta})$$
$$h_2 = 0$$

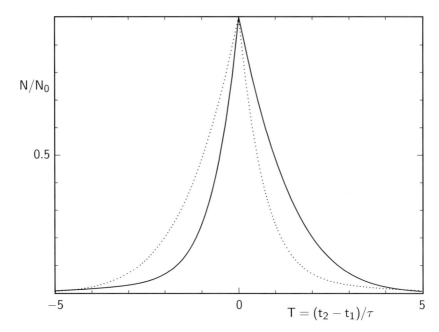

Fig. 17. Time dependent rate of $\Upsilon(4S) \to B^0 + J/\psi K_S^0$ (———) and $\Upsilon(4S) \to \overline{B}^0 + J/\psi K_S^0$ (········) for $\sin 2\beta = 0.5$ and $x = 0.7$.

which leads to a rate

$$\dot{N}(B\overline{B} \to X_1 X_2) \propto 4|A_1|^2|A_2|^2 e^{-\Gamma t_1} e^{-T} \left[1 - \xi_1 \xi_2 \big(\cos 2\phi_1 \cos 2\phi_2 + \sin 2\phi_1 \sin 2\phi_2 \cos xT\big)\right]$$

where the observable phase angles are $\phi_1 = \tilde{\phi}_1 + \tilde{\beta}$ and $\phi_2 = \tilde{\phi}_2 + \tilde{\beta}$. If CP were conserved throughout the decay chain, the eigenvalues are $\xi_1 = -\xi_2$ since the total CP eigenvalue $+1$ is achieved via another factor (-1) from the relative $L = 1$ angular momentum of both B mesons. This corresponds to

$$\dot{N}(B\overline{B} \to X_{1\pm} X_{2\mp}) \propto 4|A_1|^2|A_2|^2 e^{-\Gamma(t_1+t_2)} \left[1 + \cos 2\phi_1 \cos 2\phi_2 + \sin 2\phi_1 \sin 2\phi_2 \cos xT\right] \qquad (113)$$

while the forbidden rate, where CP of the final state is flipped to -1, is

$$\dot{N}(B\overline{B} \to X_{1\pm} X_{2\pm}) \propto 4|A_1|^2|A_2|^2 e^{-\Gamma(t_1+t_2)} \left[1 - \cos 2\phi_1 \cos 2\phi_2 - \sin 2\phi_1 \sin 2\phi_2 \cos xT\right] \qquad (114)$$

If both decays proceed via the same flavour changing transitions, the CKM angles are the same for both B mesons, $\phi_1 = \phi_2 = \phi$, and the forbidden rate is

$$\dot{N}(B\overline{B} \to X_{1\pm} X_{2\pm}) \propto 4|A_1|^2|A_2|^2 e^{-\Gamma t_1} e^{-T} \sin^2 2\phi (1 - \cos xT)$$

This rate should be 0. It does actually vanish for $T = 0$, i.e. equal decay times of both. Only if their lifetimes differ CP violation builds up in the interference term of oscillation and decay processes. For different CKM angles, even at $T = 0$ there is a CP violating rate $\propto (1 - \cos(\phi_1 - \phi_2))$ which can be called consistently a special case of direct CP violation. The interference term is here, in contrast to the example (77), not between two amplitudes for one B decay, but between two amplitudes for two different B decays, which are in a coherent $B\overline{B}$ state. All these interesting cases will, however, not be observed in the first generation experiments on the $\Upsilon(4S)$, since they involve a product of two small branching ratios which are typically below 10^{-4} and corresponds to less than one event per year at the presently envisaged luminosities.

3.5.6 The Case of a Symmetric State

A production $e^+ e^- \to \gamma(B\overline{B})_{L=0}$ would lead to a symmetric state

$$|B^0(1)\overline{B}^0(2)\rangle + |B^0(2)\overline{B}^0(1)\rangle$$

with a time evolution described by (65a). Taking the two scaled times T_1 and T_2 as the decay times of the two mesons and multiplying this state function with $\langle X(1)X(2)|\mathbf{H}(1)\mathbf{H}(2)$ one obtains an amplitude

$$\mathcal{M}_+ := e^{i\phi_0} e^{-\frac{1}{2}(T_1+T_2)} \left[C_+ \cos(x - iy) \frac{T_1 + T_2}{2} + i S_+ \sin(x - iy) \frac{T_1 + T_2}{2} \right] \tag{115}$$

where ϕ_0 is a common, unobservable phase including the imt phases of two free B mesons, and the coefficients are

$$C_+ = A_1 \overline{A}_2 + \overline{A}_1 A_2, \quad S_+ = \eta_m \overline{A}_1 \overline{A}_2 + \frac{A_1 A_2}{\eta_m}$$

There, again, can be always two non-zero amplitude factors separated, leaving coefficients like

$$\frac{C_+}{A_1 \overline{A}_2} = 1 + \frac{\overline{A}_1}{A_1} \frac{A_2}{\overline{A}_2}, \quad \frac{S_+}{A_1 \overline{A}_2} = \frac{\eta_m \overline{A}_1}{A_1} + \frac{A_2}{\eta_m \overline{A}_2}$$

The case of $B\overline{B}$ oscillation is obtained when, e.g., for two $\overline{B}$-tags one inserts $\frac{C_+}{A_1 \overline{A}_2} = 0$, $\frac{S_+}{A_1 \overline{A}_2} = \eta_m$, and for $y = 0$ finds

$$\mathcal{M}_+ = A_1 \overline{A}_2 e^{i\phi_0} e^{-\frac{1}{2}(T_1+T_2)} i\eta_m \sin x \frac{T_1 + T_2}{2}$$

$$|\mathcal{M}_+|^2 = |A_1 \overline{A}_2|^2 e^{-(T_1+T_2)} |\eta_m|^2 \sin^2 x \frac{T_1 + T_2}{2}$$

$$= |A_1 \overline{A}_2|^2 e^{-(T_1+T_2)} |\eta_m|^2 [1 - \cos x(T_1 + T_2)]$$

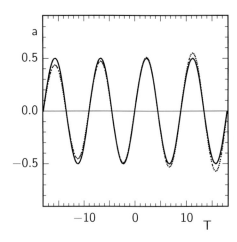

Fig. 18. Time dependent asymmetry of $B^0/\overline{B}^0 \to J/\psi K_S^0$, assuming $\Lambda_0 = 0.5$ and $y = 0$ (———) or $y = -0.01$ (·········). For the final state $J/\psi K_L^0$, these functions would just be flipped at $T = 0$, i.e. $T \longleftrightarrow -T$.

Similarly, the case of a decay to a CP eigenstate where r is just a phase factor, together with a $\overline{B}$ tag, is obtained inserting $\frac{C_+}{A_1 \overline{A}_2} = 1$, $\frac{S_+}{A_1 \overline{A}_2} = r$, and for $y = 0$ finds

$$\mathcal{M}_+ = A_1 \overline{A}_2 e^{i\phi_0} e^{-\frac{1}{2}(T_1+T_2)} \left[\cos x \frac{T_1+T_2}{2} + ir \sin x \frac{T_1+T_2}{2} \right]$$

$$|\mathcal{M}_+|^2 = |A_1 \overline{A}_2|^2 e^{-(T_1+T_2)} \left[1 - 2 \operatorname{Im} r \cos x \frac{T_1+T_2}{2} \sin x \frac{T_1+T_2}{2} \right]$$

$$= |A_1 \overline{A}_2|^2 e^{-(T_1+T_2)} \left[1 - \operatorname{Im} r \sin x (T_1+T_2) \right]$$

$$= |A_1 \overline{A}_2|^2 e^{-(T_1+T_2)} \left[1 - \operatorname{Im} r (\sin x T_1 \cos x T_2 + \cos x T_1 \sin x T_2) \right]$$

In both cases, there is an asymmetry building up as a function of the sum $T_1 + T_2$ of two lifetimes, which is one of the strange manifestations of quantum mechanics.

3.5.7 The Effects of $\Delta\Gamma$

If $y \neq 0$, the time dependent function according to (84) leads to a modulation of the asymmetry amplitude in the same way as discussed for the B_s. Since positive and negative T values are equally accessible on the $\Upsilon(4S)$, an increase and decrease of the amplitude on either side can be observed. Fig. 18 shows an example for the $J/\psi K_S^0$ channel and $y = -0.01$, which is within a factor 3 of the Standard Model expectation.

For small y values expected within the Standard model, the increase or decrease of the asymmetry amplitude will only become significant at large T values, where almost no events will be observed.

A more promising way to measure y is in the untagged decay into any common final state with small Λ_0 and Θ_0 values – corresponding to large Ω_0 numbers. Its rate is given according to (108) as

$$|\mathcal{M}|^2 + |\overline{\mathcal{M}}|^2 = 2 e^{-|T|} |A|^2 \left\{ \cosh yT + \cos \arg r \cdot \sinh yT \right\}$$

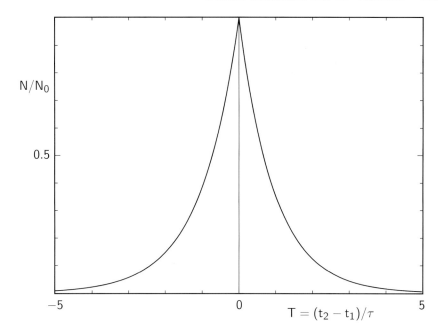

Fig. 19. Time dependent rate of untagged decays $\Upsilon(4S) \to (B^0/\overline{B}^0) + J/\psi K_S^0$ for $\sin 2\beta = 0.5$, $x = 0.7$ and $y = -0.05$. The slight asymmetry to the left yields a mean $\langle T \rangle = -0.087$. A channel with opposite CP eigenvalue, e.g. $J/\psi K_L^0$, has $\langle T \rangle = +0.087$.

$$= |A|^2 \cdot \begin{cases} \left((1+\Omega_0)e^{-\Delta t/\tau_L} + (1-\Omega_0)e^{-\Delta t/\tau_H}\right), & \Delta t > 0 \\ \left((1-\Omega_0)e^{\Delta t/\tau_L} + (1+\Omega_0)e^{\Delta t/\tau_H}\right), & \Delta t < 0 \end{cases} \quad (116)$$

The asymmetric distribution is shown in Fig. 19 for an unexpectedly high value $y = -0.05$. It has a mean value of

$$\langle T \rangle = \frac{2\Omega_0 y}{1 - y^2}$$

corresponding to

$$\langle \Delta t \rangle = \frac{\Omega_0(\tau_L - \tau_H)}{1 - y^2}$$

for the lifetime difference measurement.

3.5.8 Time Integrated Asymmetries

The total rates are proportional to $\int_0^\infty |\mathcal{M}|^2 \, dT$. Using (84) one obtains

$$\int_0^\infty |\mathcal{M}|^2 \, dT = |A|^2 \frac{1 + |r|^2}{2} \left\{ \frac{1}{1 - y^2} - \Theta_0 \frac{1}{1 + x^2} \right.$$

$$\int_0^\infty |\overline{\mathcal{M}}|^2 \, dT = \frac{|A|^2}{|\eta_m|^2} \frac{1+|r|^2}{2} \left\{ \frac{1}{1-y^2} + \Theta_0 \frac{1}{1+x^2} \right.$$
$$\left. + \Omega_0 \frac{y}{1-y^2} + \Lambda_0 \frac{x}{1+x^2} \right\}$$

leading to an asymmetry

$$a_{\text{int}} = \frac{(\Theta_0 + x\Lambda_0)(1-y^2)}{(1+y\Omega_0)(1+x^2)} \tag{117}$$

for $|\eta_m| = 1$. For $y = 0$ this simplifies to

$$a_{\text{int}} = \frac{\Theta_0 + x\Lambda_0}{1+x^2} \tag{118}$$

Therefore, a time integrated measurement is only meaningful if one of the parameters is known, e.g. for $|r| = 1$ where $\Theta_0 = 0$, the asymmetry a_{int} determines Λ_0. In this case, the asymmetry is reduced by a dilution factor

$$D_{\text{int}} = \frac{x}{1+x^2}$$

with respect to the amplitude Λ_0 of the time-dependent one, which is $D_{\text{int}} = 0.48$ for the B^0, and $D_{\text{int}} < 0.05$ for the B_s.

On the $\Upsilon(4S)$, the integration is from $-\infty$ to ∞, and e^{-T} is replaced by $e^{-|T|}$. Therefore, all odd components in $|\mathcal{M}|^2$ and $|\overline{\mathcal{M}}|^2$ lead to vanishing integrals, and the rates are

$$\int_{-\infty}^\infty |\mathcal{M}|^2 \, dT = |A|^2 \frac{1+|r|^2}{2} \left\{ \frac{1}{1-y^2} - \Theta_0 \frac{1}{1+x^2} \right\}$$
$$\int_{-\infty}^\infty |\overline{\mathcal{M}}|^2 \, dT = \frac{|A|^2}{|\eta_m|^2} \frac{1+|r|^2}{2} \left\{ \frac{1}{1-y^2} + \Theta_0 \frac{1}{1+x^2} \right\}$$

and the asymmetry is

$$a_{\text{int}} = \frac{\Theta_0}{1+x^2} \tag{119}$$

for $|\eta_m| = 1$. A measurement without decay time information can therefore be used to determine Θ_0 alone. A determination of Λ_0 in this environment requires the measurement of the lifetime difference to the second, tagging B.

In the simple case where $\Theta_0 = 0$, the integrated asymmetry is $a_{\text{int}} = 0$ and hence cannot be used as an observable to determine the physical parameter Λ_0. However, if the time order is known,

$$a_{+/-\text{int}} = \left. \frac{N(B^0 + X) - N(\overline{B}^0 + X)}{N(B^0 + X) + N(\overline{B}^0 + X)} \right|_{t_s > t_t}$$
$$= - \left. \frac{N(B^0 + X) - N(\overline{B}^0 + X)}{N(B^0 + X) + N(\overline{B}^0 + X)} \right|_{t_s < t_t} = \Lambda_0 \frac{x}{1+x^2}$$

would offer a partly time-integrated measurement with the same dilution as in (118), hence the same reduction in precision compared to the full time distribution analysis. This is of no practical use, since as the time information is available one would like to use in the most effective way.

3.5.9 Final CP Eigenstates from B^0 or B_s Decays

Weak decay amplitudes can be described by (12) and (13). For CP eigenstates with eigenvalue ξ_X their ratio is then given by

$$\frac{\overline{A}}{A} = \xi_X e^{-i(\phi_{CPB} + 2\arg V)} \tag{120}$$

where $\arg V$ is the phase angle of the CKM elements involved in A (i.e. the B^0 decay amplitude). Using the CKM representation (9) with $\eta_m = e^{i(\phi_{CPB} - 2\tilde{\beta})}$ results in

$$r = \xi_X e^{-2i(\arg V + \tilde{\beta})} \tag{121}$$

which is a convention independent phase factor of a product of four or more CKM elements. It can be transformed into an ϵ-like parameter

$$\eta_X := \frac{1-r}{1+r} = i\xi_X \tan(\arg V + \tilde{\beta}) \tag{122}$$

which is, in contrast to ϵ defined by (27), convention independent, but is specific to a final state X.

A measurement of the angle $\beta = \arg(-V_{td}^* V_{tb} V_{cd} V_{cb}^*)$ in the CKM triangle (Fig. 1a) requires a decay with $\bar{b} \to \bar{c} + c\bar{d}$. Examples for final states with this quark content are $J/\psi\pi^0$ or D^+D^-. In these decays one has $r = \xi_X e^{-2i\beta}$.

Similarly, for other final states common to B and $\overline{B}$ the rephasing invariant net product of CKM elements in the combined amplitude/mixing phase ratio r is easily extracted using (9). A summary of common final states of B^0 and $\overline{B}^0$ is given in Table 10. Since the spectator with a b quark is a $\bar{d}$ quark, the decay products of the b must have the net flavour of a d quark, accompanied by one or more quark antiquark pair.

Only the dominant contribution to the box diagram for $B^0\overline{B}^0$ mixing and to the penguin transitions $b \to d$ and $b \to s$ are considered in Table 10. Very small modifications to the phase angles ϕ_{CKM} will emerge from corrections to this approximation.

Also, many final states are reached by both tree and penguin amplitudes, specifically all direct $b \to q\bar{q}d$ final states can also be reached by the $b \to d$ penguin transition. In these cases, the true asymmetry has to be calculated using the sum of all amplitudes for A and $\overline{A}$, with additional unknown parameters that can typically not be resolved with a single asymmetry measurement.

Table 10. Examples of CP eigenstates as final states of B^0 and $\bar{B}^0$, and their sensitivity to the CKM phases. The asymmetry (98) has an amplitude $\Lambda_0 = -\xi_X \sin 2\phi_{\rm CKM}$.

b decay	$\prod$ CKM elements	angle $\phi_{\rm CKM}$	some final states
$b \to c\bar{c}d$	$V_{tb}^* V_{td} V_{cd}^* V_{cb}$	β	$\left\{ \begin{array}{c} J/\psi \\ \psi' \\ \eta_c \\ \vdots \end{array} \right\} + \left\{ \begin{array}{c} \pi^0 \\ \eta \\ \rho^0 \\ \vdots \end{array} \right\}$, $D^{(*)+}D^{(*)-}$
$b \to c\bar{c}s,\, s \to u\bar{u}d$	$V_{tb}^* V_{td} V_{cs}^* V_{cb} V_{ud}^* V_{us}$	$\beta - \phi_4 - \phi_6$	$\left\{ \begin{array}{c} J/\psi \\ \psi' \\ \eta_c \\ \vdots \end{array} \right\} + \left\{ \begin{array}{c} K_S^0 \\ K_S^0 \pi^0 \\ \vdots \end{array} \right\}$
$b \to c\bar{c}s,\, s\bar{d} \to K_L^0$	$V_{tb}^* V_{td} V_{cs}^* V_{cb} V_{cd}^* V_{cs}$	β	$\left\{ \begin{array}{c} J/\psi \\ \psi' \\ \eta_c \\ \vdots \end{array} \right\} + K_L^0$
$b \to u\bar{u}d$	$V_{tb}^* V_{td} V_{ud}^* V_{ub}$	$-\alpha$	$\pi\pi, \rho\rho \ldots$
$b \to d$	$V_{tb}^* V_{td} V_{tb}^* V_{tb}$	0	$\pi^0 \eta', \ldots$
$b \to s,\, s \to u\bar{u}d$	$V_{tb}^* V_{td} V_{ts}^* V_{tb} V_{ud}^* V_{us}$	β'	$\left\{ \begin{array}{c} K_S^0 \\ K_S^0 \pi^0 \\ \vdots \end{array} \right\} + \left\{ \begin{array}{c} \pi^0 \\ \eta \\ \eta' \\ \rho^0 \\ \vdots \end{array} \right\}$
$b \to s\bar{s}s,\, \bar{s} \to u\bar{u}\bar{d}$	$V_{tb}^* V_{td} V_{ts}^* V_{tb} V_{ud} V_{us}^*$	β'	$\left\{ \begin{array}{c} K_S^0 \\ K_S^0 \pi^0 \\ \vdots \end{array} \right\} + \left\{ \begin{array}{c} \eta \\ \eta' \\ \phi \\ \vdots \end{array} \right\}$
$b \to s,\, s\bar{d} \to K_L^0$	$V_{tb}^* V_{td} V_{ts}^* V_{tb} V_{cd}^* V_{cs}$	$\beta' + \phi_4 + \phi_6$	$K_L^0 + \left\{ \begin{array}{c} \pi^0 \\ \eta \\ \eta' \\ \rho^0 \\ \vdots \end{array} \right\}$
$b \to s\bar{s}s,\, \bar{s}d \to K_L^0$	$V_{tb}^* V_{td} V_{ts}^* V_{tb} V_{cd} V_{cs}^*$	$\beta' - \phi_4 - \phi_6$	$K_L^0 + \left\{ \begin{array}{c} \eta \\ \eta' \\ \phi \\ \vdots \end{array} \right\}$

A special case are final states with a K_S^0 or K_L^0, since these states are no CP eigenstates, and the association with an "approximate eigenstate" is by convention rather than by a physical observable (see discussion in section

3.1). However, if one decays subsequently into a CP = ±1 eigenstate, the whole system can still be taken as a CP eigenstate, and be used to extract unitarity angles. These decays have even an advantage over direct decays to the $c\bar{c}d\bar{d}$ state, since they have less contributions from penguin diagrams due to the Cabibbo-allowed transition $W \to c\bar{s}$.

In the tables, the tree level decay $s \to u + \bar{u}d$ is used to determine the CKM phase angles. There may be penguin contributions as well, which modify the asymmetry. For a c quark in the loop, the relevant unitarity angle for $B^0 \to c\bar{c}K_{S,L}$ is exactly β, for a u quark it is the same as for the tree diagram, and for a t quark it is the small angle $\phi_2 + \phi_6$.

A more precise treatment of the whole system includes oscillation and decay of the kaons as well, leading to four amplitudes which all interfere:

$$A_1 := A(B \to B|t_B) \cdot A(B \to c\bar{c}K) \cdot A(K \to K|t_K) \cdot A(K \to \pi\pi)$$
$$A_2 := A(B \to B|t_B) \cdot A(B \to c\bar{c}K) \cdot A(K \to \bar{K}|t_K) \cdot A(\bar{K} \to \pi\pi)$$
$$A_3 := A(B \to \bar{B}|t_B) \cdot \bar{A}(\bar{B} \to c\bar{c}\bar{K}) \cdot A(\bar{K} \to \bar{K}|t_K) \cdot A(\bar{K} \to \pi\pi)$$
$$A_4 := A(B \to \bar{B}|t_B) \cdot \bar{A}(\bar{B} \to c\bar{c}\bar{K}) \cdot A(\bar{K} \to K|t_K) \cdot A(K \to \pi\pi)$$

Here the oscillation amplitudes $A(K \to K)$ etc. depend on the kaon lifetime t_K. For $t_K = 0$ we have $A(\bar{K} \to K) = A(K \to \bar{K}) = 0$ and an oscillation $\Lambda_0 \sin \Delta m_B t_B$ with an amplitude $\Lambda_0 = \sin 2(\tilde{\beta} - \phi_6)$ as given in Table 10. For other times t_K, the argument $\tilde{\beta} - \phi_6 \approx \tilde{\beta}$ changes by a phase angle of $\mathcal{O}(\lambda^4)$. This is small compared to $\tilde{\beta}$, so it still measures $\tilde{\beta}$ to that precision for any kaon lifetime.

If the final kaon is a K_L^0, it will usually be detected via its strong interaction – of a strangeness flavour component – with the detector material. Since the cross section for its $\bar{K}^0$ part is considerably larger than that of the K^0 part, only two amplitudes need to be considered:

$$A_2 := A(B \to B|t_B) \cdot A(B \to c\bar{c}K) \cdot A(K \to \bar{K}|t_K \to \infty)$$
$$A_3 := A(B \to \bar{B}|t_B) \cdot \bar{A}(\bar{B} \to c\bar{c}\bar{K}) \cdot A(\bar{K} \to \bar{K}|t_K \to \infty)$$

Their ratio is

$$\frac{A_3}{A_2} = \frac{A(B \to B|t_B)}{A(B \to \bar{B}|t_B)} \cdot \frac{V_{cb}V_{cs}^*}{V_{cb}^*V_{cs}} e^{i\phi_{CPK}} \cdot \frac{1}{-\eta_{mK}}$$

The last factor for the $K^0/\bar{K}^0$ system is obtained using the limits $T \to \infty$ of (35) as

$$A(K \to \bar{K}) = e^{-imt-T/2}\left[-i\eta_{mK}\sin(x-iy)\frac{T}{2}\right]$$
$$= e^{-imt}e^{-T/2}\left[-i\eta_{mK}\frac{e^{(ix+y)T/2} - e^{(-ix-y)T/2}}{2i}\right]$$
$$\to -e^{-imt}\frac{\eta_{mK}}{2}e^{(ix+y-1)T/2} \qquad (123)$$

(the second term wins due to $y < 0$) and (36) as

$$A(\overline{K} \to \overline{K}) = e^{-imt-T/2}\left[\cos(x-iy)\frac{T}{2}\right] \to e^{-imt}\frac{1}{2}e^{(ix+y-1)T/2} \quad (124)$$

For the leading term in the box diagram we have

$$\eta_{mK} = e^{i(\phi_{\text{CP}K} + 2\arg V_{cs}^{*2}V_{cd}^2)}$$

which yields

$$\frac{A_3}{A_2} \approx \frac{A(B \to B|t_B)}{A(B \to \overline{B}|t_B)} \cdot \frac{V_{cb}V_{cd}^*}{V_{cb}^*V_{cd}} \cdot (-1)$$

i.e. it behaves indeed as a $b \to c\bar{c}d$ state with CP $= -1$ (for an $L =$ even final state). The effect of regeneration in matter will complicate the situation in a full treatment of K_L^0 interactions. Different phase changes from elastic scattering of K^0 and $\overline{K}^0$ produce a coherent K_S^0 component. This implies that the K_S^0 component is "regenerated" even at long distances in the presence of matter, and CP even $\pi\pi$ final states may occur that will dilute the asymmetry due to their opposite sign of ξ_X, if they cannot be distinguished from the inelastic $\overline{K}$ interactions.

The angles to be measured via oscillation/decay interference include all the factor $V_{tb}^*V_{td}$ from mixing, which is one side in the bd triangle (7k). Hence only the adjacent angles α and β can be measured this way. Besides the small difference of the various combinations of β and ϕ_i angles in Table 10, big differences may occur from contributions outside the Standard Model to loop (penguin) and tree decay diagrams. This may lead to different Λ_0 parameters for e.g. $J/\psi K_S^0$ and ϕK_S^0.

Similarly, interference in B_s oscillation can be used to measure the angles in the flat bs triangle (7l) given in Table 11. One of them, $\gamma' = \tilde{\gamma} - \phi_2$, is identical to an angle in the tu triangle (7h), and differs from the third angle γ in the bd triangle (7k) only at order λ^2. This would allow a test of $\alpha + \beta + \gamma = \pi$ to $\mathcal{O}(\lambda^2)$.

If the given quark level transitions are the only contributions to a final state, the asymmetry with time is a simple $\sin 2\phi_{\text{CKM}} \sin xT$ behaviour. However, many of the final states can be reached via loop graphs as well, often with different CKM elements involved. Many details can be found in a recent review [26]. In this case, both direct CP violation via the interfering amplitudes and the oscillation/decay interference lead to more complex asymmetries with the matrix elements (95) and both a $\cos xT$ and $\sin xT$ term.

Table 11. Examples of CP eigenstates as final states of B_s and $\bar{B}_s$, and their sensitivity to the CKM phases. The asymmetry (98) has an amplitude $\Lambda_0 = -\xi_X \sin 2\phi_{\text{CKM}}$.

b decay	$\prod$ CKM elements	angle ϕ_{CKM}	some final states
$b \to c\bar{c}s$	$V_{tb}^* V_{ts} V_{cs}^* V_{cb}$	$\phi_2 + \phi_6$	$\left\{ \begin{array}{c} J/\psi \\ \psi' \\ \eta_c \\ \vdots \end{array} \right\} + \left\{ \begin{array}{c} \eta \\ \phi \\ \vdots \end{array} \right\},\ D_s^{(*)+} D_s^{(*)-}$
$b \to u\bar{u}s$	$V_{tb}^* V_{ts} V_{us}^* V_{ub}$	γ'	$\phi + \left\{ \begin{array}{c} \pi^0 \\ \rho^0 \\ \vdots \end{array} \right\},\ K^{(*)+} K^{(*)-}$
$b \to u\bar{u}d,\ \bar{s} \to \bar{u}u\bar{d}$	$V_{tb}^* V_{ts} V_{ud}^* V_{ub} V_{us}^* V_{ud}$	γ'	$\left\{ \begin{array}{c} \pi^0 \\ \eta \\ \rho^0 \\ \vdots \end{array} \right\} + \left\{ \begin{array}{c} K_S^0 \\ K_S^0 \pi^0 \\ \vdots \end{array} \right\}$
$b \to u\bar{u}d,\ \bar{s}d \to K_L^0$	$V_{tb}^* V_{ts} V_{ud}^* V_{ub} V_{cs}^* V_{cd}$	$\gamma' - \phi_4 - \phi_6$	$\left\{ \begin{array}{c} \pi^0 \\ \eta \\ \rho^0 \\ \vdots \end{array} \right\} + K_L^0$
$b \to s$	$V_{tb}^* V_{ts} V_{ts}^* V_{tb}$	0	$\left\{ \begin{array}{c} \phi \\ \eta \\ \vdots \end{array} \right\} + \left\{ \begin{array}{c} \pi^0 \\ \eta' \\ \vdots \end{array} \right\}$
$b \to d,\ \bar{s} \to \bar{u}u\bar{d}$	$V_{tb}^* V_{ts} V_{td}^* V_{tb} V_{ud} V_{us}^*$	$-\beta'$	$\left\{ \begin{array}{c} K_S^0 \\ K_S^0 \pi^0 \\ \vdots \end{array} \right\} + \left\{ \begin{array}{c} \eta' \\ \vdots \end{array} \right\}$
$b \to d,\ \bar{s}d \to K_L^0$	$V_{tb}^* V_{ts} V_{td}^* V_{tb} V_{cd} V_{cs}^*$	$-\beta' - \phi_4 - \phi_6$	$K_L^0 + \left\{ \begin{array}{c} \eta' \\ \vdots \end{array} \right\}$

3.5.9.1 CP Eigenvalues of Some Final States

The most promising examples are for $\text{CP}(X) = -1$ the decay $B \to J/\psi K_S^0$ with $\Lambda_0 = \sin 2\beta$, and for $\text{CP}(X) = +1$ the decay $B \to \pi^+\pi^-$ with $\Lambda_0 = \sin 2\alpha$ up to corrections from the penguin amplitude. The CP eigenvalues of related channels can be constructed from the data listed in Table 12.

States with several possible angular momenta, like vector vector final states, are typically a mixture of CP = +1 and −1 eigenstates. Helicity 0 dominance would simplify these analyses, since it is forbidden for $L = 1$ final

Table 12. Examples of CP eigenstates relevant for B^0 decays, dependent on their relative orbital angular momentum L.

channel	L	CP	remarks
J/ψ		+1	
ψ'		+1	
χ		+1	
η_c		−1	
π^0		−1	
$K_S^0 \to \pi\pi$		+1	
K_L^0		−1	
$\rho^0 \to \pi^+\pi^-$	1	+1	
$K^{*0} \to K_S^0\pi^0$	1	+1	
$K^{*0} \to K_L^0\pi^0$	1	−1	
$J/\psi\, K_S^0$	1	−1	
$J/\psi\, (K^{*0} \to K_S^0\pi^0)$	0, 2	+1	helicities: $J_z = 0, \pm 1$
$J/\psi\, (K^{*0} \to K_S^0\pi^0)$	1	−1	helicities: $J_z = \pm 1$
$\eta_c(K^{*0} \to K_S^0\pi^0)$	1	+1	
D^+D^-	0	+1	
$D^{*+}D^{*-}$	0, 2	+1	
$D^{*+}D^{*-}$	1	−1	
$\pi^+\pi^-$	0	+1	
$\rho^+\rho^-$	0, 2	+1	
$\rho^+\rho^-$	1	−1	

states, and hence indicates a pure CP eigenstate. In general, though, these states have to be deconvoluted via a partial wave analysis.

Many final states are particle antiparticle pairs. For these, the C operator introduces only one arbitrary phase as in (11), since also $C^2 = 1$, and hence

$$C\,|X(1)\overline{X}(2)\rangle = e^{i\phi_C}e^{-i\phi_C}|X(2)\overline{X}(1)\rangle = |X(2)\overline{X}(1)\rangle$$

For pairs of spin 0 mesons, this implies

$$C(|X(1)\overline{X}(2)\rangle \pm |\overline{X}(1)X(2)\rangle) = \mp(|X(1)\overline{X}(2)\rangle \pm |\overline{X}(1)X(2)\rangle)$$
$$= P(|X(1)\overline{X}(2)\rangle \pm |\overline{X}(1)X(2)\rangle) \quad (125)$$

where 1,2 stands for $|\boldsymbol{x}\rangle, |-\boldsymbol{x}\rangle$ or $|\boldsymbol{p}\rangle, |-\boldsymbol{p}\rangle$. Therefore, eigenstates of P are always also eigenstates of C with **the same** eigenvalue, hence a CP eigenvalue of $\xi = +1$.

For pairs of mesons with spin 1 and total spin S (which is due to angular momentum conservation $S = L$), one finds

$$S = 0: \quad C(|X_\uparrow(1)\overline{X}_\downarrow(2)\rangle - |X_0(1)\overline{X}_0(2)\rangle + |X_\downarrow(1)\overline{X}_\uparrow(2)\rangle)$$

$$
\begin{aligned}
&= \mathrm{P}(|X_\uparrow(1)\bar{X}_\downarrow(2)\rangle - |X_0(1)\bar{X}_0(2)\rangle + |X_\downarrow(1)\bar{X}_\uparrow(2)\rangle) \\
S=1: \quad & \mathrm{C}(|X_\uparrow(1)\bar{X}_\downarrow(2)\rangle - |X_\downarrow(1)\bar{X}_\uparrow(2)\rangle) \\
&= -\mathrm{P}(|X_\uparrow(1)\bar{X}_\downarrow(2)\rangle - |X_\downarrow(1)\bar{X}_\uparrow(2)\rangle) \\
& \mathrm{C}(|X_\uparrow(1)\bar{X}_0(2)\rangle - |X_0(1)\bar{X}_\uparrow(2)\rangle) \\
&= -\mathrm{P}(|X_\uparrow(1)\bar{X}_0(2)\rangle - |X_0(1)\bar{X}_\uparrow(2)\rangle) \quad (126) \\
S=2: \quad & \mathrm{C}(|X_\uparrow(1)\bar{X}_\downarrow(2)\rangle + 2|X_0(1)\bar{X}_0(2)\rangle + |X_\downarrow(1)\bar{X}_\uparrow(2)\rangle) \\
&= \mathrm{P}(|X_\uparrow(1)\bar{X}_\downarrow(2)\rangle + 2|X_0(1)\bar{X}_0(2)\rangle + |X_\downarrow(1)\bar{X}_\uparrow(2)\rangle) \\
& \mathrm{C}(|X_\uparrow(1)\bar{X}_0(2)\rangle + |X_0(1)\bar{X}_\uparrow(2)\rangle) \\
&= \mathrm{P}(|X_\uparrow(1)\bar{X}_0(2)\rangle + |X_0(1)\bar{X}_\uparrow(2)\rangle) \\
& \mathrm{C}(|X_\uparrow(1)\bar{X}_\uparrow(2)\rangle) = \mathrm{P}(|X_\uparrow(1)\bar{X}_\uparrow(2)\rangle)
\end{aligned}
$$

i.e. eigenstates of P are also eigenstates of C with eigenvalue $\eta_C = \eta_P \cdot (-1)^S$, hence a CP eigenvalue of $\xi = \eta_C \eta_P = (-1)^S = (-1)^L$.

3.5.9.2 The $B \to \pi\pi$ Decay

The decay $B^0 \to \pi^+\pi^-$ can proceed via a tree diagram $b \to u\bar{u}d$, with an amplitude $A_T \propto V_{ub}^* V_{ud}$, and via a penguin type diagram $b \to d$ with an amplitude $A_P \propto V_{tb}^* V_{td}$ (see Fig. 11 in Sect. 3.3, with the s quark replaced by a d quark). The CP violating asymmetries are then determined by the ratio

$$r = \eta_m \frac{\bar{A}_T + \bar{A}_P}{A_T + A_P}$$

which has in general no simple relation to the unitarity angles [116]. This fact is often referred to as the "penguin pollution". It affects also the ratio of heavy to light B^0 eigenstates decaying into the CP even $\pi^+\pi^-$ system. The ratio B_H^0 to B_L^0 is no longer given by (73), but rather by (74) with a parameter Ω_0 determined by magnitudes and phases of $\bar{A}_T$, A_T, $\bar{A}_P$ and A_P.

The tree diagram creates a four quark system $u\bar{u}d\bar{d}$ and can be decomposed into a $\Delta I = \frac{1}{2}$ component producing a $\pi^+\pi^-$ in the isospin $I = 0$ state, and a $\Delta I = \frac{3}{2}$ component producing a $\pi^+\pi^-$ in the isospin $I = 2$ state. The penguin diagram creates only two quarks $d\bar{d}$ and can therefore produce a $\pi^+\pi^-$ only in the isospin $I = 0$ state. Both amplitudes may differ by a strong phase δ_{20} which could give rise to CP violation in the decay. The Clebsch Gordan coefficients relating these amplitudes are given in Table 13. The amplitudes in the language of the BSW model [117] are also given. The various factors $\sqrt{\frac{1}{2}}$ are from the π^0 wavefunction $|\pi^0\rangle = \sqrt{\frac{1}{2}}(|u\bar{u}\rangle + |d\bar{d}\rangle)$. The physical amplitudes may be modified by strong final state interactions. Since isospin is conserved in strong interaction, the isospin parametrization of the amplitudes is therefore more general, and in addition not limited to the tree graph.

Table 13. Spectator amplitudes (F = form factor, f = decay constant).

A	quark state	BSW model	$\Delta I = 1/2$ $I=0$	$\Delta I = 1/2$ $I=1$	$\Delta I = 3/2$ $I=1$	$\Delta I = 3/2$ $I=2$
	$B^0 \to d(\bar{u}ud)$	a_1	$\sqrt{\frac{1}{2}}$	$\sqrt{\frac{1}{2}}$	$-\sqrt{\frac{1}{2}}$	$\sqrt{\frac{1}{2}}$
T^{-+}	$\to \rho^-\pi^+$	$a_1 F_V f_P$	$\sqrt{\frac{1}{6}}$	$-\frac{1}{2}$	$\frac{1}{2}$	$\sqrt{\frac{1}{12}}$
T^{+-}	$\to \pi^-\rho^+$	$a_1 F_P f_V$	$\sqrt{\frac{1}{6}}$	$\frac{1}{2}$	$-\frac{1}{2}$	$\sqrt{\frac{1}{12}}$
$T^{\pm\mp} = \sqrt{\frac{1}{2}} \cdot$ $(T^{+-}+T^{-+})$	$\to \pi^-\pi^+$ ($L=0$)	$a_1 F_P f_P$	$\sqrt{\frac{1}{3}}$	0	0	$\sqrt{\frac{1}{6}}$
	$B^0 \to d(\bar{d}uu)$	a_2	$\sqrt{\frac{1}{2}}$	$\sqrt{\frac{1}{2}}$	$-\sqrt{\frac{1}{2}}$	$\sqrt{\frac{1}{2}}$
T^{00}	$\to \rho^0\pi^0$	$a_2\sqrt{\frac{1}{2}}F_V\sqrt{\frac{1}{2}}f_P$ $+a_2\sqrt{\frac{1}{2}}F_P\sqrt{\frac{1}{2}}f_V$	$-\sqrt{\frac{1}{6}}$	0	0	$\sqrt{\frac{1}{3}}$
	$\to \pi^0\rho^0$					
T^{00}	$\to \pi^0\pi^0$	$\sqrt{2}\cdot a_2\sqrt{\frac{1}{2}}F_P\sqrt{\frac{1}{2}}f_P$	$-\sqrt{\frac{1}{6}}$	0	0	$\sqrt{\frac{1}{3}}$
	$B^+ \to u(\bar{u}ud)$	a_1	0	1	$-\frac{1}{2}$	$\sqrt{\frac{3}{4}}$
	$B^+ \to u(\bar{d}uu)$	a_2				
T^{0+}	$\to \rho^0\pi^+$	$a_1\sqrt{\frac{1}{2}}F_V f_P$ $+a_2 F_P\sqrt{\frac{1}{2}}f_V$		$-\sqrt{\frac{1}{2}}$	$\sqrt{\frac{1}{8}}$	$\sqrt{\frac{3}{8}}$
	$\to \pi^+\rho^0$					
T^{+0}	$\to \pi^0\rho^+$	$a_1\sqrt{\frac{1}{2}}F_P f_V$ $+a_2 F_V\sqrt{\frac{1}{2}}f_P$		$\sqrt{\frac{1}{2}}$	$-\sqrt{\frac{1}{8}}$	$\sqrt{\frac{3}{8}}$
	$\to \rho^+\pi^0$					
$T^{+0}_{+0} = \sqrt{\frac{1}{2}} \cdot$ $(T^{0+}+T^{+0})$	$\to \pi^0\pi^+$ ($L=0$)	$a_1\sqrt{\frac{1}{2}}F_P f_P$ $+a_2 F_P\sqrt{\frac{1}{2}}f_P$		0	0	$\sqrt{\frac{3}{4}}$
	$\to \pi^+\pi^0$					

The weak phases involved in this CP violation from the interference of A_T and A_P and in the CP violation from the interference of oscillation and decay through A_T are both given by the factor

$$\frac{V_{ub}V_{ud}^*}{V_{tb}V_{td}^*} = -\mathrm{e}^{\mathrm{i}\alpha}\cdot\left|\frac{V_{ub}V_{ud}}{V_{tb}V_{td}}\right|$$

Therefore, the decay can be used to determine the unitarity angle α, and the amplitude ratio is

$$r = \mathrm{e}^{2\mathrm{i}\alpha}\frac{a_{T1/2}+a_{T3/2}\mathrm{e}^{\mathrm{i}\delta_{20}}/\sqrt{2}-a_P\mathrm{e}^{-\mathrm{i}\alpha}}{a_{T1/2}+a_{T3/2}\mathrm{e}^{\mathrm{i}\delta_{20}}/\sqrt{2}-a_P\mathrm{e}^{\mathrm{i}\alpha}}$$

where $T1/2$, $T3/2$, and P denote tree $\Delta I = \frac{1}{2}$, $\Delta I = \frac{3}{2}$, and penguin, respectively.

In the limit $|A_P| = 0$, the parameters are simply $r = e^{2i\alpha}$, $\Theta_0 = 0$ and $\Lambda_0 = \sin 2\alpha$.

If the phase difference δ_{20} is a multiple of π, i.e. if both isospin amplitudes of the tree process are relatively real,

$$r = e^{2i\alpha} \frac{|A_T| - |A_P|\cos\alpha + i|A_P|\sin\alpha}{|A_T| - |A_P|\cos\alpha - i|A_P|\sin\alpha}$$

and there is still $|r| = 1$, but the phase is a function of $|A_P|/|A_T|$ and α. This would be the case in a factorizing approach like the BSW model [117], which is in good agreement with exclusive two body decays from the $b \to c$ transition. However, final state interactions may be large in light quark final states, and could easily destroy this simple phase relation.

If the penguin amplitude is much smaller than the tree amplitude, the ratio

$$r = e^{2i\alpha} \frac{|A_T|e^{i\delta_{TP}} - |A_P|e^{-i\alpha}}{|A_T|e^{i\delta_{TP}} - |A_P|e^{i\alpha}}$$

can be used to determine the asymmetry amplitudes as taylor expansion in $|A_P/A_T|$ as [118]

$$\Lambda_0 = \sin 2\alpha + 2\cos 2\alpha \sin\alpha \cos\delta_{TP} \frac{|A_P|}{|A_T|} + \mathcal{O}\left(\frac{|A_P|^2}{|A_T|^2}\right)$$

$$\Theta_0 = \sin\alpha \sin\delta_{TP} \frac{|A_P|}{|A_T|} + \mathcal{O}\left(\frac{|A_P|^2}{|A_T|^2}\right)$$

$$\Omega_0 = \cos 2\alpha - 2\sin 2\alpha \sin\alpha \cos\delta_{TP} \frac{|A_P|}{|A_T|} + \mathcal{O}\left(\frac{|A_P|^2}{|A_T|^2}\right)$$

If $|A_P/A_T|$ is known, e.g. from comparison of $K\pi$ and $\pi\pi$ final state branching fractions, the two phase angles α and δ_{TP} can be determined up to discrete ambiguities from a fit to the asymmetry (96) which determines Θ_0 and Λ_0 [119].

For an arbitrary phase difference δ_{20} and penguin contribution, all parameters can be determined via additional measurements [120]. The amplitudes for $B^0 \to \pi^+\pi^-$ (A_{+-}), $\bar{B}^0 \to \pi^+\pi^-$ ($\bar{A}_{+-}$), $B^0 \to \pi^0\pi^0$ (A_{00}), $\bar{B}^0 \to \pi^0\pi^0$ ($\bar{A}_{00}$), $B^+ \to \pi^+\pi^0$ (A_{0+}), and $B^- \to \pi^-\pi^0$ ($\bar{A}_{0-}$) are related to the four amplitudes $B \to \pi\pi_{I=0}$ (A_0), $\bar{B} \to \pi\pi_{I=0}$ ($\bar{A}_0$), $B \to \pi\pi_{I=2}$ (A_2), and $\bar{B} \to \pi\pi_{I=2}$ ($\bar{A}_2$) via Clebsch Gordan coefficients:

$$A_{+-} = \sqrt{\frac{1}{3}}A_0 + \sqrt{\frac{1}{6}}A_2, \quad A_{00} = -\sqrt{\frac{1}{6}}A_0 + \sqrt{\frac{1}{3}}A_2, \quad A_{0+} = \sqrt{\frac{3}{4}}A_2$$

$$\bar{A}_{+-} = \sqrt{\frac{1}{3}}\bar{A}_0 + \sqrt{\frac{1}{6}}\bar{A}_2, \quad \bar{A}_{00} = -\sqrt{\frac{1}{6}}\bar{A}_0 + \sqrt{\frac{1}{3}}\bar{A}_2, \quad \bar{A}_{0-} = \sqrt{\frac{3}{4}}\bar{A}_2$$

leading to the relations

$$A_{+-} + \sqrt{2}\,A_{00} - \sqrt{2}\,A_{0+} = 0 \qquad (127)$$
$$\bar{A}_{+-} + \sqrt{2}\,\bar{A}_{00} - \sqrt{2}\,\bar{A}_{0-} = 0$$

which can be represented by triangles in the complex plane. The shapes of these triangles are determined by the lengths of their sides, i.e. the absolute values of the amplitudes, and the phase differences can then be calculated using triangle geometry.

The value of $|A_{0+}| = |\bar{A}_{0-}|$ is given by the B^+ branching fraction to $\pi^+\pi^0$ as

$$|A_{0+}|^2 = |\bar{A}_{0-}|^2 = f\,\frac{\tau(B^0)}{\tau(B^+)}\,\mathcal{B}(B^+ \to \pi^+\pi^0)$$

The factor f is common to all amplitudes, if we ignore the tiny difference in the phase space factor of the three channels. This decay goes totally to $I = 2$, since $I_3 = \pm 1$ and $I = 1$ is impossible for a two pion system with angular momentum $L = 0$.

The other two amplitudes can be reconstructed from the Θ_0 parameter of the asymmetry function (96),

$$\Theta_0 = \frac{|\bar{A}|^2 - |A|^2}{|\bar{A}|^2 + |A|^2}$$

and the sum of both branching fractions,

$$|A_{+-}|^2 + |\bar{A}_{+-}|^2 = f \cdot \left[\mathcal{B}(B^0 \to \pi^+\pi^-) + \mathcal{B}(\bar{B}^0 \to \pi^+\pi^-)\right]$$
$$|A_{00}|^2 + |\bar{A}_{00}|^2 = f \cdot \left[\mathcal{B}(B^0 \to \pi^0\pi^0) + \mathcal{B}(\bar{B}^0 \to \pi^0\pi^0)\right]$$

In principle, the same method works for both final states. However, since there is no lifetime measurement, the difference $|A_{00}|^2 - |\bar{A}_{00}|^2$ can only be determined at the $\Upsilon(4S)$, where the rates integrated over all time differences are

$$f \cdot \mathcal{B}(B^0_{T=0} \to \pi^0\pi^0) = (1-\chi)\,|A_{00}|^2 + \chi\,|\bar{A}_{00}|^2 \qquad (128)$$
$$f \cdot \mathcal{B}(\bar{B}^0_{T=0} \to \pi^0\pi^0) = (1-\chi)\,|\bar{A}_{00}|^2 + \chi\,|A_{00}|^2$$

and their asymmetry (119) depends only on one parameter Θ_0. The time integrated asymmetry (118) for a single B^0 or $\bar{B}^0$ produced in fragmentation depends also on Λ_0 which involves an unknown phase angle.

The analysis of the time dependent $\pi^+\pi^-$ asymmetry (96) provides $\Lambda_0 = 2\,\mathrm{Im}\,r/(1+|r|^2)$ and $\Theta_0 = (|r|^2 - 1)/(|r|^2 + 1)$ corresponding to the ratio

$$r = \eta_m\,\frac{\bar{A}_{+-}}{A_{+-}} = e^{i(2\alpha + \theta_{+-})} \cdot \frac{|\bar{A}_{+-}|}{|A_{+-}|} \qquad (129)$$

The parameter Θ_0 thus determines $|\overline{A}_{+-}|/|A_{+-}|$, which together with the total rate determines the sides $|\overline{A}_{+-}|$ and $|A_{+-}|$. This is equivalent to the extraction from the time integrated rates as described above for $\pi^0\pi^0$ via (128).

All these values can be used to construct the two triangles given by (127) from their three sides. The two amplitudes A_{0+} and $\overline{A}_{0-}$ can be made relatively real by a suitable choice of the CP phase ϕ_{CPB}. In this convention, $A_{0+)} = \overline{A}_{0-}$ and the two triangles can be drawn with this amplitude as a common base line. Then, the phase angle θ_{+-} of $\overline{A}_{+-}/A_{+-}$ is the angle between the two sides corresponding to the amplitudes $\overline{A}_{+-}$ and A_{+-}. It can be determined up to a twofold ambiguity in magnitude, since the tips of the two triangles can be either on the same side of the base line or on opposite sides, and a twofold ambiguity in its sign.

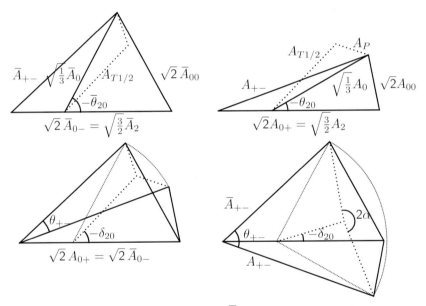

Fig. 20. Amplitude triangles for B and $\overline{B}$ to $\pi\pi$ are shown as the upper two plots. The dotted lines are the tree and penguin contributions to A_0 and $\overline{A}_0$, with a relative phase angle of $\pm\alpha$ to the tree amplitude, leading to $|A_0| = |\overline{A}_0|$. The lower left graph is the combination of the two triangles with the tips on the same side of the base line, corresponding to the dotted amplitudes shown in the upper plots. The lower right graph is the second solution with tips on different sides. It corresponds to a different composition of A_0 and $\overline{A}_0$ (not shown in the upper plots).

Fig. 20 illustrates the situation. The convention

$$\sqrt{\frac{1}{3}}A_0 = A_{T1/2} + A_P, \quad \sqrt{\frac{1}{3}}\overline{A}_0 = \overline{A}_{T1/2} + \overline{A}_P$$

is used leading to

$$A_{T+-} = A_{T1/2} + \frac{\sqrt{2}}{3}A_{0+}, \quad \sqrt{2}A_{T00} = -A_{T1/2} + \frac{2\sqrt{2}}{3}A_{0+}$$

$$\bar{A}_{T+-} = \bar{A}_{T1/2} + \frac{\sqrt{2}}{3}\bar{A}_{0-}, \quad \sqrt{2}\bar{A}_{T00} = -\bar{A}_{T1/2} + \frac{2\sqrt{2}}{3}\bar{A}_{0-}$$

and A_{0+}, $\bar{A}_{0-}$ are pure $\Delta I = \frac{3}{2}$.

The asymmetry amplitude Λ_0 can be calculated from (129) to be

$$\Lambda_0 = \frac{2}{1 + |\bar{A}_{+-}/A_{+-}|} \sin(2\alpha + \theta_{+-})$$

The amplitude ratio is given by the second fit parameter Θ_0, and four θ_{+-} values can be read from the triangles.

The ratio of B_H^0 to B_L^0 in this state is given by

$$\Omega_0 = \frac{2}{1 + |\bar{A}_{+-}/A_{+-}|} \cos(2\alpha + \theta_{+-})$$

and is in general different from that for $\pi^0\pi^0$, since both the $|\bar{A}_{00}/A_{00}|$ ratio and θ_{00} differ.

An alternative way to use the triangles is obtained if the ratio r is expressed in isospin amplitudes:

$$r = e^{2i\alpha} \frac{|A_0| + |A_2|e^{i\theta_{20}}}{|A_0| + |A_2|e^{i\bar{\theta}_{20}}}$$

The angles θ_{20} and $\bar{\theta}_{20}$ can be read from the two upper triangles in Fig. 20, up to the ambiguous sign, as well as the amplitude ratio $r_{20} := |A_2/A_0|$. The asymmetry parameter

$$\Lambda_0 = \frac{\sin\theta_{20} - \sin\bar{\theta}_{20} + r_{20}\sin(\theta_{20} - \bar{\theta}_{20})}{1 + r_{20}(\cos\theta_{20} + \cos\bar{\theta}_{20}) + r_{20}^2} r_{20}\cos(2\alpha)$$
$$+ \frac{1 + r_{20}(\cos\bar{\theta}_{20} + \cos\theta_{20}) + r_{20}^2\cos(\theta_{20} - \bar{\theta}_{20})}{1 + r_{20}(\cos\theta_{20} + \cos\bar{\theta}_{20}) + r_{20}^2}\sin(2\alpha)$$

is then a function of these three parameters and 2α.

There are several ways to present a measurement of the $\pi^+\pi^-$ asymmetry in absence of clear knowledge of the penguin contribution. One can give the amplitudes Λ_0 and Θ_0 of the $\sin xT$ and $\cos xT$ terms, or give $|r|$ and a measure of the phase $\arg r$ like $\sin 2\alpha_{\text{eff}} := \sin\arg r = \text{Im}\, r/|r|$. There are theoretical predictions [121] relating α_{eff} with α. They are, however, based on the factorization assumption which is likely to be invalid for $b \to$ light quark transitions.

A similar analysis can be used with the three pion decays $B \to \rho\pi \to \pi\pi\pi$, with common final states $\pi^+\pi^-\pi^0$ and $\pi^0\pi^0\pi^0$ to B^0 and $\overline{B}^0$ ($\rho^+\pi^-$ and $\rho^-\pi^+$ are no CP eigenstates). Here, one has five independent amplitudes from different $\rho\pi$ charge combinations, but also two more amplitudes ($\Delta I = \frac{1}{2}$ and $\Delta I = \frac{3}{2}$) leading to $I = 1$ final states. Phase differences can be calculated from the absolute values of the five amplitudes in complete analogy to the two pion case, with discrete ambiguities due to different orientation of the polygons. However, these ambiguities can be resolved from a measurement of the phase difference from the interference regions in the Dalitz plot [122]. Therefore, the three pion channels offer not only higher rates, but also a unique solution for α in the Standard Model analysis.

At a high level of precision, further corrections have to be considered. The isospin analysis is based on the assumption that penguin type amplitudes produce only isospin 0 final states. This is not true for electroweak penguin diagrams, which may lead to $I = 2$ states with four light quarks hadronizing to $\pi\pi$, and which may contribute a small fraction to the decay. There are also other quarks besides the t in the loops, which modify the phase angles. While all these effects can be considered negligible in the first generation of CP violation experiments, inconsistent results could still emerge. They would indicate CP violating effects beyond the Standard Model, where the assumption of a description via the CKM matrix phases does not hold.

3.5.10 Mixtures of CP Eigenstates

An example for a mixture of CP eigenstates is the final state $D^{*+}D^{*-}$ which is CP $= -1$ for $L = 1$ and CP $= +1$ for $L = 0$ or 2. Other vector vector decays like $B_s \to D_s^{*+}D_s^{*-}$ or $B^0 \to \rho^+\rho^-$ or vector vector decays with two CP eigenstates like $B_s \to J/\psi\phi$ or $B^0 \to J/\psi(K_S^0\pi^0)_{K^{*0}}$ show the same properties. In these cases the amplitude is

$$A = (V) \cdot (A_+ + A_-)$$

where (V) denotes the common CKM factors, and the subscripts of the residual factors are the CP eigenvalues, e.g. $A(B^0 \to D^{*+}D^{*-}, L = 1) = V_{cb}^* V_{cd} A_-$ and $A(B^0 \to D^{*+}D^{*-}, L = 0, 2) = V_{cb}^* V_{cd} A_+$. They correspond to different helicities of the vector mesons, therefore the factors A_+ and A_- have different phases changing with the angles of the decay products, e.g. $D^{*+} \to D^0\pi^+$ or $J/\psi \to l^+l^-$. Their ratio $r_A := A_-/A_+$ depends on the decay angles, conveniently described as θ_1, θ_2 and $\phi_1 - \phi_2$ in the two helicity frames. The amplitude ratio observable in CP violation experiments is

$$r = e^{i\phi}\frac{1 - r_A}{1 + r_A}$$

where ϕ is the invariant phase from the CKM elements in mixing and decay, which leads for $B^0 \to D^{*+} D^{*-}$ to

$$e^{i\phi} = \eta_m \frac{V_{cb} V_{cd}^*}{V_{cb}^* V_{cd}} = e^{-2i\beta}$$

and the coefficients in the time evolution (84) are

$$\frac{1+|r|^2}{2} = \frac{1+|r_A|^2}{1+|r_A|^2 + 2\operatorname{Re} r_A}$$

$$\frac{1-|r|^2}{2} = \frac{2\operatorname{Re} r_A}{1+|r_A|^2 + 2\operatorname{Re} r_A}$$

$$\operatorname{Re} r = |r|\cos\arg r = \frac{(1-|r_A|^2)\cos\phi + 2\operatorname{Im} r_A \sin\phi}{1+|r_A|^2 + 2\operatorname{Re} r_A}$$

$$\operatorname{Im} r = |r|\sin\arg r = \frac{(1-|r_A|^2)\sin\phi - 2\operatorname{Im} r_A \cos\phi}{1+|r_A|^2 + 2\operatorname{Re} r_A}$$

It has been emphasized in [123] that interference in mixed CP eigenstates like these can be used to observe a small phase angle ϕ in untagged B_s decays. This is possible since the sum of both initial flavours is given by (108), being proportional to

$$|\mathcal{M}|^2 + |\overline{\mathcal{M}}|^2 = 2e^{-T}|A|^2 \left\{ \frac{1+|r|^2}{2} \cosh yT - \operatorname{Re} r \sinh yT \right\}$$

and the coefficient of the $\sinh yT$ term, $\operatorname{Re} r$, has a component $\operatorname{Im} r_A \sin\phi$ which becomes dominant in regions of decay angle space where $\operatorname{Im} r_A \gg 1 - |r_A|^2$.

Since the B has total spin 0, the sum of the spins of the two vector mesons must equal the angular momentum, and since $L_z = 0$ in the flight direction of the daughter, so must be $S_{z1} + S_{z2} = 0$. The Clebsch-Gordan coefficients of the possible partial waves are

	$L, L_z:$	$0,0$	$1,0$	$2,0$
$S_z =$	$+1,-1:$	$\sqrt{\frac{1}{3}}$	$\sqrt{\frac{1}{2}}$	$\sqrt{\frac{1}{6}}$
	$0,0:$	$-\sqrt{\frac{1}{3}}$	0	$\sqrt{\frac{2}{3}}$
	$-1,+1:$	$\sqrt{\frac{1}{3}}$	$-\sqrt{\frac{1}{2}}$	$\sqrt{\frac{1}{6}}$
		CP $= +\xi$	CP $= -\xi$	CP $= +\xi$

The CP eigenvalue is given by the intrinsic CP eigenvalue of the two mesons, ξ, and a factor $(-1)^L$ from the parity of the angular momentum. The longitudinal amplitude $A_{0,0}$ is only present in even angular momenta, and hence coresponds to a CP eigenstate, the helicities ± 1 are mixtures of both eigenstates. However, states of definite transverse spin projections with respect to the decay plane of the second particle

$$A_\| = A_{+1,-1} + A_{-1,+1}, \qquad A_\perp = A_{+1,-1} - A_{-1,+1}$$

are pure CP eigenstates, too [124], and may be used to decompose the angular distribution into states with even and odd CP eigenvalue.

3.5.10.1 $B \to J/\psi K^*$

Partial wave decompositions of the decay amplitudes $B^0 \to J/\psi K^{*0}$ have been performed by several experiments. The e^+e^- colliders have also used the charged decay $B^+ \to J/\psi K^{*+}$, assuming isospin invariance. The results are listed in Table 14.

Note that the non-trivial phase for $A_\perp$ disfavours the fragmentation hypothesis that is used in many B decay calculations.

3.5.11 Observations on Non-eigenstates

CP violation in oscillation/decay interference can also be observed in final states that are not CP eigenstates, as long as they can be reached by both B and $\bar{B}$. As an example [131], the channel shown in Fig. 10 on page 103 has the decays $\bar{b} \to \bar{c}(\bar{s}u)$ leading to the final state $\bar{D}^0 K_S^0$ from a B^0 meson (a), and $b \to u(\bar{s}c)$ leading to the same final state from a $\bar{B}^0$ meson (b). The amplitudes are

$$A \propto V_{cb}^* V_{us} \approx A\lambda^3$$
$$\bar{A}' \propto V_{ub} V_{cs}^* \approx A\lambda^3(\rho - i\eta)$$

which are of the same order of magnitude. Using $V_{cs}^* V_{cd}$ for the $K^0 \to K_S^0$ amplitude and ignoring small CKM phase angles $\mathcal{O}(\lambda^4)$ the observable ratio is

$$r = \eta_m \frac{\bar{A}'}{A}$$
$$= e^{-2i\beta} \exp\left(i \arg \frac{V_{ub} V_{cd}^*}{V_{cb}^* V_{us} V_{cs}^* V_{cd}}\right) e^{-i\delta_{AA'}} \frac{|A'|}{|A|}$$
$$= e^{-i(2\beta+\gamma+\delta_{AA'})} \frac{|A'|}{|A|}$$

where $\delta_{AA'}$ is the strong interaction phase difference. Similarly, for the final state $D^0 K_S^0$

$$r = \eta_m \frac{\bar{A}}{A'} = e^{-i(2\beta+\gamma-\delta_{AA'})} \frac{|A|}{|A'|}$$

Table 14. Amplitude decomposition for $B \to J/\psi K^{*0}$. The phases are given in radian relative to A_0.

$\|A_0\|^2$ CP +1	$\|A_\perp\|^2$ CP −1	$\|A_\|\|^2$ CP +1	$\phi_\perp$	$\phi_\|$	experiment
0.97 ± .16 ± .15					ARGUS 94 [125]
0.65 ± .10 ± .04					CDF 95 [126]
0.52 ± .07 ± .04	0.16 ± .08 ± .04		−0.11 ± .46 ± .03	3.00 ± .37 ± .04	CLEO 97 [127]
0.59 ± .06 ± .01	0.13 ± .12 .09 ± .06		−0.56 ± .54	2.16 ± .47	CDF 00 [128]
0.597 ± .028 ± .024	0.160 ± .032 ± .014	0.243 ± .034 ± .017	−0.17 ± .16 ± .07	2.50 ± .20 ± .08	BABAR 01 [129]
0.60 ± .03 ± .04	0.19 ± .04 ± .04		0.01 ± .19 ± .08	2.86 ± .25 ± .05	BELLE 01 [130]
0.589 ± .025	0.165 ± .028		−0.12 ± .12	2.66 ± .14	average

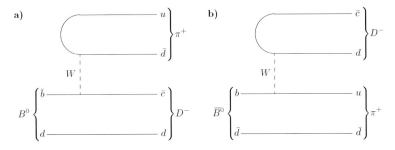

Fig. 21. Diagrams for $B^0 \to D^- \pi^+$ (a) and $\overline{B}^0 \to D^- \pi^+$ (b).

From an observation of the time-dependent asymmetry (96) both the absolute values and the phases of these numbers can be determined, and the three unknown parameters $\delta_{AA'}$, $2\beta + \gamma$ and $|A|/|A'|$ be extracted.

Measuring separately $D^0/\overline{D}^0$ decays into CP eigenstates like $\pi\pi$ or KK, where both amplitudes interfere, the angle γ can be extracted from the rate measurement, which becomes simpler if the non-mixing charged B decay ($\to D^0/\overline{D}^0 K^\pm$) [132] or the self-tagging decay $B^0 \to D^0/\overline{D}^0 (K^+\pi^-)_{K^{*0}}$ [133] are used.

A similar self-tagging decay $B_s \to D_s^+ K^-$ has as weak phase $\tilde{\gamma}+2\phi_2-\phi_6 \approx \gamma$ and has been suggested [134] to determine this angle.

3.5.11.1 The Decay $B \to D\pi$

Among the most frequent exclusive decay modes are $B^0 \to D^- \pi^+$ and $B^0 \to D^{*-} \pi^+$. These final states can be produced from the $\overline{B}^0$ by a highly suppressed decay graph, as illustrated in Fig. 21.

The amplitudes are

$$A \propto V_{cb}^* V_{ud} \approx A\lambda^2$$
$$\overline{A}' \propto V_{cd}^* V_{ub} \approx -A\lambda^4(\rho + i\eta)$$

The observable ratio is

$$r = \eta_m \frac{\overline{A}'}{A} = e^{-2i\tilde{\beta}} \exp\left(i \arg \frac{V_{cd}^* V_{ub}}{V_{cb}^* V_{ud}^*}\right) e^{-i\delta_{AA'}} \frac{|A'|}{|A|} = -e^{-i(2\beta+\gamma+\delta_{AA'})} \frac{|A'|}{|A|}$$

where $\delta_{AA'}$ is the strong interaction phase difference, and $|A'|/|A| \approx \lambda^2 \sqrt{\rho^2 + \eta^2}$. Here the phase is exactly $2\beta + \gamma$ ($= 2\tilde{\beta} + \tilde{\gamma} + \phi_4$ in the standard phase convention), and $|r| \ll 1$.

Similarly, for the final state $D^+\pi^-$

$$r = \eta_m \frac{\overline{A}}{A'} = -e^{-i(2\beta+\gamma-\delta_{AA'})} \frac{|A|}{|A'|}$$

with $|r| \gg 1$. Again, from an observation of the time-dependent asymmetry (96) both the absolute values and the phases of these numbers can be determined, and the three unknown parameters $\delta_{AA'}$, $2\beta+\gamma$ and $|A|/|A'|$ be extracted [135].

The physical dilution is

$$D_P \approx 2\lambda^2 \sqrt{\rho^2 + \eta^2} \approx 0.03$$

and the asymmetry parameters are $\Theta_0 \approx \mp 1$ indicating that the state is to a good approximation a tagging state that shows ordinary oscillation behaviour, and

$$\Lambda_0 = D_P \sin(2\beta + \gamma \pm \delta_{AA'})$$

3.6 Measurement of Time Dependent Asymmetries of Neutral B Mesons

Experimental information on CP violation of B mesons produced in jets have been obtained by the CDF collaboration at the Tevatron, and with B mesons produced in $\Upsilon(4S)$ decays at the asymmetric colliders PEP II/BABAR and KEKB/BELLE.

The asymmetry described by (98) has been measured in the distribution of the signal B meson's proper lifetime or lifetime difference via its distance of flight and the momentum reconstructed from its decay products. The measurement is illustrated using the most abundant channel $B^0 \to J/\psi K_S^0$. Here the asymmetry amplitude $\Lambda_0 = \sin 2\beta$ is related directly to the angle β in the CKM unitarity triangle, and (98) reads

$$a(T) = \Lambda_0 \sin xT$$

with $T = t/\tau$ and assuming $y = 0$. The example $J/\psi K_S^0$ can be replaced by many other final states which show a similar behaviour with the same or a different amplitude Λ_0. The integrated asymmetry is 0 at the $\Upsilon(4S)$ and diluted by the factor $x/(1+x^2) = 0.48$ in b jets.

In any experiment using the time-dependent asymmetry, the CP violating parameters are measured by the analysis of final states from a continuum or bound $b\bar{b}$ quark pair involving three main steps:
- The relevant B meson final states are reconstructed with a maximum signal-to-background ratio,
- the lifetime or lifetime difference is measured,
- and the initial beauty flavour of the B^0 (or B_s) at $T = 0$ is determined.

The latter, tagging, has already been discussed in section 2.3.1. In contrast to a pure frequency measurement as in $B\bar{B}$ oscillation, the determination of CP asymmetries includes measurements of oscillation amplitudes, which are diluted as given in (66). The bigger the dilution

3.6.1 Observed Versus True Asymmetry

Using the example intruduced above with leptons as flavour tags, the asymmetry a is in an ideal case, where the flavour B^0 or $\bar{B}^0$ at $T = 0$ is known unambiguously from the lepton charge,

$$a(T) = \frac{\dot{N}(\bar{B}^0 \to J/\psi\, K_S^0) - \dot{N}(B^0 \to J/\psi\, K_S^0)}{\dot{N}(\bar{B}^0 \to J/\psi\, K_S^0) + \dot{N}(B^0 \to J/\psi\, K_S^0)}$$

$$= \frac{\dot{N}(J/\psi\, K_S^0 + l^+) - \dot{N}(J/\psi\, K_S^0 + l^-)}{\dot{N}(J/\psi\, K_S^0 + l^+) + \dot{N}(J/\psi\, K_S^0 + l^-)}$$

In a real experiment, there are a few additional asymmetries involved at $\Upsilon(4S)$ decays, and even more at hadronic B factories using pp, $\bar{p}p$ or p nucleus collisions. The most obvious effect comes from the probability of wrong tagging of the initial flavour. For a charged lepton tag with mistag probability w, i.e. a fraction w of leptons with the opposite charge compared to a lepton from $b \to cl\nu$, the observed number of initial B^0 and $\bar{B}^0$ mesons is

$$N_1 \equiv \dot{N}(J/\psi\, K_S^0 + l^+) = (1-w) \cdot \dot{N}(\bar{B}^0 \to J/\psi\, K_S^0) + w \cdot \dot{N}(B^0 \to J/\psi\, K_S^0)$$
$$N_2 \equiv \dot{N}(J/\psi\, K_S^0 + l^-) = (1-w) \cdot \dot{N}(B^0 \to J/\psi\, K_S^0) + w \cdot \dot{N}(\bar{B}^0 \to J/\psi\, K_S^0)$$

and the observed asymmetry

$$a_{\text{obs}} = \frac{\dot{N}_1 - \dot{N}_2}{\dot{N}_1 + \dot{N}_2} = (1 - 2w) \cdot a = D \cdot a$$

is diluted as in the mixing case (66). In general we have the following diluting effects:

1) The production rates of B^0 and $\bar{B}^0$ are equal in experiments where the sum of all initial flavours is 0, as e^+e^- or $\bar{p}p$ annihilation, but different e.g. in pp collisions, due to the fact that an excess of 4 u-quarks and 2 d-quarks is present from the beginning. There are four different fragmentation probabilities:

$$f_0 = N(B^0)/N(\bar{b}) \qquad \bar{f}_0 = N(\bar{B}^0)/N(b)$$
$$f_s = N(B_s)/N(\bar{b}) \qquad \bar{f}_s = N(\bar{B}_s)/N(b)$$

This introduces an intrinsic asymmetry, which is not present at particle antiparticle colliders.

2) The second b-hadron used for tagging can have oscillated into its antiparticle with probability

$$\chi = \frac{N(\bar{b} \to l^-)}{N(\bar{b} \to l)} = f_0\chi_0 + f_s\chi_s$$

$$\bar{\chi} = \frac{N(b \to l^+)}{N(b \to l)} = \bar{f}_0\chi_0 + \bar{f}_s\chi_s$$

Here χ is the average probability for $\bar{b} \to b$, and $\bar{\chi}$ for $b \to \bar{b}$ through mixing; χ_0 and χ_s denote the mixing probabilities of B^0 and B_s, respectively. Due to the coherent antisymmetric $B\bar{B}$ state in $\Upsilon(4S)$ decays, this effect is absent at B factories operating at the $\Upsilon(4S)$.

3) The lepton can be from semileptonic charm decay in the $b \to c$ cascade. Likewise, almost all tags have a chance to occur at the "wrong" charge. Part of this effect is even due to wrong particle identification of b decay products. This mistag probability w is present at all experiments, but is reduced if determined as a function of discriminating variables, as described below. This probability can be different for b and $\bar{b}$ tags due to asymmetric background and different efficiencies for positive and negative particles.

This can be parametrized with the branching fractions B for $b \to l^- X$ and C for $b \to l^+ X$ being

$$B \approx 2\left[\mathcal{B}(b \to l^-\nu X) + 0.18\,\mathcal{B}(B \to \tau^-\nu X) + 0.08\,\mathcal{B}(B \to D_s^- X)\right] \approx 0.24$$
$$C \approx \mathcal{B}(b \to cX) \cdot 2\,\mathcal{B}(c \to l^+\nu X) \approx 0.20$$

and the charge-dependent efficiencies (including geometry, reconstruction and identification of leptons associated with a $B \to J/\psi K_S^0$ signal) $\varepsilon_{B\pm}$ for "right sign" leptons from the second b-hadron, and $\varepsilon_{C\pm}$ for "wrong sign" leptons, mainly from secondary charmed hadron decays. The given branching fractions illustrate the main contributions to both classes. Note that the resulting event numbers are about equal, if the efficiency is uniform over the whole phase space, and tagging power has to come from the use of kinematic differences of both lepton samples.

4) The lepton can be faked by π or K, with absolute multiplicities b_+ and b_- for faked positive and negative tag leptons, with

$$b_+ = N(\pi^+) \cdot \delta_\pi^+ \cdot \varepsilon_{\pi+} + N(K^+) \cdot \delta_K^+ \cdot \varepsilon_{K+}$$
$$b_- = N(\pi^-) \cdot \delta_\pi^- \cdot \varepsilon_{\pi-} + N(K^-) \cdot \delta_K^- \cdot \varepsilon_{K-}$$

Here ε_X is the kinematic tagging acceptance and δ_X the misidentification probability for hadron X.

Two sources of hadrons contribute in two different ways:

- At hadron beam experiments, the charged hadron production through fragmentation yields $b_+ \neq b_-$ due to the initial quarks from the pp or pn state, with no correlation to the b flavour.
- The charged hadron production through b-hadron decays may show a substantial charge asymmetry, which is correlated to the b flavour and therefore has also effective tagging power and mistag probability. This case is absorbed in the mistag probability w, which includes mistags from true and faked tag leptons. To include these misidentified hadrons, we just change the meaning of B and C above to represent all right- and wrong-sign tracks, and $\varepsilon_{B\pm}$ and $\varepsilon_{C\pm}$ to be the average probability that these tracks are (correctly or wrongly) identified as leptons.

5) Two b-hadrons with the same beauty may have been produced simultaneously. High rate hadronic b factories have a small chance to produce two $b\bar{b}X$ events in a single bunch crossing. In addition, single interactions like $pp \to b\bar{b}b\bar{b}X$ have a small non-zero frequency $\sim 10^{-5} \ldots 10^{-4}$ compared to all $pp \to b\bar{b}X$ events. These effects are typically below fractions of 10^{-4} which can safely be neglected.

A related background is a $c\bar{c}X$ event that occurs together with a $b\bar{b}$ event, producing additional leptons from charmed hadron decays with no relation to the beauty flavour. Their effect is the same as from misidentified hadrons, and can be included in the fractions b_+ and b_- for the analytical calculation.

For b and $\bar{b}$ flavours as tags at $T = 0$ we have two event rates $\dot{N} = \mathrm{d}N/\mathrm{d}T$ as a function of the proper scaling lifetime T

$$\dot{N}(J/\psi K_S^0 + \bar{b}) = (1+a) \cdot f_0 \cdot N$$
$$\dot{N}(J/\psi K_S^0 + b) = (1-a) \cdot \bar{f}_0 \cdot N$$

for given true asymmetry $a = a(T)$. Tagging the b or $\bar{b}$ with an electron or muon, the effects mentioned above give the following rates:

$$\begin{aligned}
\dot{N}_1 = \dot{N}(J/\psi K_S^0 + l^+) &= (1-a) \cdot f_0 \cdot [(1-\chi)\varepsilon_{B+}B + \chi\varepsilon_{C+}C + b_+] \cdot N \\
&+ (1+a) \cdot f_0 \cdot [\bar{\chi}\varepsilon_{B+}B + (1-\bar{\chi})\varepsilon_{C+}C + b_+] \cdot N \\
\dot{N}_2 = \dot{N}(J/\psi K_S^0 + l^-) &= (1+a) \cdot f_0 \cdot [(1-\bar{\chi})\varepsilon_{B-}B + \bar{\chi}\varepsilon_{C-}C + b_-] \cdot N \\
&+ (1-a) \cdot \bar{f}_0 \cdot [\chi\varepsilon_{B-}B + (1-\chi)\varepsilon_{C-}C + b_-] \cdot N
\end{aligned} \tag{130}$$

The parameters b_- and b_+ are the fake rates from misidentification of hadrons that are not correlated with the b flavour.

Double tags, e.g. by a true plus a fake lepton, have been ignored in these formulae. However, in a more general tagging procedure that incorporates multiple particles for tagging, the same calculation can be used with appropriate translation of the meaning of the individual constants.

The normalization constant is

$$N = N(b\bar{b}) \cdot \mathcal{B}(B^0 \to J/\psi K_S^0 \to l^+l^-\pi^+\pi^-) \cdot \varepsilon(T) \cdot e^{-|T|}$$

where ε is the reconstruction and trigger efficiency. It will cancel in all ratios. The observed asymmetry a_{obs} is then

$$\begin{aligned} a_{\text{obs}} &= \frac{\dot{N}_1 - \dot{N}_2}{\dot{N}_1 + \dot{N}_2} \\ &= \frac{2f_b d_b + d_\varepsilon + D_w \chi_{\text{av}} \Delta_\chi - D_w D_m d_f + a[D_w D_m - d_f(2f_b d_b + d_\varepsilon + D_w \chi_{\text{av}} \Delta_\chi)]}{1 + 2f_b + f_\varepsilon(\chi_{\text{av}} \Delta_\chi - D_m d_f) + a[D_m f_\varepsilon - d_f(1 + 2f_b + \chi_{\text{av}} \Delta_\chi f_\varepsilon)]} \end{aligned} \quad (131)$$

with

$$D_w = 1 - 2w, \quad w = \frac{\varepsilon_C C}{\varepsilon_B B + \varepsilon_C C}, \quad d_f = \frac{\bar{f}_0 - f_0}{2}$$

$$D_m = 1 - \bar{\chi} - \chi, \quad \Delta_\chi = \bar{\chi} - \chi, \quad \chi_{\text{av}} = \frac{\bar{\chi} + \chi}{2}$$

$$f_b = \frac{(b_+ + b_-)}{\varepsilon_B B + \varepsilon_C C}, \quad d_b = b_+ - b_-$$

$$\varepsilon_C = \frac{\varepsilon_{C+} + \varepsilon_{C-}}{2}, \quad \varepsilon_B = \frac{\varepsilon_{B+} + \varepsilon_{B-}}{2}$$

$$d_\varepsilon = \frac{1}{2}(\varepsilon_{B+} - \varepsilon_{B-})(1 - w) + (\varepsilon_{C+} - \varepsilon_{C-})$$

$$f_\varepsilon = \frac{1}{2}(\varepsilon_{B+} - \varepsilon_{B-})(1 - w) - (\varepsilon_{C+} - \varepsilon_{C-})$$

or approximately, ignoring terms $\mathcal{O}(a^2)$,

$$a_{\text{obs}} \approx I + D \cdot a \quad (132)$$

with an "intrinsic" asymmetry

$$I = \frac{2f_b d_b + d_\varepsilon + D_w \chi_{\text{av}} \Delta_\chi - D_w D_m d_f}{1 + 2f_b + f_\varepsilon(\chi_{\text{av}} \Delta_\chi - D_m d_f)}$$

and a "dilution factor"

$$D = D_m \cdot (1 - d_f^2) \cdot \frac{D_w(1 + 2f_b) - f_\varepsilon(d_\varepsilon + 2f_b d_b)}{(1 + 2f_b + f_\varepsilon(\chi_{\text{av}} \Delta_\chi - D_m d_f))} = D_m \cdot D_t$$

where the dilution factor can be split into a tagging component D_t and a mixing component $D_m = 1 - \chi - \bar{\chi} \approx 0.78$ for b jets and $D_m = 1$ at the $\Upsilon(4S)$. The linear approximation holds very well for all practical purposes. Nonlinear corrections to (132) occur for $f_0 \neq \bar{f}_0$, $\varepsilon_{C+} \neq \varepsilon_{C-}$ or $\varepsilon_{B+} \neq \varepsilon_{B-}$, but are small due to the smallness of these asymmetries.

The rate of reconstructed tagged events is

$$\dot{N}_{\text{tot}} = \dot{N}_1 + \dot{N}_2 = (\varepsilon_B B + \varepsilon_C C + b_+ + b_-) \cdot [f_0 + \bar{f}_0 + a(\bar{f}_0 - f_0)] \cdot N$$

On the $\Upsilon(4S)$ a $B\bar{B}$ pair is produced exclusively and $T = 0$ is the decay time of the tag-B. There are no background tracks corresponding to $b_- = b_+ = 0$, no mixing $\chi = \bar{\chi} = 0$, and equal production rate $f_0 = \bar{f}_0$ and the relation (131) simplifies considerably to

$$a_{\text{obs}} = \frac{d_\varepsilon + D_w a}{1 + f_\varepsilon a} \approx (1 - 2w)a \tag{133}$$

There is no intrinsic asymmetry unless the detector has different acceptances for positive and negative particles, there is no mixing dilution, $D_m = 1$, and the dilution factor is related to the mistag probability simply as $D = D_t = 1 - 2w$. This tagging dilution, which is also a good first approximation in more complicated jet environments, can be expressed in the simple form

$$D_t = 1 - 2w = \frac{\text{right-sign} - \text{wrong-sign}}{\text{right-sign} + \text{wrong-sign}} \tag{134}$$

where "right-sign" and "wrong-sign" refers to the number of correct and wrong tags, respectively.

As will be shown in detail below, the error on the observed asymmetry amplitude is $\sigma_a \propto 1/\sqrt{N_1 + N_2}$. Therefore, the error on the asymmetry amplitude Λ_0 is

$$\sigma(\Lambda_0) = \frac{1}{D\sqrt{N_1 + N_2}}\sqrt{1 - (D\Lambda_0)^2} \approx \frac{1}{\sqrt{\varepsilon_t D^2 N_s}}$$

where N_s is the number of signal events, ε_t is the tagging efficiency, i.e. $\varepsilon_t N_s$ is the fraction of signal events with a flavour tag, and the approximation holds for small asymmetries $(D\Lambda_0)^2 \ll 1$. The performance of tagging can therefore be defined by the factor $(\varepsilon_t D_t^2)_{\text{eff}}$ which gives the effective reduction in number regarding statistical precision. It is known under many names. One is **separation**, since it is 1 if b and $\bar{b}$ can be separated perfectly event by event, and 0 if they cannot be distinguished. It is also called **effective tagging efficiency** since it is the tagging efficiency weighted by a measure of usefullness of a tag.

3.6.2 Statistical Tagging

The correlation exploited in a flavour tag is perfect for fully reconstructed charged beauty mesons or for beauty baryons. Even neutral B mesons reveal their flavour unambiguously in most of their final states. However, a complete reconstruction of B decays is only possible in a very limited number of events. A more universal approach is to collect this information via certain characteristics of the particles which are able to identify the flavour. Tagging information can be obtained in many ways, as has been discussed in section 2.3.1. In the example above, only leptons have been used, with a charge correlated to the beauty flavour via semileptonic decays $b \to l^- \bar{\nu}c$.

This tagging method suffers from two problems:
- only about 20% of all b hadrons decay semileptonically, and
- a substantial fraction of leptons from other sources have the "wrong" charge.

In the statistical approach one considers the distributions of the variables $X_{11}, X_{12} \ldots$ of daughter particle 1 with charge Q_1, the variables $X_{21}, X_{22} \ldots$ of daughter particle 2 with charge Q_2, and so on, i.e.

$$f_{\bar{b}}(X_{11}, X_{12} \ldots, X_{21}, X_{22}, \ldots, \ldots | Q_1, Q_2 \ldots)$$

if the parent was a B meson (or, in general, a $\bar{b}$-hadron), and

$$f_b(X_{11}, X_{12} \ldots, X_{21}, X_{22}, \ldots, \ldots | Q_1, Q_2 \ldots)$$

if the parent was a $\bar{B}$ meson (or b-hadron). The mostly continuous variables X_{ij} and the charges Q_i may both be considered random variables. The flavour of this event is assigned by maximum likelihood, i.e. if $f_{\bar{b}} > f_b$ the beauty of the event is taken to be $+1$ and vice versa. This assignment is unique, and depends on the X and Q variables together. The signed dilution factor is then

$$D_t(X \ldots, Q \ldots) = \frac{f_{\bar{b}}(X \ldots | Q \ldots) - f_b(X \ldots | Q \ldots)}{f_{\bar{b}}(X \ldots | Q \ldots) + f_b(X \ldots | Q \ldots)}$$

where the sign gives the estimated beauty quantum number, and its absolute value gives the dilution factor corresponding to this flavour assignment.

A very effective discriminating variable for leptons is their momentum in a system as close as possible to the B rest system. Other discriminating variables can be e.g. the impact parameter to the beam axis or the angle to the closest track from the same B. Often information on the environment may have even better or at least complementary discriminating power, e.g. the energy of a neutrino as missing momentum in the case of a charged lepton tag.

For one-dimensional distributions, the true flavour is statistically related to Q by the densities $f_{\bar{b}}(X_j|Q)$ of tags with a $\bar{b}$ quark, and $f_B(X_j|Q)$ of tags with a b quark. The average beauty flavour of a sample with a certain value X_j and Q is then

$$\hat{B}_i = \frac{f_{\bar{b}}(X_j|Q) - f_b(X_j|Q)}{f_{\bar{b}}(X_j|Q) + f_b(X_j|Q)} \tag{135}$$

If efficiencies are independent of charge, the densities for the opposite charge are $f_{\bar{b}}(X_j| - Q) = f_b(X_j|Q)$ and $f_b(X_j| - Q) = f_{\bar{b}}(X_j|Q)$ up to negligible explicit CP violations. This reduces four functions in the general ansatz to only two functions. Instead of treating $Q = +1$ and -1 as two values of an additional random variable, the flavour of the parent b-hadron may be a

priori assigned to be Q. This assignment is arbitrary, and could be opposite, as long as it is uniquely defined. Then the flavour estimator can be written as
$$\hat{B}_i = D_t(X_j) \cdot Q \tag{136}$$

The absolute value of the signed factor
$$D_t(X_j) = \frac{f_{\bar{b}}(X_j|+) - f_b(X_j|+)}{f_{\bar{b}}(X_j|+) + f_b(X_j|+)} = \frac{f_{\bar{b}}(X_j|+) - f_{\bar{b}}(X_j|-)}{f_{\bar{b}}(X_j|+) + f_{\bar{b}}(X_j|-)}$$

is the tagging dilution factor for all tags with a value X_j. It is $|D_t| = 1$ if the flavour is perfectly correlated with Q. The sign is a flavour corrector: It is negative if the arbitrary a priori assignment from Q is more often wrong than right:
$$D_t = \frac{\text{right-sign} - \text{wrong-sign}}{\text{right-sign} + \text{wrong-sign}}$$

It is typically a smooth function of X_j, and can be obtained from a Monte Carlo simulation or from real data measuring the $B\bar{B}$ oscillation amplitude as described below.

The flavour estimator $\hat{B}$ defined in (136) is the average flavour of events in a sample with tagging particles of charge Q and discriminating variable X_j, e.g. of positive leptons at a given cms momentum:

$$\hat{B} = (+1) \cdot \frac{f_{\bar{b}}(X_j|Q)}{f_{\bar{b}}(X_j|Q) + f_b(X_j|Q)} + (-1) \cdot \frac{f_b(X_j|Q)}{f_{\bar{b}}(X_j|Q) + f_b(X_j|Q)}$$

The coefficients are the fractions (or *a posteriori* probabilities) of events from B and $\bar{B}$, respectively. This variable is itself a discriminating variable comprising all used information in the event. It can be split into its sign – used as flavour guess – and its absolute value $|\hat{B}|$, which is the tagging dilution factor
$$D_t(\hat{B}) = |\hat{B}|$$

All informations X_{ij} from one or several tags in an event are combined to estimate the flavour of the tag hadron more reliably.

A full exploitation of this tagging method is, however, impossible since the detailed information required to determine the innumerable multidimensional functions cannot be achieved within the statistical precision of any experiment. Several approximations are proposed to overcome the technical problems.

One approach is to use only one-dimensional distributions and to assume factorization
$$f_{\bar{b}}(X_{11}, X_{12}\ldots, X_{21}\ldots|Q_1, Q_2\ldots)$$
$$= f_{\bar{b}}(X_{11}|Q_1) \cdot f_{\bar{b}}(X_{12}|Q_1) \cdots f_{\bar{b}}(X_{21}|Q_2) \cdots$$

With this ansatz, one can add all flavour estimators of the same event like relativistic velocities. For two tagging particles with variables X_1, Q_1 and X_2, Q_2, inserting (136) into the result

$$\hat{B} = \hat{B}_1 \oplus \hat{B}_2 := \frac{\hat{B}_1 + \hat{B}_2}{1 + \hat{B}_1 \hat{B}_2} \tag{137}$$

leads immediately to

$$\hat{B} = \frac{f_{\bar{b}}(X_1|Q_1)f_{\bar{b}}(X_2|Q_2) - f_b(X_1|Q_1)f_b(X_2|Q_2)}{f_{\bar{b}}(X_1|Q_1)f_{\bar{b}}(X_2|Q_2) + f_b(X_1|Q_1)f_b(X_2|Q_2)}$$

Repeated addition of the estimators from all tagging particles and discriminating variables yields the total flavour estimator of the event.

Although factorization is not valid, the approximation is usually quite good and can be tested by determining $D_t(\hat{B})$. The deviation $D_t(\hat{B}) - |\hat{B}|$ is a measure of the correlations; if it is small, the approximation is a good one, if it is large, a different way to combine tagging information may improve the result.

A better approach for many discriminating variables of one particle is to **combine** them into one single variable which is an optimum representation of the flavour information including the correlations. Either a linear combination of carefully chosen variables is used in a Fisher discriminant analysis, or, even better, a neural net is trained with all discriminating variables as input. A combination of both methods, or more flexible neural networks which allow input of a variable number of tag particles and their corresponding parameters, could eventually give a maximum of information.

In a statistical tagging method, the effective performance $(\varepsilon_t D_t^2)_{\text{eff}}$ can no longer be split into two factors. In fact, in the ultimate realization $\varepsilon_t = 1$, i.e. every event is used as a tag. This is only reasonable if, instead of fitting $a_{\text{obs}}(T) = D \Lambda_0 \sin xT + I$, a fit to a two-dimensional distribution

$$a_{\text{obs}}(T, D) = D \Lambda_0 \sin xT + I \tag{138}$$

is used to obtain the parameter Λ_0. Here the mistag contribution to D is not a constant average dilution factor, but a **function** of the discriminating variables $X_1, X_2, \ldots$, and varies between -1 and $+1$. Negative values mean that in the kinematic range defined by $\{X_1, X_2, \ldots\}$ there are more "wrong-sign" tags with opposite sign of charge and flavour than "right-sign" (or same-sign) tags. In other words, the correlation between the charge and the beauty flavour has flipped. D is evaluated event by event in addition to the scaled lifetime (difference) T.

The product $(\varepsilon_t D_t^2)_{\text{eff}}$ can still be evaluated as

$$(\varepsilon_t D_t^2)_{\text{eff}} = \int_0^1 \frac{d\varepsilon_t}{ds} D_t^2(s)\, ds$$

where
$$\frac{\mathrm{d}\varepsilon_t}{\mathrm{d}s} = f(s)$$

is the probability density of $s = |\hat{B}|$. In the general concept, a flavour estimator is constructed for every event – with a value close to 0 if no good tagging information is available – and the performance number

$$(\varepsilon_t D_t^2)_{\mathrm{eff}} = \langle D_t^2 \rangle$$

is the average tagging dilution of all events.

This concept should be applied to all dilution factors, e.g. to mixing dilution as a function of the lifetime of a B meson in the second jet or to the dilution due to limited precision in the determination of the oscillation time T as a function of vertex precision. The relevant information for the fit from tagging is the dilution coefficient $D_t(X_1, X_2, \ldots)$ itself. It is given by all tag particles, i.e. particles with a nonzero charge-like quantum number Q that can be correlated to the beauty flavour. The discriminating variables for each particles have to be chosen in a way to differentiate maximally between kinematic situations with different correlation strength. An obvious variable is the momentum of the particle in the parent's rest frame, if this can be reconstructed. This is illustrated in Fig. 22, where the momentum of muons in the $\Upsilon(4S)$ rest frame – which comes close to the B rest frame – is used as discriminator: Above a value of about $1.5\,\mathrm{GeV}/c$, almost all muons are from semileptonic B decays, and show therefore an almost perfect correlation with the beauty flavour. This is indicated by a dilution factor close to 1 in Fig. 22b, which is calculated from the distributions in Fig. 22a using (134). At values below $0.7\,\mathrm{GeV}/c$ there is an opposite correlation due to $b \to c \to l^+ \nu X$, though diluted to about one fifth. The flip of the correlation is at $0.9\,\mathrm{GeV}/c$, where the mistag probability is $w = \frac{1}{2}$, and no flavour information is obtained.

The idea to calculate a flavour estimator for every event is especially promising on the $\Upsilon(4S)$, where $B^0 \overline{B}^0$ are produced exclusively. After removing the signal CP channel (e.g. $J/\psi K_S^0$), the whole residual event is from the other B, so every charged particle can be used for statistical tagging.

3.6.2.1 Some Examples

Present experiments are not fully using this concept, because a separation in categories with a single piece of information gives a better control on systematic uncertainties. Leptons provide the cleanest flavour tags, if discriminating variables are properly used to distinguish various sources of leptons in b decays (see section 2.3.1.1). For exclusive signals, some discriminating variables used are

1) the momentum p^* in the $\Upsilon(4S)$ or tag-B cms, or the transverse momentum $p_\perp$ at hadron machines,

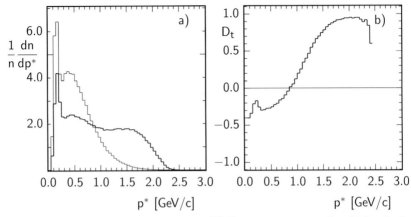

Fig. 22. Distributions of muons in the $\Upsilon(4S)$ cms momentum for "right-sign" (———) and "wrong-sign" (———) tags **(a)**, and the corresponding dilution factor **(b)**. The simulation has been done for the BABAR detector; the peak at low momentum is due to K and π decays and particle misidentification.

2) the isolation in the tag-B cms, e.g. the angular distance to the next neighbouring track, and/or the energy E_{90} of all other detected particles in the same hemisphere, i.e. with less than 90° to the tag lepton,
3) the charged multiplicity of the tag-B decay,
4) the angle between the lepton and the next charged track from the tag,
5) vertex information from the lepton, e.g. the impact parameter or the closest radial distance between the track and the beam axis,
6) the recoil mass m_r, i.e. the invariant mass of all other particles in the event assigned to the tag,
7) the missing momentum, energy and mass in an $\Upsilon(4S)$ event.

The best discriminator is the lepton momentum in the tag-B centre of mass system, which can be determined on the $\Upsilon(4S)$ from the known cms momentum and the reconstructed signal-B. Using the nominal $\Upsilon(4S)$ system instead is, however, a good approximation, and can be used even with incomplete signal-B final states. Fig. 22 shows the right-sign and wrong-sign distributions and the tagging dilution used to calculate $\hat{B}$. This is from an early simulation before the start of the BABAR experiment, but will suffice to illustrate statistical tagging. For leptons with $p^* > 1.4\,\text{GeV}/c$, there is negligible improvement from additional variables, but they can help for low momentum leptons. The figure includes misidentified hadrons which contribute their share to the overall performance.

The recoil mass is a powerful supplement to the momentum, showing peaks of reconstructed D and D^* mesons, which indicate a direct semileptonic B decay. Vertex information could also be useful, but the charm lifetimes being much shorter than the B lifetimes, the efficiency to reconstruct separate vertices is very low.

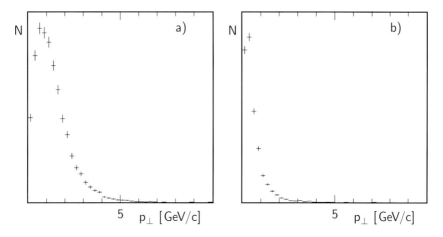

Fig. 23. Shape of the $p_\perp$ distribution of muons from semileptonic b-decays (a) and from $b \to c \to l$ decays (b) from a Monte Carlo simulation for a fixed-target experiment at LHC.

The dilution and performance of most of the mentioned variables are significantly better in discriminating primary from secondary leptons, than from discriminating right-sign from wrong-sign. This is mainly due to the fact that a substantial fraction of secondary leptons – as from D_s meson decays via process (d) – has the "right" sign of charge.

The performance of lepton tagging alone is for an ideal detector $(\varepsilon_t D_t^2)_\mathrm{eff} = 0.24$. Within the acceptance of BABAR, it is still $(\varepsilon_t D_t^2)_\mathrm{eff} = 0.20$ for perfect particle identification.

The actual experiment used only high-momentum leptons and used an average dilution. Thus they achieved a performance $(\varepsilon_t D_t^2)_\mathrm{eff} = 0.081$ [5]. However, part of the information from poorly identified leptons is regained in a second tagging class using a neural net to employ more discriminating variables.

3.6.2.2 Lepton Tags at b Jets

At hadron experiments, the prime discriminating variable to distinguish direct tag leptons from wrong-sign secondaries is the transverse momentum $p_\perp$ with respect to the beam axis. For illustration we present simulation results for a planned fixed-target experiment LHB at LHC [136]. Multiplicities for a $p_\perp$ cut on right-sign and wrong-sign charged muons have been derived from Monte Carlo distributionsand are compared to those of charged pions and kaons in Table 15.

Misidentification of hadrons as leptons is a possible source of dilution and intrinsic asymmetry. Typical misidentification probabilities at high momenta are in the order of $\delta_K \approx 2\%$ and $\delta_\pi \approx 1\%$.

Table 15. Multiplicity of various particles in $pBe \to bB^0X$ (7 TeV protons with 0.68 minimum bias event overlap) for different $p_\perp$ cuts, excluding the decay products of the B^0 from a PYTHIA 5.7 / JETSET 7.4 / FRITIOF 7.02 / ARIADNE 4.02 Monte Carlo simulation.

$p_\perp >$	$n(e^-)$	$n(e^+)$	$n(\mu^-)$	$n(\mu^+)$	$n(\pi^-)$	$n(\pi^+)$	$n(K^-)$	$n(K^+)$
0.00	0.380	0.342	0.128	0.096	17.85	18.63	1.91	1.74
0.25	0.146	0.094	0.119	0.072	11.20	11.80	1.53	1.36
0.50	0.109	0.043	0.102	0.041	5.28	5.54	0.89	0.75
0.75	0.087	0.022	0.085	0.022	2.71	2.82	0.458	0.352
1.00	0.069	0.0122	0.068	0.0125	1.675	1.716	0.235	0.159
1.25	0.053	0.0071	0.053	0.0066	1.230	1.242	0.126	0.075
1.50	0.039	0.0039	0.040	0.0041	0.994	0.997	0.073	0.039
1.75	0.028	0.0025	0.030	0.0024	0.835	0.835	0.042	0.021
2.00	0.020	0.0015	0.021	0.0015	0.714	0.709	0.027	0.012

In a real experiment, these have to be determined from the data. A sample of clean $K^\pm$ and $\pi^\pm$ can be obtained from the decay $D^{*+} \to D^0 \pi^+_{\text{slow}}$, $D^0 \to K^-\pi^+$, where the K is the particle with opposite charge to the slow π. The purity is obtained by the secondary vertex of $D^0 \to K^-\pi^+$ and the small mass difference $m_{D^*} - m_D$. Counting the number of muon and electron candidates passing this selection as K or π, the misidentification probabilities can be determined in the experiment.

In a multiparticle environment of hadronic interactions, the transverse momentum may be the only useful discriminating variable. Since the acceptance of hadrons drops more rapidly with $p_\perp$ than that of leptons (Table 15), a cut on $p_\perp$ can remove systematic uncertainties and will cost only a small reduction in the performance factor $(\varepsilon_t D_t^2)_{\text{eff}}$. A cut in this variable may also be required as additional tool for background rejection. To optimize this cut with respect to tagging performance, the value of $(\varepsilon_t D_t^2)_{\text{eff}}$ has been calculated for different $p_\perp$ cuts as shown in Fig. 24 for electrons and muons. No contributions from misidentified hadrons are included here. The maximum $(\varepsilon_t D_t^2)_{\text{eff}} = 0.039$ for both is close to $p_\perp = 1\,\text{GeV}/c$, suggesting a cut at this value, with a total $(\varepsilon_t D_t^2)_{\text{eff}} = 0.08$. Including estimated contributions from hadron misidentification, the total performance of lepton tagging reduces $(\varepsilon_t D_t^2)_{\text{eff}} = 0.06$. Statistical tagging without cut corresponds to $(\varepsilon_t D_t^2)_{\text{eff}} = 0.046$ for electrons and $(\varepsilon_t D_t^2)_{\text{eff}} = 0.054$ for muons. However, most of this advantage will be lost if misidentified pions with substantial intrinsic charge asymmetry are added.

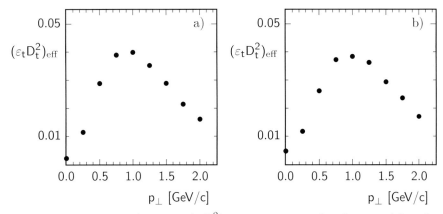

Fig. 24. Tagging performance $(\varepsilon_t D_t^2)_{\text{eff}}$ versus $p_\perp$ cut for electrons **(a)** and muons **(b)** in $p\text{Be} \to B^0 bX$ events at LHB. The maximum is close to $1\,\text{GeV}/c$.

3.6.2.3 Other Tags

Charged kaon tags are the most efficient. It is therefore desirable to investigate correlations with other information in the event in order to optimize their use. The signature provided by charged kaons is the sign of strangeness. This signature is weakened, if other strange particles are present in the same B decay. While the fraction of Λ baryons is negligible, events with other kaons are relatively abundant. Therefore, it is useful to divide kaon tags into two classes:

1) Events with exactly one charged kaon and no K_S^0, or with two (or more) charged kaons of the same sign. This class is the main contributor of beauty flavour information. The remaining events with kaons add only a little to the overall performance:

2) Events with one charged kaon and at least either a reconstructed K_S^0 (Fig. 26) or another charged kaon of different sign. The latter are usually not taken for tagging.

In $\Upsilon(4S)$ factories, kaons are partly identified via dE/dx in the main tracker. However, their tagging power can be only fully exploited with a Cherenkov detector, e.g. a DIRC (BABAR) or aerogel counters (BELLE). The performance of pure charged kaon tagging at BABAR is $(\varepsilon_t D_t^2)_{\text{eff}} = 0.14$ [5].

Hadron B factories rely typically on gas RICH systems for kaon identification. The use of discriminating variables has been explored for the LHB experiment using the event generators PYTHIA 5.7, JETSET 7.4 [137], FRITIOF 7.02 [138] and ARIADNE 4.02. A main vertex is defined as the one the reconstructed B points at, which is in most cases the true origin of the $b\bar{b}$ jets. For all tracks, an impact parameter d_{0i} with respect to the main

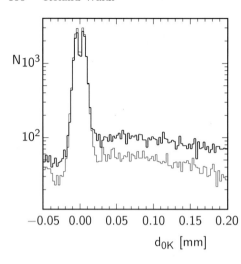

Fig. 25. Simulated impact parameter distribution for a fixed target experiment at LHC of "right-sign" (———) and "wrong-sign" (———) charged kaons.

vertex is computed. The selection cuts ensure tracking in the detector and identification in the RICH ($40\,\text{GeV}/c < p_K < 350\,\text{GeV}/c$).

Primary vertex kaons are reduced by a cut on the impact parameter to the primary vertex $d_{0K} > 20\,\mu\text{m}$ The importance of a cut on d_0 is demonstrated in Fig. 25, which shows that the tagging power of charged kaons vanishes in the peak close to $d_0 = 0$. This condition also removes kaons from minimum bias pileups occurring before the main vertex. Pileup events occurring after the main vertex contribute to the kaon tag sample, and are included in the numbers given below. With a detailed study of the different event vertices, we will be able to reduce this contribution, and thus improve the performance of kaon tagging with respect to this simulation.

The charge correlation of the K with the B has been adjusted to agree with the ARGUS measurement [139]. If more than one kaon is found in an event, the one with highest impact parameter d_{0K} is taken as tag. Including secondary interactions in the silicon, this yields a tag rate of $\varepsilon_K = 0.51$ per bB event, and a dilution factor $D_K = 0.36\cdot$. If kaons from secondary interactions in the silicon counter could be rejected completely, the performance would improve to $\varepsilon_K = 0.34$ at a dilution $D_K = 0.42$. From the pulse height measurement, it is expected to identify at least 50% of these kaons. For a conservative estimate, kaons are not used if a tag lepton is detected. This reduces the fraction of kaons from other sources more than the fraction of kaons from b decays, thus improving the dilution factor. Including these two factors, the final kaon tag rate and dilution are

$$\varepsilon_K = 0.36, \quad D_K = 0.45, \quad (\varepsilon_t D_t^2)_{\text{eff}} = 0.073$$

The dilution may improve considerably if we can assign the charged kaon to a secondary vertex. This has been shown by the SLD collaboration [44], who used kaons from a b-hadron vertex to measure $B\overline{B}$ oscillations, with an estimated dilution factor $D_K \approx 0.8$.

A $p_\perp$-cut which is the main tool to purify tag leptons does not help, since the majority of the kaons is from the cascade charmed hadron decay.

A typical B event consists of many more tags, which can be combined using the prescription in (137). All particles which can be cleanly associated with a b-hadron either via a common vertex or at the $\Upsilon(4S)$ are potential tags. In addition to leptons and kaons, these are charged pions with a significant flavour correlation at the high end of the momentum spectrum due to two-body decays with a charged pion or ρ meson, and at the low end due to pions from $D^{*+} \to \pi^+ D^0$.

The momentum spectrum in the $\Upsilon(4S)$ cms is shown in Fig. 26. As examples for other discriminating variables, distributions in missing mass and recoil mass for charged pions are shown in the following figures.

Some less important tags are protons and Λ hyperons from baryonic B decays. It should be emphasized, that the simple diquark antidiquark pair creation model in Monte Carlo simulations may be wrong in reproducing baryon-flavour correlations, so the "predictions" should be taken with reservation. Assuming a vertex finding efficiency of 90% and the geometrical acceptance of the BABAR detector, the reconstruction efficiency of $\Lambda \to p\pi^-$ is $\sim 60\%$. The contribution to tagging performance has been derived from a BABAR Monte Carlo simulation. It is $(\varepsilon_t D_t^2)_{\text{eff}} = 0.005$, making it a negligible fraction. Only recently, experimental information of baryon flavour correlations have become available from CLEO [140]. They find an average dilution factor for Λ tags of $D_\Lambda = 0.40 \pm 0.11$ (for a mixture of neutral and charged B mesons). This is significantly worse than the Monte Carlo prediction of $D_\Lambda \approx 0.70$. A better tag would be the full reconstruction of a Λ_c baryon, with $D_{\Lambda_c} = 0.68 \pm 0.19$.

At hadron machines, tagging with pions and protons is expected to give even less information, since these particles are abundant in fragmentation, and show intrinsic asymmetries which reduce further their usefulness as flavour tags. This is confirmed in a simulation for LHB, where Λ hyperons add negligible information due to $(\varepsilon_t D_t^2)_{\text{eff}} = 0.003$.

3.6.3 Determining I and D from Data

An experimental determination of the dilution factor D is essential to control the systematic error of an asymmetry amplitude determination. The dilution factor can only be guessed from Monte Carlo simulations, and has to be determined from data. In addition, the intrinsic asymmetry I, although it could be obtained from the fit of the time dependent function

$$a_{\text{obs}}(T) = I + D \cdot A_0 \sin xT$$

alone, can profit from additional independent statistical information. Both parameters are obtained using a self-tagging decay channel with no or

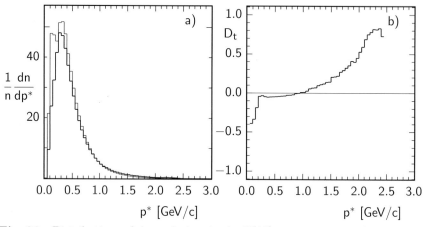

Fig. 26. Distributions of charged pions in the $\Upsilon(4S)$ cms momentum for "right-sign" (———) and "wrong-sign" (———) tags **(a)**, and the corresponding dilution factor **(b)**.

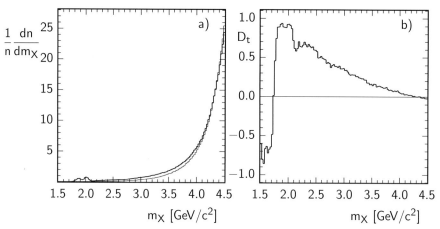

Fig. 27. Distributions of the missing mass from the decay $B \to \pi^\pm X$ for "right-sign" (———) and "wrong-sign" (———) tags **(a)**, and the corresponding dilution factor **(b)**.

negligible CP asymmetry[14] and small combinatoric background. Its time dependent tagged rate asymmetry is due to $B^0 \bar{B}^0$ oscillation

$$a_{\text{obs}}(T) = I + D \cdot \cos xT$$

which has a physical amplitude of 1, up to negligible corrections $\mathcal{O}(\delta_\epsilon)$.

---

[14] analysing many different channels separately can help to avoid stepping into this trap

Combining, for example, a reconstructed $B \to D^*\pi$ with a tag lepton, the following rates can be be measured as a function of the B proper lifetime in full analogy with equation (130):

$$\dot{N}_1 = \dot{N}(D^{*+}\pi^- l^+) = (1+\tilde{a}) \cdot \bar{f}_0 \cdot [(1-\chi)\varepsilon_{B+}B + \chi\varepsilon_{C+}C + b_+] \cdot \frac{N}{2}$$
$$+ (1-\tilde{a}) \cdot f_0 \cdot [\bar{\chi}\varepsilon_{B+}B + (1-\bar{\chi})\varepsilon_{C+}C + b_+] \cdot \frac{N}{2}$$

$$\dot{N}_1' = \dot{N}(D^{*-}\pi^+ l^+) = (1-\tilde{a}) \cdot \bar{f}_0 \cdot [(1-\chi)\varepsilon_{B+}B + \chi\varepsilon_{C+}C + b_+] \cdot \frac{N}{2}$$
$$+ (1+\tilde{a}) \cdot f_0 \cdot [\bar{\chi}\varepsilon_{B+}B + (1-\bar{\chi})\varepsilon_{C+}C + b_+] \cdot \frac{N}{2}$$

$$\dot{N}_2 = \dot{N}(D^{*-}\pi^+ l^-) = (1+\tilde{a}) \cdot f_0 \cdot [(1-\bar{\chi})\varepsilon_{B-}B + \bar{\chi}\varepsilon_{C-}C + b_-] \cdot \frac{N}{2}$$
$$+ (1-\tilde{a}) \cdot \bar{f}_0 \cdot [\chi\varepsilon_{B-}B + (1-\chi)\varepsilon_{C-}C + b_-] \cdot \frac{N}{2}$$

$$\dot{N}_2' = \dot{N}(D^{*+}\pi^- l^-) = (1-\tilde{a}) \cdot f_0 \cdot [(1-\bar{\chi})\varepsilon_{B-}B + \bar{\chi}\varepsilon_{C-}C + b_-] \cdot \frac{N}{2}$$
$$+ (1+\tilde{a}) \cdot \bar{f}_0 \cdot [\chi\varepsilon_{B-}B + (1-\chi\varepsilon_{C-}C) + b_-] \cdot \frac{N}{2}$$

where $\tilde{a}(T) = \cos xT$ is the true asymmetry. This implies

$$\dot{N}_1 + \dot{N}_1' = f_0 \cdot [(1-\bar{\chi})\varepsilon_{B-}B + \bar{\chi}\varepsilon_{C-}C + b_-] \cdot \frac{N}{2}$$
$$+ \bar{f}_0 \cdot [\chi\varepsilon_{B-}B + (1-\chi)\varepsilon_{C-}C + b_-] \cdot \frac{N}{2}$$

$$\dot{N}_2 + \dot{N}_2' = \bar{f}_0 \cdot [(1-\chi)\varepsilon_{B+}B + \chi\varepsilon_{C+}C + b_+] \cdot \frac{N}{2}$$
$$+ f_0 \cdot [\bar{\chi}\varepsilon_{B+}B + (1-\bar{\chi})\varepsilon_{C+}C + b_+] \cdot \frac{N}{2}$$

Using the integrated numbers $N_i = \int \dot{N}_i \, dT$, the intrinsic asymmetry, which is constant in time, can be determined very precisely as the ratio

$$I = \frac{(N_1 + N_1') - (N_2 + N_2')}{N_1 + N_1' + N_2 + N_2'} \tag{139}$$

The dilution factor can be obtained from a fit of $a_{\text{obs}}(T) = I + D \cdot \tilde{a}(T)$ to

$$a_{\text{obs}}(T) = \frac{\dot{N}_1 - \dot{N}_2'}{\dot{N}_1 + \dot{N}_2'}$$

and of $a_{\text{obs}}(T) = I - D \cdot \tilde{a}(T)$ to

$$a'_{\text{obs}}(T) = \frac{\dot{N}_1' - \dot{N}_2}{\dot{N}_1' + \dot{N}_2}$$

The dilution for a single tag charge can be investigated from the ratio

$$a_{\text{obs}+}(T) = \frac{\dot{N}_1 - \dot{N}_1'}{\dot{N}_1 + \dot{N}_1'} = D_+ \cdot \tilde{a}(T), \quad a_{\text{obs}-}(T) = \frac{\dot{N}_2 - \dot{N}_2'}{\dot{N}_2 + \dot{N}_2'} = D_- \cdot \tilde{a}(T)$$

with

$$D = \frac{D_+ + D_-}{2}(1 - I^2)$$

Regarding the fact, that constants as f_0 and χ depend actually on momentum and proper lifetime, respectively, a channel with very similar response to cuts is best to determine I and D. For the decay $J/\psi K_S^0$ the decay $B^0 \to J/\psi K^{*0}$ is well suited in this respect, since the B flavour can be identified if $K^{*0} \to K^+\pi^-$. The only pre-condition is a good K/π separation, as is desired by any B factory experiment. The branching fraction $\mathcal{B}(B^0 \to J/\psi K^{*0}) \cdot \mathcal{B}(K^{*0} \to K^+\pi^-)$ is a factor 2.9±0.5 larger than the signal channel $\mathcal{B}(B^0 \to J/\psi K_S^0) \cdot \mathcal{B}(K_S^0 \to \pi^+\pi^-)$ [9]. For tagging methods relying on the other b-hadron the dilution factor could even be determined from charged $B^+ \to J/\psi K^+$ decays, which are about a factor 3 more copious than $B^0 \to J/\psi K^0$, $K^0 \to K_S^0 \to \pi^+\pi^-$.

With more channels e^+e^- colliders can compensate for their lower numbers of B mesons and their need to study a function $D_t(\hat{B})$ or D_t in several tagging classes instead of just one number. The BABAR experiment has used in its first study of CP violation [141] a pure sample of 4600 hadronic B decays[15] to $D^{(*)-}\pi^+(\pi^0, \pi^+\pi^-)$ and $J/\psi(K^+\pi^-)_{K^{*0}}$ and a sample of 15200 events from $B^0 \to D^{*-}l^+\nu$ in a selection with 80% purity

Fully reconstructed charged B mesons are a perfect tool to study opposite-jet tagging in $b\bar{b}$ events. Of course, they are not useful on the $\Upsilon(4S)$, since here the tag is not provided by the fragmentation products of a b quark, but solely by B^0 and $\bar{B}^0$ mesons. In fact, an experimental determination of mistag rates via a fully reconstructed B^0 or $\bar{B}^0$ has to be corrected for two effects from $B^\pm$ background: the fact that a charged B does not mix, and the different tagging dilution for charged and neutral B mesons.

3.6.4 Effects of Vertex Resolution

The discussion of vertex resolution effects in section 2.3.3.1 applies also to CP violating asymmetry oscillations. Fig. 28 shows one of the distributions

$$f(T) = e^{-|T|}(1 \pm \Lambda \sin xT)$$

describing decay rates to CP eigenstates on the $\Upsilon(4S)$ and its smearing from a convolution with a Gaussian. The asymmetry calculated from the smeared

---

[15] all decay channels are for B^0 but represent the charge-conjugate decay of $\bar{B}^0$, too.

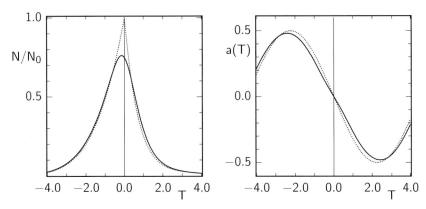

Fig. 28. Original (dotted) and smeared (solid line) distribution of $T = \Delta t/\tau$ and the corresponding asymmetry $a(T)$ assuming a Gaussian error in T with $\sigma_T = 0.4$, and the parameters $x = 0.7$, $\Lambda = 0.5$. The dilution factor from the approximation (70) is $D_r = 0.9616$ corresponding very precisely to the true reduced amplitude $\Lambda = 0.4808$.

distribution is diluted and distorted. As an example, the dilution factor from (70) is $D_r > 0.995$ for the fully reconstructed $B^0 \to J/\psi K_S^0$ events at CDF. The asymmetric e^+e^- collider experiments are operating at moderate boosts much smaller than that of a B produced in a high energy jet, and are therefore more sensitive to resolution effects than jet experiments.

BABAR [141] uses a parametrization of their vertex resolution with a sum of three Gaussians to describe the statistical uncertainty as well as the bias from sequential charm decays of the tag-B. The bias varies for different tagging categories, and the parameters – nine in total – are determined from a large sample of fully reconstructed tagging modes that exhibit the oscillation asymmetry (59).

These include a minor correction also specific to the $\Upsilon(4S)$ experiments: The B production vertex (x_0, y_0, z_0) is not known, so they cannot determine the signal B lifetime $t_s = \frac{z_s - z_0}{\beta_{zs}\gamma_s}$ and the tag-B lifetime $t_t = \frac{z_t - z_0}{\beta_{zt}\gamma_t}$ separately, in order to calculate the relevant time difference $\Delta t = t_s - t_t$. The easiest approximation is to use $\beta_{z(t)} \approx \beta_{z(s)} \approx \beta$, the boost of the $\Upsilon(4S)$, and

$$\Delta t = \frac{z_s - z_t}{\beta\gamma}$$

$$\sigma(\Delta t) = \frac{\Delta t}{\beta\gamma}\sigma(\beta\gamma) \oplus \frac{\sigma(\Delta z)}{\beta\gamma}$$

For $\beta\gamma = 0.55$ (as at PEP II), taking into account the exponential distribution of Δt and the angular distribution of the pseudoscalar B mesons $dn/d\cos\theta \propto \sin^2\theta$, the movement of the B mesons in the $\Upsilon(4S)$ centre of mass system infers an rms error on the lifetime difference of 0.22 ps (a flat distribution would increase the width to 0.30 ps). This component is increasing with Δt, but on average is a negligible contribution to the total error of $\sigma(\Delta t) = 1.1\,\text{ps}$ (rms).

This small effect can be further reduced, because the signal B is fully reconstructed, hence its full momentum vector is known. This can be exploited, assuming the first decaying B to have lived its average lifetime, which is $\tau/2$, to estimate the location of z_0, yielding the following recipe where $v_s := \beta_{zs}\gamma_s = p_z/m_B$ of the $\pi^+\pi^-$ combination, $v := \beta\gamma$, and $r_m := m(\Upsilon(4S))/m(B^0) \approx 2.00$:

$$\Delta t = \begin{cases} \dfrac{z_s - z_t}{v_s} + \tau \dfrac{r_m v - 2v_s}{2v_s} & z_s \geq z_t \\ \dfrac{z_s - z_t}{2v - v_s} + \tau \dfrac{r_m v - 2v_s}{2(r_m v - v_s)} & z_s \leq z_t \end{cases}$$

The resolution is improved considerably, with an rms of $0.09\,\text{ps}$ and essentially no bias.

The tag vertex has a correctable bias of a few μm due to the cascade charmed particle decay, and a resolution ranging from $100\,\mu\text{m}$ to $200\,\mu\text{m}$. This corresponds to an error in the scaled lifetime difference $\sigma_T = 0.3\ldots 0.5$ which is still substantially smaller than half an oscillation period of $\pi/x \approx 4.5$. Fortunately, the better resolutions are achieved in events which also have a good flavour tag, hence they contribute more to the overall fit than the events with less precise vertex information. Therefore, the effective dilution factor D_r from the time measurement is around 0.95.

The effect of non-zero time resolution is in reality different from this simplified approximation, since the distributions that are actually smeared are the exponential decay distributions. Due to their slope, more events are shifted to larger times than vice versa. The effect of Gaussian smearing with $\sigma_T = 0.3$ is shown in Fig. 29.

3.6.5 Fitting CP Asymmetries

In the limit of large data samples the fit of CP parameters to data asymmetry distributions can be performed analytically. These calculations give useful limits to the precision that can be reached in any real experiment. The following sections give a tool for quick estimates of this precision in certain combinations of free parameters. It is always assumed that the B^0 lifetime τ and the mixing parameter x are known with higher precision than the CP violation parameters.

A maximum of information is obtained from the time-dependent measurement of the CP asymmetry, where the lifetime t_s of the signal B or on the $\Upsilon(4S)$ also the difference of its lifetime $t_s - t_t$ to the tag-B is measured. The observed asymmetry function is in the most simple case

$$a_{\text{obs}}(T) = \frac{N_1 - N_2}{N_1 + N_2} = D \cdot \Lambda_0 \sin xT$$

where the dimensionless time variable is either $T = t_s/\tau$ for $b\bar{b}$ production, or $T = (t_s - t_t)/\tau$ for coherent $B\bar{B}$ production on the $\Upsilon(4S)$.

For large numbers of events, i.e. in the Gaussian approximation, the error is

$$\delta a = \sqrt{\frac{4N_1 N_2}{(N_1 + N_2)^3}} = \frac{1}{\sqrt{N_1 + N_2}} \cdot \sqrt{1 - a^2}$$

For $\Lambda^2 = D^2 \Lambda_0^2 \ll 1$, the factor $\sqrt{1-a^2}$ can be neglected. Therefore, for estimates of the error it is sufficient to use the approximation

$$\delta a \approx \frac{1}{\sqrt{N_1 + N_2}} \qquad (140)$$

The fluctuations of a_{obs} are thus proportional to

$$\delta a_{\text{obs}}(T) \propto \frac{1}{\sqrt{e^{-|T|}}}$$

and the significance varies as

$$S(T) = \frac{|a_{\text{obs}}(T)|}{\delta a_{\text{obs}}(T)} \propto |\sin xT| \cdot e^{-|T|/2} \qquad (141)$$

as shown in Fig. 29. The correct expression for not too small asymmetries

$$S(T) \propto \frac{|\sin xT| \cdot e^{-|T|/2}}{\sqrt{1 + \Lambda^2 \sin^2 xT}}$$

is also shown for $\Lambda = 0.5$ and demonstrates that the approximation is reasonable. The most sensitive region for $x \approx 0.7$ is in the range $T = 0.3 \ldots 3.0$, and a cut $T > 0.3$ to reject background will remove only a very insensitive subset from the sample.

The situation is completely different for fits where a term

$$a(T) = \Theta \cos xT, \quad a_{\text{obs}}(T) = D\Theta \cos xT$$

is present. Here, the most sensitive region to the amplitude is at $T = 0$ (dotted line in Fig. 29). In situations, where both terms appear with amplitudes Λ and $\Theta \neq 0$, the region near $T = 0$ requires special attention, since background tends to strike in this region most violently.

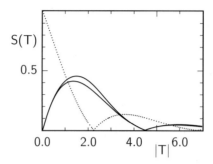

Fig. 29. Significance of the observed CP asymmetry ($\propto \sin xT$) versus decay time for $x = 0.70$, $\Lambda \approx 0$ (lower curve) and $\Lambda = 0.5$ (upper curve). The dotted line is a pure $\cos xT$ term (as in the mixing asymmetry) in the limit of vanishing amplitude $\Theta \to 0$.

3.6.5.1 Fit to the Time Dependent Asymmetry

A fit of the observed amplitude Λ to the experimental asymmetry values $a_1 \ldots a_n$ obtained at n scaled B lifetime or lifetime difference intervals centred at $T_1 \ldots T_n$ for small bin sizes ΔT can be performed by minimizing

$$\chi^2 = \sum_{i=1}^{n} \frac{(a_i - \Lambda \sin xT_i)^2}{\sigma(a_i)^2} = N_0 \Delta T \sum_{i=1}^{n} (a_i - \Lambda \sin xT_i)^2 e^{-|T_i|} \quad (142)$$

where the error on a_i is approximated by $\sigma(a_i) \approx 1/\sqrt{N_1(i) + N_2(i)} \approx 1/\sqrt{N_0 e^{-|T_i|} \Delta T}$. Here N_0 denotes the number of events for all lifetimes, while $N_{\text{tot}} = N_1 + N_2$ is the number of events within the selected lifetime intervals, hence the actual sample size. N_1 and N_2 are the subsamples with apparent, i.e. tagged, B and $\bar{B}$ mesons, respectively, and $N_1(i), N_2(i)$ their contribution to bin i. Assuming constant acceptance in the time interval investigated, N_0 is related to N_{tot} via

$$N_{\text{tot}} = N_0 \int_{T_1 \min}^{T_n \max} e^{-|T|} \, dT = N_0 \Delta T \sum_{i=1}^{n} e^{-|T_i|}$$

where N_1, N_2 are the numbers of B and $\bar{B}$ tagged events, respectively.

The best fit value $\hat{\Lambda}$ is obtained from $\frac{\partial \chi^2}{\partial \Lambda} = 0$ as

$$\hat{\Lambda} = \frac{\sum a_i e^{-|T_i|} \sin xT_i}{\sum e^{-|T_i|} (\sin xT_i)^2} \quad (143)$$

The error can be calculated from Fisher's information function $\mathcal{I} = \frac{1}{2} \frac{\partial^2 \chi^2}{\partial \Lambda^2}$ (Rao-Cramér bound $\sigma^2 \geq 1/\langle \mathcal{I} \rangle$):

$$\sigma_\Lambda^2 = \left(\frac{1}{2} \frac{\partial^2 \chi^2}{\partial \Lambda^2} \bigg|_{\Lambda = \hat{\Lambda}} \right)^{-1} = \frac{1}{N_0 \Delta T \sum e^{-|T_i|} (\sin xT_i)^2}$$

For all possible fits we have $\mathcal{I} = S_0^2 \cdot N_{\text{tot}} = \tilde{S}_0^2 \cdot N_0$, where the coefficient

$$\tilde{S}_0 = \sqrt{\Delta T \sum \mathrm{e}^{-|T_i|}(\sin xT_i)^2}$$

describes the sensitivity on Λ and may be used as figure of merit for a sample of given size N_0 with no lifetime dependent acceptance or lifetime cuts. Correspondingly,

$$S_0 = \sqrt{\frac{N_{\text{tot}}}{N_0}}\tilde{S}_0 = \frac{\sqrt{N_{\text{tot}}}}{\sigma_\Lambda} \qquad (144)$$

is the sensitivity coefficient for a remaining sample of given size N_{tot} after cuts in proper lifetime or a selection with non-uniform acceptance in this variable. The sensitivity coefficients can be used to determine the error from the accepted number of events N_{tot} (or the available number N_0 at all lifetimes) as

$$\sigma(\Lambda) = \frac{1}{S_0 \sqrt{N_{\text{tot}}}} = \frac{1}{\tilde{S}_0 \sqrt{N_0}} \qquad (145)$$

The latter is especially interesting when the effects of time cuts are investigated.

The squares of these figures are $S_0^2 = \mathcal{I}/N_{\text{tot}}$ and $\tilde{S}_0^2 = \mathcal{I}/N_0$ and give the information (*sensu* Fisher) per event on the parameter Λ. In the same spirit, $(S_0 D)^2$ is the **information per event** on the undiluted physical parameter Λ_0.

The following sections will discuss various cases of one- and two-parameter fits in the large N_0 limit. This treatment gives simple formulae which are useful to compare different cases, but is not recommended for a real data analysis, since the approximations are not good for small data samples and since no information from individual events is used.

The method described above can be simplified by defining the following quantities, which have to be calculated from the asymmetries a_i at scaled time T_i:

$$F_{00} = \Delta T \sum \mathrm{e}^{-|T_i|} \qquad G_{10} = \Delta T \sum a_i \mathrm{e}^{-|T_i|} \sin xT_i$$
$$F_{20} = \Delta T \sum \mathrm{e}^{-|T_i|} \sin^2 xT_i \qquad (146)$$

For the more general shape

$$a_{\text{obs}}(T) = \Lambda \sin xT + \Theta \cos xT + I$$

we need in addition:

$$F_{10} = \Delta T \sum \mathrm{e}^{-|T_i|} \sin xT_i \qquad G_{00} = \Delta T \sum a_i \mathrm{e}^{-|T_i|}$$
$$F_{01} = \Delta T \sum \mathrm{e}^{-|T_i|} \cos xT_i \qquad G_{01} = \Delta T \sum a_i \mathrm{e}^{-|T_i|} \cos xT_i$$
$$F_{02} = \Delta T \sum \mathrm{e}^{-|T_i|} \cos^2 xT_i$$
$$F_{11} = \Delta T \sum \mathrm{e}^{-|T_i|} \sin xT_i \cos xT_i$$

$$(147)$$

Here $\Delta T \ll \tau/x$ is the sampling width. For large intervals ΔT, the function values should be replaced by integrals, e. g.

$$G_{10} = \sum_i a_i \int_{T_{i\,\min}}^{T_{i\,\max}} e^{-|T|} \sin xT \, dT$$

$$= \sum_i a_i \frac{e^{-|T_{i\,\min}|}(x \cos xT_{i\,\min} + \sin xT_{i\,\min}) - e^{-|T_{i\,\max}|}(x \cos xT_{i\,\max} + \sin xT_{i\,\max})}{(1+x^2)}$$

$$G_{01} = \sum_i a_i \frac{e^{-|T_{i\,\min}|}(\cos xT_{i\,\min} - x \sin xT_{i\,\min}) - e^{-|T_{i\,\max}|}(\cos xT_{i\,\max} - x \sin xT_{i\,\max})}{(1+x^2)}$$

while the a_i-free F-terms can always be replaced by integrals over the whole T interval, from $T_{\min} = T_{1\,\min}$ to $T_{\max} = T_{n\,\max}$, like

$$\begin{aligned}
F_{11} &= \int_{T_{\min}}^{T_{\max}} e^{-|T|} \sin xT \cos xT \, dT \\
&= \frac{e^{-|T_{\min}|}(2x \cos 2xT_{\min} + \sin 2xT_{\min})}{2(1+4x^2)} \\
&\quad - \frac{e^{-|T_{\max}|}(2x \cos 2xT_{\max} + \sin 2xT_{\max})}{2(1+4x^2)} \\
F_{20} &= \frac{(e^{-|T_{\min}|} - e^{-|T_{\max}|})}{2} \\
&\quad + \frac{e^{-|T_{\min}|}(2x \sin 2xT_{\min} - \cos 2xT_{\min})}{2(1+4x^2)} \\
&\quad - \frac{e^{-|T_{\max}|}(2x \sin 2xT_{\max} - \cos 2xT_{\max})}{2(1+4x^2)} \\
F_{02} &= \frac{(e^{-|T_{\min}|} - e^{-|T_{\max}|})}{2} \\
&\quad + \frac{e^{-|T_{\min}|}(-2x \sin 2xT_{\min} + \cos 2xT_{\min})}{2(1+4x^2)} \\
&\quad - \frac{e^{-|T_{\max}|}(-2x \sin 2xT_{\max} + \cos 2xT_{\max})}{2(1+4x^2)}
\end{aligned} \qquad (148)$$

For the full time range $T = 0 \ldots \infty$ and constant acceptance, the F integrals can be expressed in terms of x as

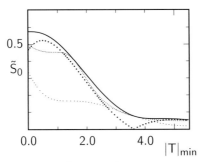

Fig. 30. Sensitivity coefficient $\tilde{S}_0 = \frac{1}{\sigma_\Lambda \sqrt{N_0}}$ for a constant number of $B^0\bar{b}$ events versus lower bound on $|T|$, i.e. for the interval $T_{min}\ldots\infty$, for the one-parameter fit (———), the two-parameter fit of Λ and I ($\cdots\cdots$), the two-parameter fit of Λ and Θ (— — —), and the time-integrated measurement ($\cdot\cdot\cdot\cdot\cdot\cdot$), assuming a mixing parameter $x = 0.70$.

$$\begin{aligned}
F_{00} &= \int dT\, e^{-|T|} = 1 \\
F_{10} &= \int dT\, e^{-|T|} \sin xT = \frac{x}{1+x^2} \\
F_{20} &= \int dT\, e^{-|T|} \sin^2 xT = \frac{2x^2}{1+4x^2} \\
F_{01} &= \int dT\, e^{-|T|} \cos xT = \frac{1}{1+x^2} \\
F_{02} &= \int dT\, e^{-|T|} \cos^2 xT = \frac{1+2x^2}{1+4x^2} \\
F_{11} &= \int dT\, e^{-|T|} \sin xT \cos xT = \frac{x}{1+4x^2}
\end{aligned} \quad (149)$$

3.6.5.2 Fit of Λ

A one parameter fit of Λ with non-vanishing intrinsic asymmetry I minimizes

$$\chi^2 = \text{const} - 2\Lambda G_{10} + \Lambda^2 F_{20} + 2\Lambda I F_{10} \quad (150)$$

and yields (with $\Theta = 0$ and I known) a best fit value

$$\hat{\Lambda} = \frac{G_{10} - I F_{10}}{F_{20}} \quad (151)$$

or $\hat{\Lambda} = G_{10}/F_{20}$ for $I = 0$. The error is given by

$$\sigma_\Lambda^2 = \frac{1}{N_0 F_{20}}$$

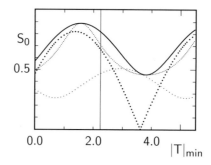

Fig. 31. Sensitivity coefficient $S_0 = \frac{1}{\sigma_\Lambda \sqrt{N_{\text{tot}}}}$ for a constant number of accepted $B^0 b$ events versus lower bound on $|T|$, for the one-parameter fit (———), the two-parameter fit of Λ and I (········), the two-parameter fit of Λ and Θ (— — —), and the time-integrated measurement (········), assuming a mixing parameter $x = 0.70$. N_{tot} is the number of events in the interval $T_{\min} \ldots \infty$, whereas N_0 in Fig. 30 is the number of events from 0 to ∞. The value of $S_{0\Theta} = \frac{1}{\sigma_\Theta \sqrt{N_{\text{tot}}}}$ can also be read from this figure, using a shift of $\frac{\pi}{2x} = 2.24$.

and the sensitivity coefficients are

$$S_0 = \sqrt{\frac{F_{20}}{F_{00}}}, \quad \tilde{S}_0 = \sqrt{F_{20}} \qquad (152)$$

The error is independent of Λ and I to the extent of approximation (140). For the whole interval $T = 0 \ldots \infty$ we have $N_0 = N_{\text{tot}}$ and

$$\sigma_\Lambda = \frac{1}{S_0 \sqrt{N_{\text{tot}}}}, \quad S_0 = \tilde{S}_0 = \sqrt{\frac{2x^2}{1 + 4x^2}} = 0.59 \qquad (153)$$

using $x = 0.75$. Table 16 gives numbers for more intervals and x values. The effect of a limited lifetime interval for a fixed integrated luminosity is described by the $\tilde{S}_0$ values, which vary below 1% for reasonable intervals and all possible x values. Only very hard restrictions to $0.3 \ldots 3.0$ lifetimes shown in the last column give a 3 to 4% reduction in sensitivity. It can also be seen that a larger x value helps to increase the experimental precision.

Table 16. Sensitivity coefficients for a one-parameter fit of Λ

$|T|$ interval:	$0 \ldots \infty$	$0.2 \ldots 6.0$		$0.3 \ldots \infty$		$0.3 \ldots 3.0$	
	$S_0 = \tilde{S}_0$	S_0	$\tilde{S}_0$	S_0	$\tilde{S}_0$	S_0	$\tilde{S}_0$
$x = 0.70$	0.575	0.634	0.573	0.665	0.572	0.669	0.556
$x = 0.75$	0.588	0.648	0.586	0.680	0.585	0.688	0.572

The sensitivity coefficient $\tilde{S}_0 = \sqrt{F_{20}}$ for a general interval $T = T_{\min} \ldots \infty$ is shown as solid line in Fig. 30 as a function of $T_{\min}$, and S_0 is shown in Fig. 31.

3.6.5.3 Fit of Λ and I

An intrinsic asymmetry can be determined simultaneously. This **two parameter** fit of Λ and I with $\Theta = 0$ can again be performed analytically. The

$$\chi^2 = \text{const} + I^2 F_{00} - 2IG_{00} - 2\Lambda G_{10} + \Lambda^2 F_{20} + 2\Lambda I F_{10} \tag{154}$$

is minimized by

$$\hat{\Lambda} = \frac{F_{00}G_{10} - F_{10}G_{00}}{F_{20}F_{00} - F_{10}^2}, \quad \hat{I} = \frac{F_{20}G_{00} - F_{10}G_{10}}{F_{20}F_{00} - F_{10}^2} \tag{155}$$

The covariance matrix

$$\mathbf{C}(\Lambda, I) = \frac{1}{N_0} \begin{pmatrix} \frac{F_{00}}{F_{20}F_{00} - F_{10}^2} & \frac{F_{10}}{F_{20}F_{00} - F_{10}^2} \\ \frac{F_{10}}{F_{20}F_{00} - F_{10}^2} & \frac{F_{20}}{F_{20}F_{00} - F_{10}^2} \end{pmatrix} \tag{156}$$

is independent of Λ and I, leading to

$$\sigma_\Lambda^2 = \frac{1}{N_0(F_{20} - F_{10}^2/F_{00})}$$

For the whole interval $T = 0 \ldots \infty$ the sensitivity coefficient for Λ is then

$$S_0 = \tilde{S}_0 = \sqrt{\frac{x^2(1 + 2x^4)}{(1 + 4x^2)(1 + x^2)^2}} = 0.34 \tag{157}$$

for $x = 0.75$. Table 17 gives numbers for more intervals and x values. The sensitivity coefficient for a general interval $T = T_{\min} \ldots \infty$ is shown in Fig. 30 as a function of $T_{\min}$ (wide-dotted curve).

Table 17. Sensitivity coefficients for Λ in a two-parameter fit of Λ and I

$|T|$ interval:	$0 \ldots \infty$	$0.2 \ldots 6.0$		$0.3 \ldots \infty$		$0.3 \ldots 3.0$	
	$S_0 = \tilde{S}_0$	S_0	$\tilde{S}_0$	S_0	$\tilde{S}_0$	S_0	$\tilde{S}_0$
$x = 0.70$	0.332	0.291	0.263	0.288	0.248	0.255	0.212
$x = 0.75$	0.340	0.301	0.272	0.298	0.257	0.250	0.208

While for known I the region near $T = 0$ can be safely dropped, here this region is very important to fix I. Also, the sensitivities are significantly lower than for the one parameter fit, which makes it highly desirable to determine the intrinsic asymmetry independently using $B^0\bar{B}^0$ oscillation events.

If this measurement gives $I = I_m \pm \sigma_I$, a fit of Λ and I with a constraint added to (154),

$$\chi^2 + \left(\frac{I - I_m}{\sigma_I}\right)^2$$

leads to the same equations as above with the replacements

$$F_{00} \to F_{00} + \frac{1}{\sigma_I^2}$$

$$G_{00} \to G_{00} + \frac{I_m}{\sigma_I^2}$$

The solutions are

$$\hat{\Lambda} = \frac{(F_{00} + 1/\sigma_I^2)G_{10} - F_{10}(G_{00} + I_m/\sigma_I^2)}{F_{20}(F_{00} + 1/\sigma_I^2) - F_{10}^2},$$

$$\hat{I} = \frac{I_m/\sigma_I^2 + G_{00} - F_{10}G_{10}/F_{20}}{1/\sigma_I^2 + F_{00} - F_{10}^2/F_{20}} \qquad (158)$$

The covariance matrix is

$$\mathbf{C}(\Lambda, I) = \frac{1}{N_0} \frac{1}{F_{20}(F_{00} + 1/\sigma_I^2) - F_{10}^2} \begin{pmatrix} F_{00} + 1/\sigma_I^2 & F_{10} \\ F_{10} & F_{20} \end{pmatrix} \qquad (159)$$

leading to

$$\sigma_\Lambda^2 = \frac{1}{N_0 \left(F_{20} - \frac{F_{10}^2}{F_{00} + 1/\sigma_I^2}\right)}$$

For the whole interval $T = 0 \ldots \infty$ the sensitivity coefficient is

$$S_0 = \tilde{S}_0 = \sqrt{\frac{2x^2}{1 + 4x^2} - \frac{x^2}{(1 + x^2)^2} \frac{\sigma_I^2}{1 + \sigma_I^2}} = 0.59 \sqrt{1 - 0.67 \frac{\sigma_I^2}{1 + \sigma_I^2}} \qquad (160)$$

for $x = 0.75$.

3.6.5.4 Fit of Θ

The fit of Θ (with I known and $\Lambda = 0$), which is performed for a control measurement to determine the dilution factor $D = \Theta$, leads to the solution

$$\hat{\Theta} = \frac{G_{01} - IF_{01}}{F_{02}} \qquad (161)$$

or $\hat{\Theta} = G_{01}/F_{02}$ for $I = 0$, with error

$$\sigma_\Theta^2 = \frac{1}{N_0 F_{02}}$$

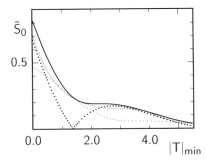

Fig. 32. Sensitivity coefficient for the fit of a $\cos xT$ contribution, as in a mixing measurement or in final states with more than one strong amplitudes. The plots show $\tilde{S}_{0\Theta} = \frac{1}{\sigma_\Theta \sqrt{N_0}}$ for a constant total number of $B^0 b$ events versus lower bound on $|T|$, for the one-parameter fit (———), the two-parameter fit of Θ and I (········), the two-parameter fit of Λ and Θ (·········), and the time-integrated measurement (·········), assuming a mixing parameter $x = 0.70$. The value of $S_{0\Theta} = \frac{1}{\sigma_\Theta \sqrt{N_{\text{tot}}}}$ can be read from Fig. 30, with a phase shift of $\pi/2$, i.e. $T_{\Theta\,\text{min}} = 0$ is at $T_{\Lambda\,\text{min}} = \frac{\pi}{2x} = 2.24$.

and sensitivity coefficients

$$S_{0\Theta} = \sqrt{\frac{F_{02}}{F_{00}}}, \quad \tilde{S}_{0\Theta} = \sqrt{F_{02}} \tag{162}$$

For the whole interval $T = 0 \ldots \infty$ this yields

$$S_{0\Theta} = \tilde{S}_{0\Theta} = \sqrt{\frac{1 + 2x^2}{1 + 4x^2}} = 0.81 \tag{163}$$

Table 18 gives numbers for more intervals and x values. Fig. 32 shows $\tilde{S}_{0\Theta}$ versus a lower time cut (solid line). The sensitivity is maximum for $T = 0$, hence in contrast to the Λ measurement $S_{0\Theta}$ drops sharply when a cut on low T is applied.

Table 18. Sensitivity coefficients for a one-parameter fit of Θ

$|T|$ interval:	$0 \ldots \infty$	$0.2 \ldots 6.0$		$0.3 \ldots \infty$		$0.3 \ldots 3.0$	
	$S_{0\Theta} = \tilde{S}_{0\Theta}$	$S_{0\Theta}$	$\tilde{S}_{0\Theta}$	$S_{0\Theta}$	$\tilde{S}_{0\Theta}$	$S_{0\Theta}$	$\tilde{S}_{0\Theta}$
$x = 0.70$	0.818	0.773	0.699	0.747	0.643	0.744	0.618
$x = 0.75$	0.809	0.761	0.688	0.734	0.631	0.726	0.603

3.6.5.5 Fit of Θ and I

The combined fit of Θ and I with $\Lambda = 0$, which is performed for the control measurement to determine the dilution $D = \Theta$ and the intrinsic asymmetry I, leads to the solutions

$$\hat{\Theta} = \frac{F_{01}G_{01} - F_{02}G_{00}}{F_{01}^2 - F_{02}F_{00}}, \quad \hat{I} = \frac{F_{01}G_{00} - F_{00}G_{01}}{F_{01}^2 - F_{02}F_{00}} \tag{164}$$

The covariance matrix

$$\mathbf{C}(\Theta, I) = \frac{1}{N_0} \begin{pmatrix} \frac{F_{00}}{F_{00}F_{02} - F_{01}^2} & \frac{F_{01}}{F_{00}F_{02} - F_{01}^2} \\ \frac{F_{01}}{F_{00}F_{02} - F_{01}^2} & \frac{F_{02}}{F_{00}F_{02} - F_{01}^2} \end{pmatrix} \tag{165}$$

is independent of Θ and I. For the whole interval $T = 0\ldots\infty$ this yields

$$S_{0\Theta} = \tilde{S}_{0\Theta} = \frac{x^2}{1+x^2}\sqrt{\frac{5+2x^2}{1+4x^2}} = 0.49 \tag{166}$$

and

$$S_{0I} = \tilde{S}_{0I} = \frac{x^2}{1+x^2}\sqrt{\frac{5+2x^2}{1+2x^2}} = 0.60 \tag{167}$$

for $x = 0.75$. Table 19 gives numbers for more intervals and x values. Fig. 33 shows $\tilde{S}_{0\Theta}$ versus a lower time cut (wide-dotted line).

Table 19. Sensitivity coefficients for Θ and I in a two-parameter fit of Θ and I

$|T|$ interval:	$0\ldots\infty$		$0.2\ldots6.0$		$0.3\ldots\infty$		$0.3\ldots3.0$	
	$S_{0\Theta} = \tilde{S}_{0\Theta}$		$S_{0\Theta}$	$\tilde{S}_{0\Theta}$	$S_{0\Theta}$	$\tilde{S}_{0\Theta}$	$S_{0\Theta}$	$\tilde{S}_{0\Theta}$
$x = 0.70$	0.467		0.487	0.440	0.496	0.427	0.357	0.297
$x = 0.75$	0.494		0.514	0.464	0.521	0.448	0.394	0.327
	$S_{0I} = \tilde{S}_{0I}$		S_{0I}	$\tilde{S}_{0I}$	S_{0I}	$\tilde{S}_{0I}$	S_{0I}	$\tilde{S}_{0I}$
$x = 0.70$	0.572		0.630	0.569	0.664	0.572	0.480	0.399
$x = 0.75$	0.611		0.675	0.610	0.710	0.611	0.542	0.451

3.6.5.6 Fit of Λ and Θ

The combined fit of Λ and Θ, which is required in more complicated decays like $B^0 \to \pi\pi$, leads to the solutions

$$\hat{\Lambda} = \frac{I(F_{10}F_{02} - F_{01}F_{11}) + G_{10}F_{11} - G_{10}F_{02}}{F_{11}^2 - F_{20}F_{02}}$$

$$\hat{\Theta} = \frac{I(F_{20}F_{01} - F_{10}F_{11}) + G_{10}F_{11} - G_{01}F_{20}}{F_{11}^2 - F_{20}F_{02}} \tag{168}$$

The covariance matrix

$$\mathbf{C}(\Lambda, \Theta) = \frac{1}{N_0} \begin{pmatrix} \frac{F_{02}}{F_{20}F_{02}-F_{11}^2} & \frac{F_{11}}{F_{20}F_{02}-F_{11}^2} \\ \frac{F_{11}}{F_{20}F_{02}-F_{11}^2} & \frac{F_{20}}{F_{20}F_{02}-F_{11}^2} \end{pmatrix} \quad (169)$$

is independent of Λ and Θ. For the whole interval $T = 0 \ldots \infty$ this yields

$$S_0 = \tilde{S}_0 = \sqrt{\frac{x^2}{1+2x^2}} = 0.51 \quad (170)$$

for $x = 0.75$, which is 86% of the precision reached in a fit with only one unknown Λ. The corresponding

$$S_{0\Theta} = \tilde{S}_{0\Theta} = \sqrt{\frac{1}{2}} = 0.71 \quad (171)$$

is valid for any x. Table 20 gives numbers for more intervals and x values. Fig. 30 shows $\tilde{S}_0$ for many fits with Λ and Fig. 32 shows $\tilde{S}_{0\Theta}$ versus a lower time cut (narrow-dotted lines).

Table 20. Sensitivity coefficients for Λ and Θ in a two-parameter fit of Λ and Θ

$|T|$ interval:	$0\ldots\infty$	$0.2\ldots6.0$		$0.3\ldots\infty$		$0.3\ldots3.0$	
	$S_0 = \tilde{S}_0$	S_0	$\tilde{S}_0$	S_0	$\tilde{S}_0$	S_0	$\tilde{S}_0$
$x = 0.70$	0.497	0.525	0.474	0.545	0.469	0.507	0.421
$x = 0.75$	0.514	0.545	0.492	0.567	0.488	0.540	0.449
	$S_{0\Theta} = \tilde{S}_{0\Theta}$	$S_{0\Theta}$	$\tilde{S}_{0\Theta}$	$S_{0\Theta}$	$\tilde{S}_{0\Theta}$	$S_{0\Theta}$	$\tilde{S}_{0\Theta}$
$x = 0.70$	0.707	0.641	0.579	0.612	0.527	0.564	0.469
$x = 0.75$	0.707	0.641	0.579	0.612	0.527	0.570	0.474

3.6.6 Using Time Integrated Numbers

The time-integrated asymmetry with a cut $|T| > T_{\min}$ is at hadron colliders given by (118) as

$$a_{\text{int}} = F_{10}\Lambda + F_{01}\Theta + I$$

Only one parameter can be determined. For Λ (with known I and $\Theta = 0$) the sensitivity coefficients

$$S_0 = F_{10} \quad \tilde{S}_0 = \frac{F_{10}}{\sqrt{F_{00}}} \quad (172)$$

are shown as function of $T_{\min}$ in Figs. 30 and 31 (thick dots), and listed for several intervals and values of x in Table 21. They are smaller than the ones for a time-dependent fit of Λ (Table 16) by about 20%.

For a time-integrated dilution measurement (I known and $\Lambda = 0$, $D = \Theta$) the sensitivity coefficient $\tilde{S}_{0\Theta} = F_{01}/\sqrt{F_{00}}$ is shown in Fig. 32 (thick dots).

Table 21. Sensitivity coefficients for a one-parameter determination of Λ from rates integrated over two time intervals

$|T|$ interval:	$0\ldots\infty$	$0.2\ldots6.0$		$0.3\ldots\infty$		$0.3\ldots3.0$	
	$S_0 = \tilde{S}_0$	S_0	$\tilde{S}_0$	S_0	$\tilde{S}_0$	S_0	$\tilde{S}_0$
$x = 0.70$	0.470	0.509	0.460	0.516	0.444	0.514	0.427
$x = 0.75$	0.480	0.519	0.469	0.526	0.452	0.532	0.443

3.6.6.1 Background

Most backgrounds to a time-integrated sample contribute equally to the B and $\bar{B}$ signal. If we denote the tagged B events by b and $\bar{b}$, and the background by c, the numbers

$$N_1 = N_b + \tfrac{1}{2}N_c, \quad N_2 = N_{\bar{b}} + \tfrac{1}{2}N_c$$

lead to an asymmetry

$$a = \frac{N_1 - N_2}{N_1 + N_2} = \frac{N_b - N_{\bar{b}}}{N_b + N_{\bar{b}}} \cdot (1-c) \tag{173}$$

using the background fraction

$$c = \frac{N_c}{N_b + N_{\bar{b}} + N_c} = \frac{N_c}{N_\text{tot}}$$

This corresponds to an additional dilution factor

$$D_c = 1 - c = \frac{1}{1+c'} = \frac{N_b + N_{\bar{b}}}{N_b + N_{\bar{b}} + N_c}$$

where

$$c' = B:S = \frac{N_c}{N_b + N_{\bar{b}}}$$

is the background normalized to the number of signal events. Since the number of events increases at the same time by a factor $(1+c') = 1/D_c$, the error on the undiluted parameters Λ_0 or Θ_0 increases by the factor $1/\sqrt{D_c}$, e. g.

$$\sigma_{\Lambda_0} = \frac{1}{D_x D_c} \sigma_\Lambda = \frac{1}{D_x D_c S_0} \frac{1}{\sqrt{N_\text{tot}}} = \frac{1}{D_x \sqrt{D_c} S_0} \frac{1}{\sqrt{N_b + N_{\bar{b}}}}$$

where D_x summarizes all other dilution factors.

For the time dependent fit, this is only a first approximation, since the time dependent background fraction $cf_c(T)$ drops sharply with increasing $|T|$. The real loss in precision is there smaller in Λ_0, but larger in Θ_0.

3.6.7 Fit Procedure for Small Data Samples

For small data samples, instead of fitting the asymmetry histogram, a maximum likelihood fit to the single events has to be performed. To derive the likelihood function, we go back to the original probability densities

$$f_b(T) = \frac{1}{2} e^{-|T|} (1 + D\Lambda_0 \sin xT) \qquad (174)$$
$$f_{\bar{b}}(T) = \frac{1}{2} e^{-|T|} (1 - D\Lambda_0 \sin xT)$$

where no intrinsic asymmetry is assumed ($I = 0$). An event i with $\hat{B}_i$ estimating the flavour of its tag partner, i.e. $-\hat{B}_i$ estimates the flavour of the signal B at $T = 0$, will have the log-likelihood

$$L_i(\Lambda_0) = \ln\left(1 + \hat{B}_i D_i \Lambda_0 \sin xT_i\right) = \ln(1 + \mathcal{K}_i \Lambda_0)$$

where all terms independent of Λ_0 have already been dropped. The likelihood obviously depends only on the observables

$$\mathcal{K}_i = \hat{B}_i D_i \sin xT_i$$

where D_i is a product of dilution factors, including a correction to the tagging dilution if $D_t(\hat{B}) \neq |\hat{B}|$. This coefficient can be written as $\hat{B} D = \operatorname{sign} \hat{B} \cdot D_t(\hat{B}) \cdot D_r \cdot D_m$. For each event i, the individual lifetimes and dilution factors can be transformed into the optimum variable $\mathcal{K}_i$.

In real experiments, a more detailed description of the distortions and dilutions is used, which is accomplished by a convolution of these effects and the functions (174). The final function is then used for the calculation of L and of $\mathcal{K}$.

The total log-likelihood is $L(\Lambda_0) = \sum L_i(\Lambda_0|\mathcal{K}_i)$, and the best estimate of Λ_0 is determined by its maximum. Instead of a max L fit, there is another easy way to obtain Λ_0 from the distribution of the $\mathcal{K}$ observable [142], making use of the fact that its distribution can be written as

$$f(\mathcal{K}|\Lambda_0) = g(\mathcal{K})(1 + \Lambda_0 \mathcal{K})$$

and $g(\mathcal{K}) = f(\mathcal{K}|0)$ is symmetric about 0, i.e. $g(\mathcal{K}) = g(-\mathcal{K})$ and $f(\mathcal{K})$ is symmetric if $\Lambda_0 = 0$. This can be seen from the fact that on the $\Upsilon(4S)$ the distribution $f_b(T) = f_{\bar{b}}(T) = f(|T|)$ is symmetric in T if $\Lambda_0 = 0$, while the variable $\mathcal{K}$ is an odd function of T, and that the distribution of $\hat{B}$ is symmetric in $\hat{B}$, since there are equal numbers of B and $\bar{B}$ tags (ignoring small effects of direct CP violation), while $\mathcal{K}$ is also an odd function of $\hat{B}$. One of these conditions is sufficient to guarantee the symmetry of $g(\mathcal{K})$. The first condition is lost at incoherent $b\bar{b}$ production, where only non-negative values of T occur, but the second may still hold and then implies $I = 0$.

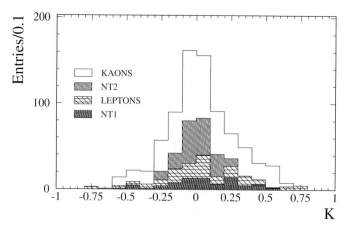

Fig. 33. Distribution of $\mathcal{K}$ for the events used in the BABAR $\sin 2\beta$ analysis [5]. The distribution is shown for four different tagging categories. The smallest dilution, i.e. biggest dilution factor, qis obtained with lepton tags, hence their distribution shows the maximum deviation from 0.

The moments of the $\mathcal{K}$ distribution are then

$$\langle \mathcal{K}^{2n-1} \rangle = \int \mathcal{K}^{2n-1} g(\mathcal{K}) \, d\mathcal{K} + \Lambda_0 \int \mathcal{K}^{2n} g(\mathcal{K}) \, d\mathcal{K} = \Lambda_0 \int \mathcal{K}^{2n} g(\mathcal{K}) \, d\mathcal{K}$$

$$\langle \mathcal{K}^{2n} \rangle = \int \mathcal{K}^{2n} g(\mathcal{K}) \, d\mathcal{K} + \Lambda_0 \int \mathcal{K}^{2n+1} g(\mathcal{K}) \, d\mathcal{K} = \int \mathcal{K}^{2n} g(\mathcal{K}) \, d\mathcal{K}$$

$$\langle \mathcal{K}^{2n-1} \rangle = \Lambda_0 \langle \mathcal{K}^{2n} \rangle \tag{175}$$

especially:

$$\langle \mathcal{K} \rangle = \int \mathcal{K} g(\mathcal{K}) \, d\mathcal{K} + \Lambda_0 \int \mathcal{K}^2 g(\mathcal{K}) \, d\mathcal{K} \qquad = \Lambda_0 \int \mathcal{K}^2 g(\mathcal{K}) \, d\mathcal{K}$$

$$\langle \mathcal{K}^2 \rangle = \int \mathcal{K}^2 g(\mathcal{K}) \, d\mathcal{K} + \Lambda_0 \int \mathcal{K}^3 g(\mathcal{K}) \, d\mathcal{K} \qquad = \int \mathcal{K}^2 g(\mathcal{K}) \, d\mathcal{K}$$

$$\langle \mathcal{K} \rangle = \Lambda_0 \langle \mathcal{K}^2 \rangle \tag{176}$$

Unbiased estimators for moments of a distribution are the arithmetic means, $\langle \mathcal{K}^n \rangle = \frac{1}{N} \sum \mathcal{K}_i^n$, leading to an estimator for Λ_0

$$\hat{\Lambda}_0 = \frac{\sum \mathcal{K}_i}{\sum \mathcal{K}_i^2} \tag{177}$$

The $\mathcal{K}$ distribution from the BABAR analysis [5] is shown in Fig. 33.

A visual verification can obtained dividing the experimental distribution by the Monte Carlo distribution with $\Lambda_{\mathrm{MC}} = 0$. If the $\mathcal{K}_i$ are computed correctly, this gives a straight line

$$\frac{N_{\mathrm{MC}}}{N_{\mathrm{tot}}} \frac{dn/d\mathcal{K}}{dn_{\mathrm{MC}}/d\mathcal{K}} = 1 + \Lambda_0 \mathcal{K}$$

with slope Λ_0. This is a powerful verification of the Monte Carlo model used to describe tagging and T measurement. Also the asymmetry of the $\mathcal{K}$ distribution

$$a(\mathcal{K}) = \frac{f(\mathcal{K}) - f(-\mathcal{K})}{f(\mathcal{K}) + f(-\mathcal{K})} = 1 + \Lambda_0 \mathcal{K}$$

is a straight line with slope Λ_0, if $\mathcal{K}$ is calculated properly.

The error from (177) is approximately

$$\delta \hat{\Lambda}_0 \approx \frac{1}{\sqrt{N \cdot \langle \mathcal{K}^2 \rangle}} \sqrt{1 - \Lambda_0^2 \frac{\langle \mathcal{K}^4 \rangle}{\langle \mathcal{K}^2 \rangle}} < \frac{1}{\sqrt{N \cdot \langle \mathcal{K}^2 \rangle}} \qquad (178)$$

approaching the upper limit for $\Lambda_0 \to 0$. Hence the sensitivity coefficient is $S_0 = \sqrt{\langle \mathcal{K}^2 \rangle}$, which can be extracted from Monte Carlo studies. If we use $\mathcal{K} = \sin xT$ to extract the diluted parameter Λ, it agrees with the high-statistics value for S_0 given in (153) above.

Maximizing the log-likelihood

$$L(\Lambda_0) = \sum \ln(1 + \Lambda_0 \mathcal{K}_i)$$

is equivalent to solving

$$\frac{\mathrm{d}L}{\mathrm{d}\Lambda_0} = \sum \frac{\mathcal{K}_i}{1 + \Lambda_0 \mathcal{K}_i} = 0 \qquad (179)$$

The best accuracy that can be obtained in any fit is given by the Rao–Cramér bound

$$\sigma^2(\Lambda_0) \geq \frac{1}{N \cdot \left\langle \frac{\mathcal{K}^2}{(1 + \Lambda_0 \mathcal{K})^2} \right\rangle} \approx \frac{1}{N \cdot \langle \mathcal{K}^2 \rangle} \left[1 - \Lambda_0^2 \frac{\langle \mathcal{K}^4 \rangle}{\langle \mathcal{K}^2 \rangle} \right] \qquad (180)$$

The approximation for small Λ_0 values is the error (178) obtained by the simple estimator $\hat{\Lambda}_0$ of (177), leaving little room for improvement in fit accuracy. The taylor expansion of (179)

$$\frac{\mathrm{d}L}{\mathrm{d}\Lambda_0} = \sum \mathcal{K}_i - \Lambda_0 \sum \mathcal{K}_i^2 + \mathcal{O}(\Lambda_0^2)$$

leads to the same solution as (177) for small Λ_0.

The general case, a simultaneous determination of Λ_0 and Θ_0 at the presence of a non-vanishing intrinsic asymmetry I, is solved accordingly. The log-likelihood

$$L_i(\Lambda_0, \Theta_0) = \ln(1 + \operatorname{sign} \hat{B}_i \cdot I + \mathcal{K}_i \Lambda_0 + \vartheta_i \Theta_0)$$

is used in an unbinned maximum likelihood fit, where $\vartheta_i = \hat{B}_i D_i \cos xT_i$, solving a set of coupled equations

$$\frac{\mathrm{d}L}{\mathrm{d}\Lambda_0} = \sum \frac{\mathcal{K}_i}{1 + \operatorname{sign} \hat{B}_i \cdot I + \Lambda_0 \mathcal{K}_i + \vartheta_i \Theta_0} = 0;$$

$$\frac{\mathrm{d}L}{\mathrm{d}\Theta_0} = \sum \frac{\vartheta_i}{1 + \operatorname{sign} \hat{B}_i \cdot I + \Lambda_0 \mathcal{K}_i + \vartheta_i \Theta_0} = 0 \tag{181}$$

where I enters as a constant, since it has been determined precisely using (139), or it is $I = 0$ at the $\Upsilon(4S)$. In this general case, $\mathcal{K}$ and ϑ are highly correlated and the fit cannot be replaced by simple estimators for Λ_0 and Θ_0 using moments.

3.6.7.1 Background

The fit function including background is in the most simple case of CP asymmetry

$$f_b(T) = (1-c)\mathrm{e}^{-|T|}\left[1 + \hat{B}D\Lambda_0 \sin xT\right] + c f_c(T) \tag{182}$$

where $f_c(T)$ is the background distribution in apparent scaled oscillation time T, while the estimated background fraction c_i can be evaluated event by event in a similar way as the flavour estimator $\hat{B}_i$, using other event properties but ignoring T. Here we assume that background events show no asymmetry in their apparent flavour distribution.

The fit strategy is the same as in the previous section for the background-free case, maximizing a log-likelihood, using again an optimum variable

$$\mathcal{K}_i = \frac{\hat{B}_i D_i}{1 + \frac{c_i}{1-c_i} f_c(T_i)\mathrm{e}^{|T_i|}} \sin xT_i = \hat{B}_i D_i D_{c,i} \sin xT_i$$

where the whole coefficient can be considered a time-dependent dilution factor, with contribution

$$D_c = \frac{1}{1 + \frac{c}{1-c} f_c(T)\mathrm{e}^{|T|}}$$

from background. For the one parameter fit, a symmetry in $\hat{B}$ or T still allows to use the estimator (177) from the moments of this variable.

A conservative background model is one where the background rate drops with T not faster than the signal, i.e. with $f_c(T) \sim \mathrm{e}^{-|T|}$. In this case, the background dilution factor is $D_c = 1 - c$.

From the previous chapters one can conclude that without loss of information one can drop all events with $T \lesssim 0.2$, where the background is probably large and – due to systematic uncertainties – most violent. This cut, in the optimum variable formalism, removes, however, only events with small $\mathcal{K}$ values, since the factor $\sin xT$ is small, while $f_c(T)$ is large, independent of the event signature, which determines $\hat{B}$ and c. These events are thereby guaranteed to contribute only a negligible amount of information.

3.7 Experimental Data on $B \to J/\psi K_S^0$

If we assume that the Standard Model explanation for CP violation is the correct and only one, then the channel $B \to J/\psi K_S^0$ where the K_S^0 decays to a CP = +1 eigenstate offers a clean measurement of $\sin 2\beta$. Results have become available since 1999. A summary of all present experimental information on $\sin 2\beta$ is given in Table 22. The statistical error is limited by $\sigma(\Lambda_0) > 1.7/\sqrt{N_{\text{sig}} \cdot (\varepsilon_t D_t^2)_{\text{eff}}}$, and typically larger due to background, time-resolution and the statistical error on the dilution.

Table 22. Results on the asymmetry amplitude $\Lambda_0 = \sin 2\beta$ from $B \to J/\psi K_S^0$ and related channels. The number of signal events and the effective number $(\varepsilon_t D_t^2)_{\text{eff}} N_{\text{sig}}$ is given to compare the sensitivity of the measurements.

0.0 0.5 1.0	Λ_0	$N_{\text{sig}} \cdot (\varepsilon_t D_t^2)_{\text{eff}} / N_{\text{sig}}$	experiment
	$(3.2 \pm^{1.8}_{2.0} \pm 0.5)$	$1.6/14 \pm 2$	OPAL 98 [143]
	$(1.8 \pm 1.1 \pm 0.3)$	$12/198 \pm 17$	CDF 98 [144]
	$0.79 \pm^{0.38}_{0.41} \pm 0.16$	$25/397 \pm 31$	CDF 99 [145] *
	$(0.84 \pm^{0.82}_{1.04} \pm 0.16)$	$4.4/17$	ALEPH 00 [146]
	$0.34 \pm 0.20 \pm 0.05$	$140/520$	BABAR 01 [141]
	$0.58 \pm^{0.32}_{0.34} \pm^{0.09}_{0.10}$	$70/260$	BELLE 01 [147]
	$0.59 \pm 0.14 \pm 0.05$	$280/1080$	BABAR 01 [5]
	$0.99 \pm 0.14 \pm 0.06$	$280/1030$	BELLE 01 [6]
	$0.75 \pm 0.09 \pm 0.04$	$540/2200$	BABAR 02 prel. [148]*
	$0.82 \pm 0.12 \pm 0.05$	$380/1380$	BELLE 02 prel. [149] *
	0.79 ± 0.10	$\chi^2 = 3.5/2df$	average 2001
	0.78 ± 0.08	$\chi^2 = 0.2/2df$	average of (*)

The first measurement with non-zero information content on the CP-violating parameter $\sin 2\beta$ was performed by the CDF experiment and published 1999 [145]. They found a value of $\sin 2\beta = 0.79 \pm^{0.41}_{0.44}$ (stat. and syst. errors combined). The golden final state $\mu^+\mu^-\pi^+\pi^-$ was reconstructed. A total sample of 395 ± 31 signal events is divided into a subsample of 202 ± 18 events with both muons measured in the silicon vertex detector, and the remainder without lifetime information. This latter sample is used to measure the time integrated asymmetry, while the subsample with vertex information is submitted to a fit of the asymmetry function (98).

The flavour was tagged combining information from soft leptons and a jet charge variable of the opposite jet, and same-side tagging with a charged particle (taken as pion) close to the reconstructed B^0 meson in direction and rapidity. From several tracks, the one with the smallest transverse momentum relative to the $B\pi$ system is selected. The opposite-jet tags have been investigated using fully reconstructed charged B mesons, and the same-jet tags using $D^{(*)}lX$ candidates. The overall effective tagging performance is $(\varepsilon_t D_t^2)_{\text{eff}} = 0.063 \pm 0.017$.

3.7.1 More Final States at the Υ(4S)

The year 2000 saw first results from the e^+e^- B factories at SLAC and KEK which produce $B^0\bar{B}^0$ pairs in Υ(4S) decays. These experiments have much lower background than experiments using B^0 mesons in b jets, and can therefore reconstruct many additional decay channels with lower branching fraction or worse resolution.

At the end of 2000, the BABAR sample [141] was the largest sample of B mesons, comprising 23 Million B meson pairs from Υ(4S) decays. Both BABAR and BELLE were accumulating more data rapidly. They could significantly establish CP violation in B mesons in 2001 with the samples given in Table 23.

Signals are reconstructed for $(J/\psi \to l^+l^-) + (K_S^0 \to \pi^+\pi^-, \pi^0\pi^0)$, $(\psi(2S) \to l^+l^-, J/\psi\pi^+\pi^-) + (K_S^0 \to \pi^+\pi^-)$, $(\chi_{c1} \to J/\psi\gamma) + (K_S^0 \to \pi^+\pi^-)$, $(J/\psi \to l^+l^-) + (K^{*0} \to K_S^0\pi^0)$, and $(J/\psi \to l^+l^-) + K_L^0$. Each event with a signal candidate is assigned a B^0 or $\bar{B}^0$ tag if the rest of the event (with the daughter tracks of the signal B removed) satisfies the criteria for one of several tagging categories. The BABAR experiment uses only events with a minimum tag discrimination, thus reducing the number of tagged events to 68% of all reconstructed events. The total statistical tagging performance is $(\varepsilon_t D_t^2)_{\text{eff}} = 0.261 \pm 0.012$.

BELLE [6] has reconstructed additional final states, $(\eta_c \to K^+K^-\pi^0, K_S^0 K^\pm \pi^\mp) + (K_S^0 \to \pi^+\pi^-)$, from a sample about the same size as the BABAR sample. The channel $J/\psi\pi^0$ has been used in their first analysis [147], but has been dropped in [6] since it is a $c\bar{c}d\bar{d}$ final state without intermediate neutral kaon, and may have some additional phases from penguin diagrams. They tag all events including those with little or no flavour discrimination. Their total statistical tagging performance is $(\varepsilon_t D_t^2)_{\text{eff}} = 0.270 \pm 0.012$ and agrees with that of BABAR.

Both experiments fix the B^0 lifetime $\tau_{B^0} = 1.548\,\text{ps}$ and the oscillation (angular) frequency $\Delta m = 0.472\,\text{ps}^{-1}$ to the 2000 world average [9]. There is a slight correlation between A_0 and these data that is absorbed in the systematic error.

Table 23. Data sample composition of the BABAR [5] and BELLE [6] measurements of $\sin 2\beta$.

channel	signal	tagged signal + bgr.	purity (tagged)
BABAR: CP = −1	29/fb		
$J/\psi K_S^0$ ($K_S^0 \to \pi^+\pi^-$)	461 ± 22	316	98%
$J/\psi K_S^0$ ($K_S^0 \to \pi^0\pi^0$)	113 ± 12	64	94%
$\psi(2S) K_S^0$ ($K_S^0 \to \pi^+\pi^-$)	86 ± 17	67	98%
$\chi_{c1} K_S^0$ ($K_S^0 \to \pi^+\pi^-$)	44 ± 8	33	97%
total low background	704 ± 31	480	96%
CP = ±1			
$J/\psi K^{*0}$ ($K^{*0} \to K_S^0 \pi^0$)	64 ± 10	50	74%
CP = +1			
$J/\psi K_L^0$	257 ± 24	273	51%
channel	signal + bgr.		purity
BELLE: CP = −1	29/fb		
$J/\psi K_S^0$ ($K_S^0 \to \pi^+\pi^-$)	457		97%
$J/\psi K_S^0$ ($K_S^0 \to \pi^0\pi^0$)	76		88%
$\psi(2S) K_S^0$ ($K_S^0 \to \pi^+\pi^-$)	85		96%
$\chi_{c1} K_S^0$ ($K_S^0 \to \pi^+\pi^-$)	24		90%
$\eta_c K_S^0$ ($K_S^0 \to \pi^+\pi^-$)	64		61%
total low background	706		93%
CP = ±1			
$J/\psi K^{*0}$ ($K^{*0} \to K_S^0 \pi^0$)	41		84%
CP = +1			
$J/\psi K_L^0$	569		61%

The $J/\psi K^{*0}$ CP eigenstates have contributions with CP = +1 ($L = 0, 2$) and CP = −1 ($L = 1$) eigenvalues. Therefore, a partial wave analysis has been performed to disentangle both contributions. While BABAR use their result [150] to apply a physical dilution factor $D_P = 0.65 \pm 0.07$, BELLE [6] and an updated BABAR analysis [148] include the transversity angle [124] of these events in the fit to improve the separation of the two eigenstates.

Tha raw asymmetries varying as $\sin xT = \sin \Delta m \Delta t$ are clearly visible in Fig. 34, which shows the analysis with the largest sample (at present).

The $J/\psi K_L^0$ channel suffers from high background, including crosstalk from $B \to J/\psi K^{*0} \to J/\psi K_L^0 \pi^0$. This component can be included with an average CP eigenvalue in the fit, and a conservative range for this parameter

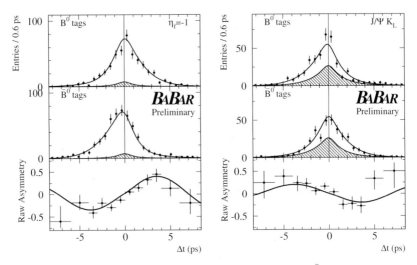

Fig. 34. Number of candidates from the $\{J/\psi, \psi(2S), \chi_{c1}\}K_S^0$ (left, CP = −1) and $J/\psi K_L^0$ (right, CP = +1) final states in the signal region with B^0 tag (top) and with $\bar{B}^0$ tag (middle), together with the raw asymmetry (bottom). Overlaid is the result of the fit for the diluted asymmtry, assuming equal numbers of B^0 and $\bar{B}^0$ tags [148].

has been taken to estimate its systematic error. Other systematic errors arise from uncertainties in input parameters to the maximum likelihood fit, incomplete knowledge of the time resolution function, uncertainties in the mistag fractions, and possible limitations in the analysis procedure. The sum of all systematic uncertainties is still considerably smaller than the statistical error (see Table 22).

A two-parameter fit of the subset of channels with low background has been performed by BABAR [5,148]. They find $|r| = 0.92 \pm 0.06 \pm 0.02$, compatible with the value of 1 if there were no direct CP violation present. The value of Λ_0 is not changed by the two-parameter fit in the subsample. The fit results correspond to

$$\Lambda_0 = 0.76 \pm 0.10, \quad \Theta_0 = -0.08 \pm 0.07$$

3.7.2 Interpretation

The average value from Table 22 is $\Lambda_0 = 0.82 \pm 0.08$. In the Standard Model, this parameter is related to the tree-level ratio

$$r_{c\bar{c}d\bar{d}} = \eta_m \frac{A(\bar{B}^0 \to c\bar{c}d\bar{d})}{A(B^0 \to c\bar{c}d\bar{d})}$$

and the corresponding "ϵ"-parameter

as
$$\eta_{c\bar{c}d\bar{d}} = \frac{A(B_{\rm H} \to c\bar{c}d\bar{d})}{A(B_{\rm L} \to c\bar{c}d\bar{d})}$$

$$\Lambda_0 = \frac{2\,{\rm Im}\, r_{c\bar{c}d\bar{d}}}{1+|r_{c\bar{c}d\bar{d}}|^2} = -\frac{2\,{\rm Im}\, \eta_{c\bar{c}d\bar{d}}}{1+|\eta_{c\bar{c}d\bar{d}}|^2}$$

Using the assumption $|r_{c\bar{c}d\bar{d}}|=1$ or equivalently ${\rm Re}\,\eta_{c\bar{c}d\bar{d}} = 0$, which is only justified within the Standard Model, we have

$${\rm Im}\, r_{c\bar{c}d\bar{d}} = \sin 2\beta = 0.78 \pm 0.08$$

From there we can calculate the two solutions $\beta = (26\pm4)°$ and $\beta = (64\pm4)°$. Taking the lower value which is compatible with the Standard Model unitary CKM matrix, we obtain

$$\eta_{c\bar{c}d\bar{d}} = -{\rm i}\tan\beta = (-0.48 \pm 0.09){\rm i}$$

Including the BABAR result on Θ_0 [148], the complex parameter is

$$\eta_{c\bar{c}d\bar{d}} = (0.05 \pm 0.04) - (0.48 \pm 0.09){\rm i}$$

with its real part compatible with 0.

The experimental result establishes the existence of CP violation in the B^0 system and is in good agreement with the Standard Model prediction. This is demonstrated in an update of a recent overall fit to the CKM matrix including the $\sin 2\beta$ measurements [151]. The allowed region for the tip of the unitarity triangle in the $\bar{\rho}, \bar{\eta}$-plane is shown in Fig. 35.

However, if the Standard Model description of this CP asymmetry is not complete, we have to consider different Λ_0 and Θ_0 parameters for different classes of channels, since e.g. $B \to J/\psi\pi^0$ may suffer more from loop diagrams than $B \to J/\psi K_{\rm S}^0$.

3.7.3 Other Related Channels

For cross checks like this, more final states are presently investigated by BABAR and BELLE. Among these are several with the same asymmetry as $J/\psi K_{\rm S}^0$ from the leading tree diagrams. One class is $B^0 \to D^+D^-$ or $D^{*+}D^{*-}$. Both are CP eigenstates – the latter a mixture of both eigenvalues. Here, penguin diagrams contribute with different phases and a different asymmetry, including possibly a $\cos xT$ term from direct CP violation, is expected.

Another class is $B^0 \to \eta' K_{\rm S}^0$ or $\phi K_{\rm S}^0$, where the angle β' of the tu unitarity triangle is measured via an asymmetry with $\Lambda_0 = \sin 2\beta'$. Since the two triangles shown in Fig. 1, are very similar, again a value close to the one for $J/\psi K_{\rm S}^0$ is expected.

All experimental results [103,149] on these states are, unfortunately, at present not yet sensitive to observe differences.

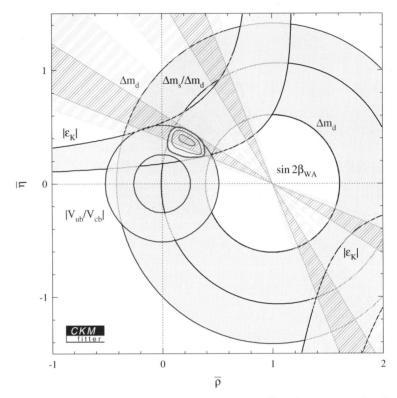

Fig. 35. Constraints on the CKM unitarity triangle (8). The contours for the tip of the triangle show the parameters at 90%, 32% and 5% significance level of a combined chisquare test to all measurements. The theoretical parameters are taken freely within allowed ranges to minimize this χ^2 without assigning any weight (probability) to them [151].

3.8 Experimental Data on $B \to \pi\pi$

The difficulties of the interpretation of the $B^0 \to \pi^+\pi^-$ asymmetry has been discussed in detail in section 3.5.9.2. Due to crossfeed between the channels, it is always investigated experimentally together with the penguin-dominated channel $B \to K^+\pi^-$. The asymmetry in the latter, self-tagging final state is a sign of direct CP violation; experimental limits are given in Table 7.

Experimental results on branching fractions are given in Table 6 in section 3.3. They strongly indicate a substantial penguin contribution to the $\pi^+\pi^-$ final state. Therefore, as long as this channel alone is investigated, experiments can only determine the coefficients Λ_0 and Θ_0 which allow no direct translation into α or $\sin 2\alpha$. The results are given in Table 24. The latest numbers from BABAR and BELLE for both coefficients differ by more than two standard deviations: while the BELLE result suggests a substantial

CP asymmetry, BABAR's result indicates a very small deviation from the completely CP symmetric decay.

Table 24. Experimental results on the coefficients of the CP violating asymmetry $a(T) = \Theta_0 \cos xT + \Lambda_0 \sin xT$ for the $B \to \pi^+\pi^-$ decay.

Λ_0	Θ_0	experiment
$0.03 \pm {}^{0.53}_{0.56} \pm 0.11$	$0.25 \pm {}^{0.45}_{0.47} \pm 0.14$	BABAR 01 [102]
$-0.01 \pm 0.37 \pm 0.07$	$0.02 \pm 0.29 \pm 0.07$	BABAR 02 prel. [103]
$-1.21 \pm {}^{0.38}_{0.27} \pm {}^{0.16}_{0.13}$	$0.94 \pm {}^{0.25}_{0.31} \pm 0.09$	BELLE 02 prel. [149]

A full analysis requires a flavour-tagged measurement of the decay $B \to \pi^0\pi^0$. For this channel, up to now no significant signal has been observed.

4 Outlook

B^0 meson oscillation is by now well measured. There are, however, more unknown parameters: the difference in width $\Delta\Gamma$ of the two B^0 eigenstates, and the oscillation frequency and difference in width of the two B_s eigenstates, where only limits are available.

The first parameter describing CP violation in B^0 meson decays, the value of $\sin 2\beta$ in the Standard Model, is now known to ± 0.08, and demonstrates a significant CP violation in the neutral B meson system.

The Standard Model description of CP violation has more parameters to check. The asymmetry amplitude $\sin 2\alpha$ is within reach after a few more years of B factory running via several final states of the $b \to u\bar{u}d$ decay. Also the discrepancy in the CP asymmetry of the $\pi^+\pi^-$ final state will be settled soon with more data available.

The third angle γ (or γ') in the unitarity triangle is difficult to access, and will need very large data samples of B^0 and B_s mesons before the constraint $\alpha + \beta + \gamma = 180°$ can be used to give limits on contributions to CP violation outside the Standard Model.

Acknowledgements

Many of the concepts and methods described in this paper have been clarified in fruitful discussions with other colleagues, among them Michael Beyer, Bob Cahn, Giovanni Carboni, Robert Fleischer, Joachim Graf, Frank Krauss, Christof Kreuter, Otto Nachtmann, Yossi Nir, Sibylle Petrak, Helen Quinn, Henning Schröder, Klaus R. Schubert, Bernhard Spaan, Jörg Urban, and many members of the ARGUS, LHB and BABAR collaborations. I also acknowledge a very pleasing environment and inspiring atmosphere provided by this School at Prerow.

References

1. T. D. Lee, C. N. Yang, Phys. Rev. **104**, 254 (1956).
2. C. S. Wu et al., Phys. Rev. **105**, 1413 (1957); R. L. Garvin et al., Phys. Rev. **105**, 1415 (1957); J. J. Friedman, V. L. Telegdi, Phys. Rev. **105**, 1681 (1957).
3. J. H. Christenson, J.W. Cronin, V.L. Fitch, R. Turlay, Phys. Rev. Lett. **13**, 138 (1964).
4. recent textbooks on "CP Violation" are
   G. C. Branco, L. Lavoura, J. P. Silva, Clarendon Press, Oxford (1999); I. I. Bigi, A. I. Sanda, Cambridge University Press (2000).
5. BABAR Collab., Phys. Rev. Lett. **87**, 091801 (2001).
6. BELLE Collab., Phys. Rev. Lett. **87**, 091802 (2001).
7. N. Cabibbo, Phys. Rev. Lett. **10**, 531 (1963).
8. M. Kobayashi, T. Maskawa, Progr. Theor. Phys. **49**, 652 (1973).
9. The Particle Data Group, Eur. Phys. J. **C15**, 1 (2000).
10. L.-L. Chau, W.-Y. Keung, Phys. Rev. Lett. **53**, 1802 (1984).
11. H. Harari, M. Leurer, Phys. Lett. **B181**, 123 (1986).
12. A. J. Buras, M. E. Lautenbacher, G. Ostermaier, Phys. Rev. **D50**, 3433 (1994).
13. L. Wolfenstein, Phys. Rev. Lett. **51**, 1945 (1984).
14. Z. Z. Xing, Phys. Rev. **D51**, 3958 (1995).
15. M. Kobayashi, Progr. Theor. Phys. **92**, 287 (1994).
16. K. R. Schubert, Proc. of the Int. Europhysics Conf. on High Energy Physics, Uppsala, Sweden, ed. by O. Botner, June 1987, vol. II, p. 791.
17. C. Hamzaoui, J. L. Rosner, A. I. Sanda, Proc. of the Workshop on High Sensitivity Beauty Physics at Fermilab, Batavia IL, USA, eds. N. Lockyerd, J. Slaughter, Nov. 1987, p. 215.
18. C. Jarlskog, Phys. Rev. Lett. **55**, 1039 (1985); Z. Phys. **C29**, 491 (1985).
19. see e. g. R. Waldi, Fortschr. Phys. **47**, 707 (1999).
20. C. Jarlskog, "Mysteries in the Standard Model", Proc. of the Int. Symp. on Production and Decay of Heavy Flavours, Heidelberg 1996, p. 331.
21. The Large Hadron Collider, Conceptual Design, CERN/AC/95-05;
    C. H. Llewellyn Smith, Proc. of the 17th Int. Symp. on Lepton-Photon Interactions, Beijing 1995, eds. Z.-P. Zheng, H.-S. Chen, World Scientific Publ., Singapore 1996, p. 370.
22. T. T. Wu, C. N. Yang, Phys. Rev. Lett. **13**, 380 (1964).
23. G. Lüders, Dan. Mat. Fys. Medd. **28**, 5 (1954); Ann. Phys. N.Y. **2**, 1 (1957).
24. e. g. E. D. Commins, P. H. Bucksbaum, "Weak Interactions of Leptons and Quarks", Cambridge University Press, 1983.
25. T. Nakada, PSI-PR-91-02 (1991);
    T. Nakada, Proc. of the XVIth Int. Symp. on Lepton-Photon Interactions, Ithaca 1993, eds. P. S. Drell, D. L. Rubin, AIP, New York 1994, p. 425.
26. A. J. Buras, R. Fleischer, TUM-HEP-275/97, in Heavy Flavours II, ed. by A. J. Buras, M. Lindner, World Scientific 1997, p. 65.

27. V. F. Weisskopf, E. P. Wigner, Z. Phys. **63**, 54 (1930) and Z. Phys. **65**, 18 (1930). A detailed discussion is in O. Nachtmann, "Elementarteilchenphysik – Phänomene und Konzepte" (in German), Vieweg, Braunschweig 1986; see also [24,25].
28. A. J. Buras, W. Słominski, H. Steger, Nucl. Phys. **B245**, 369 (1984).
29. T. Inami, C. S. Lim, Progr. Theor. Phys. **65**, 297 (1981), and erratum Progr. Theor. Phys. **65**, 1772 (1981).
30. A. J. Buras, M. Jamin, P. H. Weisz, Nucl. Phys. **B347**, 491 (1990).
31. for recent reviews see e.g. V. Giménez, G. Martinelli, C. T. Sachrajda, Nucl. Phys. Proc. Suppl. **53**, 365 (1997);
C. T. Sachrajda, Nucl. Instr. and Meth. **A462**, 23 (2001).
32. A. Ali, D. London, Nucl. Phys. Proc. Suppl. **54A**, 297 (1997)
33. H. Wittig, Int. J. Mod. Phys. **A12**, 4477 (1997).
34. I. Bigi et al., CERN-TH-7132/94 (1994), published in "B Decays", ed. by S. Stone, World Scientific Publ., 1995.
35. R. Aleksan, Phys. Lett. **B316**, 567 (1993).
36. M. Beneke, G. Buchalla, I. Dunietz, Phys. Rev. **D54**, 4419 (1996).
37. M. Beneke, G. Buchalla, C. Greub, A. Lenz, U. Nierste, Phys. Lett. **B459**, 631 (1999).
38. I. Dunietz, Phys. Rev. **D52**, 3048 (1995);
T. Browder, S. Pakvasa, Phys. Rev. **D52**, 3123 (1995).
39. Y. Nir, Phys. Lett. **B327**, 85 (1994).
40. The LEP B Oscillation Working Group, CERN-EP-2000-096, and updates on the World Wide Web, URL=http://www.cern.ch/LEPBOSC/, Mar. 2001.
41. ARGUS Collab., Phys. Lett. **B192**, 245 (1987).
42. ALEPH Collab., Phys. Lett. **B313**, 498 (1993).
43. CLEO Collab., Phys. Rev. Lett. **80**, 1150 (1998).
44. SLD Collab., SLAC-PUB-7230 (1996).
45. M. J. Teper, Phys. Lett. **90B**, 443 (1980);
C. J. Maxwell, M. J. Teper, Z. Phys. **C10**, 175 (1981).
46. DELPHI Collab., Z. Phys. **C76**, 579 (1997).
47. DELPHI Collab., Phys. Lett. **B345**, 598 (1995).
48. OPAL Collab., Z. Phys. **C66**, 19 (1995).
49. ALEPH Collab., Z. Phys. **C69**, 393 (1996).
50. CDF Collab., Phys. Rev. Lett. **80**, 2057 (1998);
CDF Collab., Phys. Rev. **D59**, 032001 (1999).
51. CLEO Collab., Phys. Lett. **B490**, 36 (2001).
52. CLEO Collab., Phys. Rev. Lett. **62**, 2233 (1989).
53. ARGUS Collab., Z. Phys. **C55**, 357 (1992).
54. CLEO Collab., Phys. Rev. Lett. **71**, 1680 (1993).
55. ARGUS Collab., Phys. Lett. **B324**, 249 (1994).
56. ARGUS Collab., Phys. Lett. **B375**, 256 (1996).
57. ALEPH Collab., Z. Phys. **C75**, 397 (1997).

58. OPAL Collab., Z. Phys. **C72**, 377 (1996).
59. OPAL Collab., Z. Phys. **C76**, 401 (1997).
60. OPAL Collab., Z. Phys. **C76**, 417 (1997).
61. OPAL Collab., Phys. Lett. **B493**, 266 (2000).
62. L3 Collab., Eur. Phys. J. **C5**, 195 (1998).
63. SLD Collab., SLAC-PUB-7228 (1996).
64. SLD Collab., SLAC-PUB-7229 (1996).
65. K. Takikawa (CDF), FERMILAB-Conf-98/054-E.
66. CDF Collab., Phys. Rev. **D60**, 072003 (1999).
67. CDF Collab., Phys. Rev. **D60**, 051101 (1999).
68. CDF Collab., Phys. Rev. **D60**, 112004 (1999).
69. BELLE Collab., Phys. Rev. Lett. **86**, 3228 (2001).
70. H. Tajima (BELLE), Proc. of the 36th Recontres de Moriond on QCD and High Energy Hadronic Interactions, Bourg-Saint-Maurice, 2001.
71. BABAR Collab., Phys. Rev. Lett. **88**, 221803 (2002).
72. BABAR Collab., SLAC-PUB-8530 (2000).
73. BABAR Collab., Phys. Rev. Lett. **88**, 221802 (2002).
74. C. Gay, Ann. Rev. Nucl. Part. Sci. **50**, 577 (2000).
75. UA1 Collab., Phys. Lett. **B186**, 247 (1987).
76. ALEPH Collab., Phys. Lett. **B322**, 441 (1994).
77. ALEPH Collab., Phys. Lett. **B356**, 409 (1995).
78. ALEPH Collab., Phys. Lett. **B377**, 205 (1996).
79. DELPHI Collab., Phys. Lett. **B414**, 382 (1997).
80. The LEP B Oscillation Working Group, CERN-EP-2001-050, and updates on the World Wide Web, URL=http://www.cern.ch/LEPBOSC/, Mar. 2002.
81. DELPHI Collab., Eur. Phys. J. **C18**, 229 (2000).
82. The LEP $\Delta\Gamma_s$ Working Group, P. Coyle et al., on the World Wide Web, URL=http://lepbosc.web.cern.ch/LEPBOSC/deltagamma_s/, Apr. 2001.
83. L3 Collab., Phys. Lett. **B438**, 417 (1998).
84. DELPHI Collab., Eur. Phys. J. **C16**, 555 (2000).
85. OPAL Collab., Phys. Lett. **B426**, 161 (1998).
86. CDF Collab., Phys. Rev. **D59**, 32004 (1999).
87. ALEPH Collab., Phys. Lett. **B486**, 286 (2000).
88. CDF Collab., contributed paper to the EPS-HEP Conference, Tampere, Finland (1999).
89. see e.g. H. Quinn, SLAC-PUB-7053 (1995).
90. an experiment to measure CP violation in Ξ^- decays is E-871: J. Antos et al., FERMILAB Proposal P-871 (1994).
91. for a review on neutrino oscillations see e.g. A. Yu. Smirnov, talk given at the 28th Int. Conf. on High Energy Physics, Warsaw 1996, e-print hep-ph/9611465 (1996);

CP and T violation is discussed in M. Tanimoto, Phys. Rev. **D55**, 322 (1997); J. Arafune, J. Sato, Phys. Rev. **D55**, 1653 (1997).
92. U. Kilian et al., Z. Phys. **C62**, 413 (1994); S. Y. Choi et al., Phys. Rev. **D52**, 1614 (1995); J. H. Kühn, E. Mirkes, TTP-96-43 (1996); Y. S. Tsai, Phys. Lett. **B378**, 272 (1996).
93. for reviews see e.g. W. Grimus, Fortschr. Phys. **36**, 201 (1988); W. Bernreuther, M. Suzuki, Rev. Mod. Phys. **63**, 313 (1991); J. Bernabéu et al., FTUV/95-15 and Proc. of the Ringberg Workshop on Perspectives for Electroweak Interactions in e^+e^- Collisions, Feb 1995, ed. by B. A. Kniehl, World Scientific, 1995.
94. CLEO Collab., CLNS 99/1650.
95. BABAR Collab., Phys. Rev. Lett. **87**, 151802 (2001).
96. BELLE Collab., Phys. Rev. Lett. **87**, 101801 (2001).
97. M. Gronau et al., Phys. Rev. **D50**, 4529 (1994); Phys. Rev. **D52**, 6374 (1994); Phys. Lett. **B333**, 500 (1994); Phys. Rev. Lett. **73**, 21 (1994);
L. Wolfenstein, Phys. Rev. **D52**, 537 (1995);
N. G. Deshpande, X. G. He, Phys. Rev. Lett. **75**, 1705 (1995); Phys. Rev. Lett. **75**, 3064 (1995);
A. J. Buras, R. Fleischer, Phys. Lett. **B360**, 138 (1995); Phys. Lett. **B365**, 390 (1996);
G. Kramer, W. F. Palmer, Phys. Rev. **D52**, 6411 (1995);
B. Grinstein, R. F. Lebed, Phys. Rev. **D53**, 6344 (1996).
98. M. Beneke, G. Buchalla, M. Neubert, C. T. Sachrajda, Nucl. Phys. **B606**, 245 (2001).
99. CLEO Collab., Phys. Rev. Lett. **84**, 5940 (2000).
100. BELLE Collab., BELLE-CONF-0108, contributed to the 20th International Symposium on Lepton and Photon Interactions at High Energies, Rome, Italy, 23-28 Jul 2001.
101. CLEO Collab., Phys. Rev. Lett. **85**, 525 (2000).
102. BABAR Collab., Phys. Rev. **D65**, 051502 (2002).
103. BABAR Collab., presented at the XXXVII Recontres de Moriond, March 2002.
104. BELLE Collab., Phys. Rev. **D64**, 071101 (2001).
105. BABAR Collab., Phys. Rev. **D65**, 051101 (2002).
106. CLEO Collab., Phys. Rev. Lett. **84**, 5283 (2000).
107. CLEO Collab., Phys. Rev. Lett. **86**, 5661 (2001).
108. CLEO Collab., Phys. Rev. Lett. **86**, 5000 (2001).
109. BABAR Collab., Phys. Rev. Lett. **88**, 231801 (2002).
110. DELPHI Collab., M. Feindt et al., DELPHI 97-98 CONF 80, contributed paper to the 18th Int. Symp. on Lepton-Photon Interactions, Hamburg 1997.
111. OPAL Collab., Eur. Phys. J. **C12**, 609 (2000).
112. ALEPH Collab., Eur. Phys. J. **C20**, 431 (2001).
113. A. B. Carter, A. I. Sanda, Phys. Rev. Lett. **45**, 952 (1980); Phys. Rev. **D23**, 1567 (1981).

114. I. I. Bigi, A. I. Sanda, Nucl. Phys. **B193**, 85 (1981); Nucl. Phys. **B281**, 41 (1987).
115. Z.-Z. Xing, Phys. Rev. **D53**, 204 (1996)
116. M. Gronau, Phys. Rev. Lett. **63**, 1451 (1989); D. London, R. D. Peccei, Phys. Lett. **B223**, 257 (1989); B. Grinstein, Phys. Lett. **B229**, 280 (1989); H. Lipkin, Phys. Lett. **B357**, 404 (1995).
117. M. Bauer, B. Stech, M. Wirbel, Z. Phys. **C34**, 103 (1987).
118. M. Gronau, Phys. Lett. **B300**, 163 (1993);
119. R. Waldi, Nucl. Instr. and Meth. **A351**, 161 (1994).
120. M. Gronau, D. London, Phys. Rev. Lett. **65**, 3381 (1990); M. Gronau, Phys. Lett. **B265**, 389 (1991); H. J. Lipkin, Y. Nir, H. R. Quinn, A. E. Snyder, Phys. Rev. **D44**, 1454 (1991); N. G. Deshpande, X.-G. He, Phys. Rev. Lett. **74**, 26 and 4099 (1995); M. Gronau et al., Phys. Rev. **D52**, 6374 (1995); A. J. Buras, R. Fleischer, MPI-PhT/95-72; B. F. L. Ward, Phys. Rev. **D51**, 6253 (1995).
121. A. I. Sanda, K. Ukai, Prog. Theor. Phys. **107**, 421 (2002).
122. A. E. Snyder, H. R. Quinn, Phys. Rev. **D48**, 2139 (1993).
123. R. Fleischer, I. Dunietz, Phys. Lett. **B387**, 361 (1996); Phys. Rev. **D55**, 259 (1997).
124. I. Dunietz et al., Phys. Rev. **D43**, 2193 (1993).
125. ARGUS Collab., Phys. Lett. **B340**, 217 (1994).
126. CDF Collab., Phys. Rev. Lett. **75**, 3068 (1995).
127. CLEO Collab., Phys. Rev. Lett. **79**, 4533 (1997).
128. CDF Collab., Phys. Rev. Lett. **85**, 4668 (2000).
129. BABAR Collab., Phys. Rev. Lett. **87**, 151802 (2001).
130. BELLE Collab., Phys. Rev. **D64**, 071101 (2001).
131. M. Gronau, D. London, Phys. Lett. **B253**, 483 (1991).
132. M. Gronau, D. Wyler, Phys. Lett. **B265**, 172 (1991).
133. I. Dunietz, Phys. Lett. **B270**, 75 (1991).
134. R. Aleksan, I. Dunietz, B. Kayser, Z. Phys. **C54**, 653 (1992).
135. I. Dunietz, J. L. Rosner, Phys. Rev. **D34**, 1404 (1986); D. Du, I. Dunietz, D. D. Wu, Phys. Rev. **D34**, 3414 (1986); F. Buccella et al., Z. Phys. **C59**, 437 (1993); D. S. Du, X. L. Li, Z. J. Xiao, Phys. Rev. **D51**, 279 (1995); Z.-Z. Xing, Phys. Lett. **B364**, 55 (1995).
136. The LHB Collab., K. Kirsebom et al., CERN/LHCC/93-45 (Letter of Intent, 1993) and CERN/LHCC/94-11 (Addendum, 1994); see also [119].
137. T. Sjöstrand, PYTHIA 5.6 and JETSET 7.3 Physics and Manual, CERN-TH.6488/92 (1992).
138. Hong Pi, LU TP 91-28 (1992).
139. ARGUS Collab., Z. Phys. **C62**, 371 (1994).
140. CLEO Collab., Phys. Rev. **D55**, 13 (1997).
141. BABAR Collab., Phys. Rev. Lett. **86**, 2515 (2001).
142. S. Henrot-Versillé, F. Le Diberder, hep-ex/0007025.
143. OPAL Collab., Eur. Phys. J. **C5**, 379 (1998).
144. CDF Collab., Phys. Rev. Lett. **81**, 5513 (1998).

145. CDF Collab., Phys. Rev. **D61**, 72005 (2000).
146. ALEPH Collab., Phys. Lett. **B492**, 259 (2000).
147. BELLE Collab., Phys. Rev. Lett. **86**, 2509 (2001).
148. BABAR Collab., SLAC-PUB-9153 (2002).
149. BELLE Collab., presented at the XXXVII Recontres de Moriond, March 2002.
150. BABAR Collab., Phys. Rev. Lett. **87**, 241801 (2001).
151. A. Höcker, H. Lacker, S. Laplace and F. Le Diberder, Eur. Phys. J. **C21**, 225 (2001).

CP Asymmetries in Neutral Kaon and Beon Decays

Klaus R. Schubert

Institut für Kern- und Teilchenphysik,
Technische Universität Dresden, D 01062 Dresden, Gemany

Abstract. This article presents the description of CP asymmetries which is commonly used in B meson physics, including its application to K meson physics where many textbooks and review articles use a more historical description.

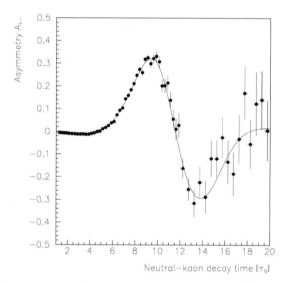

Fig. 1. The CPLEAR result [1] on the decay asymmetry of K^0 and $\overline{K}^0$ in the $\pi^+\pi^-$ mode. The abscissa is $\Gamma_S \cdot t$. Note, that the graph shows $-a$, the negative of Eq. 2.

Let us start the presentation with the CPLEAR result [1] on the asymmetry in the decay modes $K^0 \to \pi^+\pi^-$ and $\overline{K}^0 \to \pi^+\pi^-$ as shown in Fig. 1. In Kaon language, with the usual definition

$$\eta_{+-} = \frac{A(K^0_L \to \pi^+\pi^-)}{A(K^0_S \to \pi^+\pi^-)} , \qquad (1)$$

the asymmetry is given by

$$a(t) = \frac{\dot{N}(K^0 \to \pi^+\pi^-) - \dot{N}(\overline{K}^0 \to \pi^+\pi^-)}{\dot{N}(K^0 \to \pi^+\pi^-) + \dot{N}(\overline{K}^0 \to \pi^+\pi^-)}$$

$$= \frac{2\,|\eta_{+-}|\,\mathrm{e}^{-(\Gamma_{\mathrm{S}}+\Gamma_{\mathrm{L}})\,t/2}\cos(\Delta m_K \cdot t - \varphi_{+-})}{\mathrm{e}^{-\Gamma_{\mathrm{S}}t} + |\eta_{+-}|^2\,\mathrm{e}^{-\Gamma_{\mathrm{L}}t}} \, . \quad (2)$$

In this expression, t is the time between decay and production in the Kaon rest frame, the rates $\dot{N}$ are the observed decay rates for states which are pure K^0, resp. pure $\overline{K}^0$, at the time of their production, Γ_{S} is the total decay rate of the state K^0_{S}, Γ_{L} that of K^0_{L}, Δm_K is the mass difference $m(K^0_{\mathrm{L}}) - m(K^0_{\mathrm{S}})$, and φ_{+-} is the phase of the complex amplitude ratio η_{+-}.

Before coming to a discussion of η_{+-}, let me show in Fig. 2 the expectation for an analogous CP asymmetry in neutral Beon decays,

$$a(t) = \frac{\dot{N}(B^0 \to J/\Psi K^0_{\mathrm{S}}) - \dot{N}(\overline{B}^0 \to J/\Psi K^0_{\mathrm{S}})}{\dot{N}(B^0 \to J/\Psi K^0_{\mathrm{S}}) + \dot{N}(\overline{B}^0 \to J/\Psi K^0_{\mathrm{S}})} = -\sin 2\beta \cdot \sin(\Delta m_B\, t) \, , (3)$$

where β is a CKM matrix parameter to be discussed later and Δm_B is the mass difference $m(B^0_{\mathrm{H}}) - m(B^0_{\mathrm{L}})$.

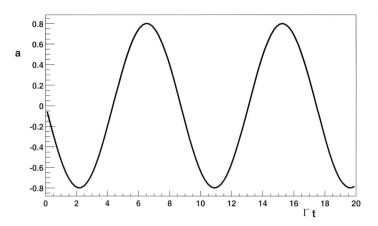

Fig. 2. Standard Model expectation for the decay asymmetry of B^0 and $\overline{B}^0$ in the $J/\Psi K^0_{\mathrm{S}}$ mode. Note that the amplitude of this oscillation, $\sin 2\beta$, is not yet well known.

The two curves in Figs. 1 and 2 look very different, but they are solutions of the same pair of coupled differential equations. Their difference originates in the fact that the parameters in these equations have very different values for neutral Kaons and Beons. This will be demonstrated in the following.

A stable particle state in its rest frame has a time dependence given by

$$\mathrm{i}\,\partial\psi/\partial t = E\psi = m\psi \, ,$$

$$\psi(t) = \psi(0) \cdot e^{-imt} , \tag{4}$$

where m is its rest mass. A decaying particle state is effectively described by

$$i\partial\psi/\partial t = \mu\psi ,$$

with a complex scalar μ and

$$m = (\mu + \mu^*)/2 , \quad \Gamma/2 = (\mu - \mu^*)/2i , \quad i\partial\psi/\partial t = (m - i\Gamma/2)\psi ,$$

$$\psi(t) = \psi(0) \cdot e^{-imt - \Gamma t/2} , \quad |\psi(t)|^2 = |\psi(0)|^2 e^{-\Gamma t} . \tag{5}$$

The total width Γ summarizes all couplings of ψ to final states, the more complete description of which would require a system of coupled differential equations. Next comes a decaying two-component state like the $K^0\overline{K}^0$ or the $B^0\overline{B}^0$ system. Its time dependence is given by [2]

$$i\frac{\partial}{\partial t}\begin{pmatrix}\psi_1\\\psi_2\end{pmatrix} = \begin{pmatrix}\mu_{11} & \mu_{12}\\\mu_{21} & \mu_{22}\end{pmatrix}\begin{pmatrix}\psi_1\\\psi_2\end{pmatrix} ,$$

with a non-Hermitian matrix μ_{ij} and

$$m_{ij} = (\mu_{ij} + \mu_{ij}^\dagger)/2 , \quad \Gamma_{ij}/2 = (\mu_{ij} - \mu_{ij}^\dagger)/2i ,$$

$$i\frac{\partial}{\partial t}\begin{pmatrix}\psi_1\\\psi_2\end{pmatrix} = \left[\begin{pmatrix}m_{11} & m_{12}\\m_{21} & m_{22}\end{pmatrix} - \frac{i}{2}\begin{pmatrix}\Gamma_{11} & \Gamma_{12}\\\Gamma_{21} & \Gamma_{22}\end{pmatrix}\right]\begin{pmatrix}\psi_1\\\psi_2\end{pmatrix} . \tag{6}$$

For discussing this equation, we assume $\psi = \psi_1 K^0 + \psi_2 \overline{K}^0$, but exactly the same arguments and results hold if $\psi = \psi_1 B^0 + \psi_2 \overline{B}^0$. The matrix μ_{ij}, decomposed into two Hermitian parts in Eq. 6, contains eight real parameters. Mass and width of K^0 and $\overline{K}^0$ have the same meaning as in Eq. 5, so we understand m_{11}, m_{22}, Γ_{11}, and Γ_{22}. The CPT theorem requires

$$m_{11} = m_{22} = m , \quad \Gamma_{11} = \Gamma_{22} = \Gamma , \tag{7}$$

leaving six real parameters. The parameters m_{12} and Γ_{12} describe the transition from $\overline{K}^0$ to K^0, as predicted by Gell-Mann and Pais in 1955 [3] and found by Good et al in 1961 [4]; m_{12} describes the dispersive part and Γ_{12} the absorptive part of this transition. In the Standard Model, both parts have their origin in second order weak interactions. The parameters m_{21} and Γ_{21} describe the CP mirror transition from K^0 to $\overline{K}^0$. Because of the hermiticity of both m_{ij} and Γ_{ij}, we have

$$m_{21} = m_{12}^* , \quad \Gamma_{21} = \Gamma_{12}^* . \tag{8}$$

Because of an unobservable phase in the definition of the states K^0 and $\overline{K}^0$, only five of the six parameters in μ_{ij} are observables. Rotations of the state

K^0 into $K^0 \cdot e^{i\phi}$ and of $\overline{K}^0$ into $\overline{K}^0 \cdot e^{-i\phi}$ have no observable effects. They lead to $m_{jj} \to m_{jj}$, $\Gamma_{jj} \to \Gamma_{jj}$ $(j = 1, 2)$, but to

$$m_{12} \to m_{12} \cdot e^{-2i\phi} \, , \quad \Gamma_{12} \to \Gamma_{12} \cdot e^{-2i\phi} \, . \tag{9}$$

Note that the two phases in Eq. 9 are the same, i. e. the ratio Γ_{12}/m_{12} is invariant in modulus and phase. The five observable parameters of μ_{ij} are m, Γ, $|m_{12}|$, $\text{Re}(\Gamma_{12}/m_{12})$, and $\text{Im}(\Gamma_{12}/m_{12})$. If CP symmetry holds in $K^0\overline{K}^0$ oscillations, the number of observables reduces to four. The imaginary part of Γ_{12}/m_{12} has then to be zero, as will be shown in Eqs. 14.

We know since 1964 [5] that CP symmetry is broken. If the violation takes place in $K^0\overline{K}^0$ transitions, as confirmed a few years later, then Γ_{12}/m_{12} is not real, and the time dependence in Eq. 6 is governed by the five parameters m, Γ, $|m_{12}|$, $\text{Re}(\Gamma_{12}/m_{12})$, and $\text{Im}(\Gamma_{12}/m_{12})$. Given these five parameters, Eq. 6 has a well-defined solution for each initial state $\psi(0)$. There are two and only two states which do not change their flavor content with time, these are the eigensolutions of Eq. 6,

$$K_S^0 = p\,K^0 + q\,\overline{K}^0 \, , \quad K_L^0 = p\,K^0 - q\,\overline{K}^0 \, ,$$
$$K_S^0(t) = K_S^0(0) \cdot e^{-\gamma_S \cdot t} \text{ with } \gamma_S = i\,m_S + \Gamma_S/2 \, ,$$
$$K_L^0(t) = K_L^0(0) \cdot e^{-\gamma_L \cdot t} \text{ with } \gamma_L = i\,m_L + \Gamma_L/2 \, . \tag{10}$$

The two eigenstates K_S^0 and K_L^0 are normalized, i. e. $|p|^2 + |q|^2 = 1$. The five observables m_S, m_L, Γ_S, Γ_L and $|p/q|$ are given by the five parameters in Eq. 6. The explicit relations between the five observables and the five parameters can be found in the literature [6]. One example is

$$\frac{q}{p} = \frac{-2m_{12}^* + i\,\Gamma_{12}^*}{2|m_{12}| + i\,|\Gamma_{12}|} \, . \tag{11}$$

Note, that p/q is not an observable, its phase is undetermined because of the arbitrariness in defining the states K^0 and $\overline{K}^0$.

We now discuss the time evolution of states ψ_K and $\psi_{\overline{K}}$ which are pure K^0 or $\overline{K}^0$ at $t = 0$; i. e. $\psi_K(0) = K^0$ and $\psi_{\overline{K}}(0) = \overline{K}^0$:

$$\psi_K(t) = \frac{1}{2p}\left[K_S^0(t) + K_L^0(t)\right] = \frac{1}{2p}\left[(pK^0 + q\overline{K}^0)\,e^{-\gamma_S t} + (pK^0 - q\overline{K}^0)\,e^{-\gamma_L t}\right]$$

$$= \frac{e^{-\gamma_S t} + e^{-\gamma_L t}}{2}\,K^0 + \frac{q}{p}\,\frac{e^{-\gamma_S t} - e^{-\gamma_L t}}{2}\,\overline{K}^0 \, ,$$

$$\psi_{\overline{K}}(t) = \frac{e^{-\gamma_S t} + e^{-\gamma_L t}}{2}\,\overline{K}^0 + \frac{p}{q}\,\frac{e^{-\gamma_S t} - e^{-\gamma_L t}}{2}\,K^0 \, . \tag{12}$$

From this time evolution of the states, we easily obtain the time evolution of the transition rates $P(t)$:

$$P(K^0 \to K^0) = P(\overline{K}^0 \to \overline{K}^0) = \left|\frac{e^{-\gamma_S t} + e^{-\gamma_L t}}{2}\right|^2$$

$$= \frac{1}{4}\left[e^{-\Gamma_S t} + e^{-\Gamma_L t} + 2e^{-(\Gamma_S+\Gamma_L)t/2}\cos\Delta mt\right], \quad (13)$$

$$P(K^0 \to \overline{K}^0) = \frac{1}{4}\left|\frac{q}{p}\right|^2\left[e^{-\Gamma_S t} + e^{-\Gamma_L t} - 2e^{-(\Gamma_S+\Gamma_L)t/2}\cos\Delta mt\right],$$

$$P(\overline{K}^0 \to K^0) = \frac{1}{4}\left|\frac{p}{q}\right|^2\left[e^{-\Gamma_S t} + e^{-\Gamma_S t} - 2e^{-(\Gamma_S+\Gamma_L)t/2}\cos\Delta mt\right]. \quad (14)$$

The rates for $K^0 \to K^0$ and $\overline{K}^0 \to \overline{K}^0$ are equal because of CPT symmetry. The equality of the rates for $K^0 \to \overline{K}^0$ and $\overline{K}^0 \to K^0$ depends on the value of $|p/q|$. CP symmetry in $K^0\overline{K}^0$ oscillations is conserved if and only if $|p/q| = 1$. From Eq. 11, we can easily derive that this condition is equivalent with $\text{Im}(\Gamma_{12}/m_{12}) = 0$.

Experimentally, the derivation of $|p/q|$ from unity is small. Therefore, we write since 1964:

$$\epsilon_m = \frac{p-q}{p+q}, \quad p = \frac{1+\epsilon_m}{\sqrt{2+2|\epsilon_m|^2}}, \quad q = \frac{1-\epsilon_m}{\sqrt{2+2|\epsilon_m|^2}}, \quad \frac{p}{q} = \frac{1+\epsilon_m}{1-\epsilon_m},$$

$$\left|\frac{p}{q}\right| = 1 + \frac{2\text{Re}(\epsilon_m)}{1+|\epsilon_m|^2}. \quad (15)$$

The index m in ϵ_m stands for mixing, another word for oscillations. Only the real part of ϵ_m is an observable, the imaginary part is arbitrary and unobservable. For keeping Eq. 15 and the following formulae simple, we choose the imaginary part to be small, i. e. $|\epsilon_m| \ll 1$, $|p/q| = 1 + 2\text{Re}(\epsilon_m)$.

CP symmetry is violated if and only if $\text{Im}(\Gamma_{12}/m_{12}) \neq 0$. CP is conserved if the phase between the two complex matrix elements Γ_{12} and m_{12} is 0 or π. Since $\text{Re}(\Gamma_{12}/m_{12})$ is negative, we define the small CP violating phase ϕ_m as

$$\Gamma_{12}/m_{12} = -|\Gamma_{12}/m_{12}|\, e^{-i\phi_m}. \quad (16)$$

Starting from the explicit solution for q/p in Eq. 11 and the approximations

$$e^{-i\phi_m} = 1 - i\phi_m,$$

$$\Delta m = 2|m_{12}|, \quad \Delta\Gamma = -2|\Gamma_{12}| = -2\Gamma = -\Gamma_S, \quad (17)$$

it can easily be derived that

$$\text{Re}(\epsilon_m) = \frac{\phi_m}{2}\frac{2\Delta m/\Gamma_S}{1+(2\Delta m/\Gamma_S)^2} \approx \frac{\phi_m}{4}. \quad (18)$$

In the next step, we describe decays of K^0 and $\overline{K}^0$ mesons into final states which are CP eigenstates, e. g. $\pi^+\pi^-$ with $CP = +1$. Without taking care of proper normalisations for the decay matrix elements, we define

$$< f|\mathcal{T}|K^0 > = A \, , \quad < f|\mathcal{T}|\overline{K}^0 > = \overline{A} \, . \tag{19}$$

The time dependent decay rates from states which are pure K^0 or $\overline{K}^0$ at $t = 0$ are obtained from Eqs. 12;

$$\dot{N}(t) = \left| \frac{e^{-\gamma_S t} + e^{-\gamma_L t}}{2} A + \frac{q}{p} \frac{e^{-\gamma_S t} - e^{-\gamma_L t}}{2} \overline{A} \right|^2 ,$$

$$\dot{\overline{N}}(t) = \left| \frac{e^{-\gamma_S t} + e^{-\gamma_L t}}{2} \overline{A} + \frac{p}{q} \frac{e^{-\gamma_S t} - e^{-\gamma_L t}}{2} A \right|^2 . \tag{20}$$

By introducing

$$r = \frac{q\overline{A}}{pA} , \tag{21}$$

we obtain

$$\dot{N}(t) = \frac{|A|^2 |1+r|^2}{4} \left| e^{-\gamma_S t} + \frac{1-r}{1+r} e^{-\gamma_L t} \right|^2 ,$$

$$\dot{\overline{N}}(t) = \frac{|A|^2 |1+r|^2}{4} \left| \frac{p}{q} \right|^2 \left| e^{-\gamma_S t} - \frac{1-r}{1+r} e^{-\gamma_L t} \right|^2 , \tag{22}$$

if $r \neq -1$. For $r = -1$ we can easily find similar expressions. We now introduce

$$\frac{1-r}{1+r} = \epsilon_f \iff \frac{1-\epsilon_f}{1+\epsilon_f} = r \, , \tag{23}$$

where we have applied the index f to the asymmetry parameter ϵ since it depends — as A, $\overline{A}$, and r — on the final state f. We use the approximations $|p/q| \approx 1$, $|\epsilon_f| \ll 1$, and obtain for the asymmetry $a = (\dot{N} - \dot{\overline{N}})/(\dot{N} + \dot{\overline{N}})$:

$$a(t) = \frac{2 e^{-(\Gamma_S + \Gamma_L)t/2} \left[\text{Re}(\epsilon_f) \cos(\Delta m t) + \text{Im}(\epsilon_f) \sin(\Delta m t) \right]}{e^{-\Gamma_S t} + |\epsilon_f|^2 e^{-\Gamma_L t}} . \tag{24}$$

This is the known result for the asymmetries introduced in Eqs. 1 and 2. For the K meson case with $f = \pi^+\pi^-$ we have $\epsilon_f = \eta_{+-} = 0.00229 \cdot e^{i \cdot 43°}$, $\Delta m = \Delta m_K = m(K_L^0) - m(K_S^0) = 0.53 \cdot 10^{10}/s$, $\Gamma_S = 1.12 \cdot 10^{10}/s$, and $\Gamma_L = 1.93 \cdot 10^7/s$ [7]. In the B meson case, the two eigenstates (p,q) and $(p,-q)$ cannot be called short- and long-living since both have the same life time in very good approximation. Because of their mass difference, they are called heavy and light, B_H^0 and B_L^0, with $\Delta m = \Delta m_B = m(B_H^0) - m(B_L^0) = 0.46 \cdot 10^{12}/s$ and $\Gamma_H = \Gamma_L = \Gamma = 0.64 \cdot 10^{12}/s$ [7]. For $f = J/\Psi K_S^0$, ϵ_f is expected to be $\epsilon_f = -i \cdot \cot \beta$ with $\beta = \arg(V_{td}^* V_{ts} V_{cs}^* V_{cd})$. Both values for the asymmetries, $\epsilon_f = \eta_{+-}$ and $\epsilon_f = -i \cot \beta$ will be discussed later in this article.

In both meson systems K and B, the two asymmetry parameters ϵ_m and ϵ_f should not be confused with each other. The CP asymmetry ϵ_m in oscillations is defined by Eq. 15,

$$\mathrm{Re}(\epsilon_m) = \frac{1}{2}\left|\frac{p}{q}\right| - \frac{1}{2},$$

its imaginary part is unobservable. The CP asymmetry ϵ_f in decays into CP eigenstates f is defined by Eq. 23,

$$\epsilon_f = \frac{1-r}{1+r}, \quad r = \frac{q\overline{A}}{pA}.$$

The arbitrary phases in defining the states K^0, $\overline{K}^0$, B^0, and $\overline{B}^0$ cancel in the definition of r. Therefore, both r and ϵ_f are observable in modulus and phase.

Oscillations have only one mechanism to break CP symmetry, this is a phase difference between m_{12} and Γ_{12} in Eq. 6. Decays into CP eigenstates can have three different mechanisms. CP is conserved if and only if $r = +1$ or $r = -1$. Since

$$r = \left|\frac{q}{p}\right| \cdot \left|\frac{\overline{A}}{A}\right| \cdot e^{-2i\phi}, \tag{25}$$

we can obtain $r \neq \pm 1$ for three reasons:

1. $|q/p| \neq 1$; CP violation in decays because of CP violation in oscillations,
2. $|\overline{A}/A| \neq 1$; called "direct CP violation", and
3. $\phi \neq n\pi/2$; called interference between oscillations and decay.

We now discuss the two decay asymmetries for $K^0, \overline{K}^0 \to \pi^+\pi^-$ and $B^0, \overline{B}^0 \to J/\Psi K_S^0$ separately and start with the second one. In $B^0 \overline{B}^0$ oscillations,

$$\mathrm{Re}(\epsilon_m)/(1+|\epsilon_m|^2) = \mathcal{O}(10^{-3}). \tag{26}$$

The first contribution can, therefore, be neglected for the large asymmetry in Fig. 2; $|q/p| = 1$. There is also no direct CP violation for $f = J/\Psi K_S^0$ since this decay mode is dominated by a single tree graph and even the next contribution, the largest penguin part, has the same CKM phase as the tree; $|\overline{A}/A| = 1$. But in the Standard weak interaction, there is a large phase between q/p and $\overline{A}/A$;

$$r = -e^{-2i\phi} \neq \pm 1. \tag{27}$$

The negative sign in front of the phase factor is the CP eigenvalue of the $J/\Psi K_S$ state. With this value of r, we obtain

$$\epsilon_f = \frac{1-r}{1+r} = \frac{1+e^{-2i\phi}}{1-e^{-2i\phi}} = \frac{1}{i \cdot \tan\phi},$$

$$\mathrm{Re}(\epsilon_f) = 0, \quad \mathrm{Im}(\epsilon_f) = -\cot\phi, \tag{28}$$

and Eq. 24, with $\Gamma_\mathrm{H} = \Gamma_\mathrm{L}$, becomes

$$a = \frac{-2\cot\phi \cdot \sin \Delta mt}{1 + \cot^2 \phi} = -\sin 2\phi \cdot \sin \Delta mt \,. \qquad (29)$$

In the Standard Model, $\phi = \beta = \arg(V_{\mathrm{td}}^* V_{\mathrm{ts}} V_{\mathrm{cs}}^* V_{\mathrm{cd}})$. The derivation can be found in textbooks [6].

The situation in the K sector is quite different. In contrast to negligible deviations of $|q/p|$ and $|\overline{A}/A|$ from unity and a large phase difference ϕ in the example above, all three sources of CP violation contribute to $a(K^0, \overline{K}^0 \to \pi^+\pi^-)$, and all three are very small. Let me start with the simpler asymmetry $a(K^0, \overline{K}^0 \to \pi\pi, I = 0)$ where the two pions are in the isospin zero state. Here we have only two contributions since

$$|<\pi\pi, I=0|\mathcal{T}|K^0>| = |A_0| = |\overline{A}_0| = |<\pi\pi, I=0|\mathcal{T}|\overline{K}^0>|\,. \qquad (30)$$

The reason for this equality is the common final state interaction phase in the tree and the penguin graph. We remain with

$$r_0 = \left|\frac{q}{p}\right| \cdot e^{-2i\phi_0} = [(1 - 2\,\mathrm{Re}(\epsilon_\mathrm{m}))]\,[1 - 2i\phi_0] \qquad (31)$$

since both $\mathrm{Re}(\epsilon_\mathrm{m})$ and ϕ_0 are very small. The resulting asymmetry parameter ϵ_0 with index $f = 0$ for isospin 0 is

$$\epsilon_0 = \frac{1 - r_0}{1 + r_0} = \mathrm{Re}(\epsilon_\mathrm{m}) + i\phi_0 \,,$$

$$\mathrm{Re}(\epsilon_0) = \mathrm{Re}(\epsilon_\mathrm{m}) \,, \quad \mathrm{Im}(\epsilon_0) = \phi_0 \,. \qquad (32)$$

Inserting this result into Eq. 24 and using

$$\mathrm{Re}(\epsilon_0)\cos(\Delta mt) + \mathrm{Im}(\epsilon_0)\sin(\Delta mt) = |\epsilon_0|\cos[\Delta mt - \phi(\epsilon_0)] \,, \qquad (33)$$

leads to Eq. 2, if $\eta_{+-} = \epsilon_0$. This holds to very good approximation since $\epsilon'/\epsilon_0 \ll 1$. In this approximation,

$$\eta_{+-} = \frac{<2\pi, I=0|\mathcal{T}|K_\mathrm{L}^0>}{<2\pi, I=0|\mathcal{T}|K_\mathrm{S}^0>} = \frac{pA_0 - q\overline{A}_0}{pA_0 + q\overline{A}_0} = \frac{1 - r_0}{1 + r_0} = \epsilon_0 \,. \qquad (34)$$

Including $\pi\pi$ contributions with isospin 2 and defining A_2 as A_0 in Eq. 30,

$$A_2 = <\pi\pi, I=2|\mathcal{T}|K^0> \,, \quad \overline{A}_2 = <\pi\pi, I=2|\mathcal{T}|\overline{K}^0> \,, \qquad (35)$$

we obtain

$$A = <\pi^+\pi^-|\mathcal{T}|K^0> = \frac{1}{\sqrt{3}}(A_2 + \sqrt{2}\,A_0) \,,$$

$$\overline{A} = <\pi^+\pi^-|\mathcal{T}|\overline{K}^0> = \frac{1}{\sqrt{3}}(\overline{A}_2 + \sqrt{2}\,\overline{A}_0) \ . \tag{36}$$

In these relations, $|\overline{A}_0/A_0| = 1$ and $|\overline{A}_2/A_2| = 1$, but there is a CP violating phase difference between the two ratios. With

$$\omega = |A_2/A_0| \tag{37}$$

we have

$$\frac{A_2}{A_0} = \omega\, e^{i\phi_2}\, e^{i(\delta_2-\delta_0)}\ , \quad \frac{\overline{A}_2}{\overline{A}_0} = \omega\, e^{-i\phi_2}\, e^{i(\delta_2-\delta_0)}\ , \tag{38}$$

where ϕ_2 is this new CP violating phase and $\delta_2 - \delta_0$ is the difference of the two strong interaction final state phases. With $\delta_2 \neq \delta_0$ and $\phi_2 \neq 0$ we have $|\overline{A}/A| \neq 1$. The asymmetry $a(K^0, \overline{K}^0 \to \pi^+\pi^-)$ has all three sources of CP violation in Eq. 25, $|p/q| \neq 1$, $\phi \neq 0$, and $|\overline{A}/A| \neq 1$. The first two are on the 10^{-3} level and the third one on the 10^{-6} level. With the definition

$$\epsilon' = \frac{i\omega}{\sqrt{2}}\, e^{i(\delta_2-\delta_0)} \cdot \phi_2 \tag{39}$$

we easily derive

$$r = r(\pi^+\pi^-) = \frac{q\overline{A}}{pA} = \frac{q\overline{A}_0(1-\epsilon')}{pA_0(1+\epsilon')} = r_0\,(1-2\epsilon')\ , \quad \left|\frac{\overline{A}}{A}\right| = 1 - 2\,\mathrm{Re}(\epsilon')\ , \tag{40}$$

$$\eta_{+-} = \frac{1-r}{1+r} = \frac{1-r_0+2r_0\epsilon'}{1+r_0} = \epsilon_0 + \epsilon'\ , \tag{41}$$

using the approximation $r_0 \approx 1$. The results in Eq. 32 are replaced by

$$\mathrm{Re}(\eta_{+-}) = \mathrm{Re}(\epsilon_m) + \mathrm{Re}(\epsilon')\ , \quad \mathrm{Im}(\eta_{+-}) = \phi_0 + \mathrm{Im}(\epsilon')\ . \tag{42}$$

The time dependence of the asymmetry becomes that of Eq. 2. The corresponding calculation for the other combination of isospins 2 and 0 gives

$$r(\pi^0\pi^0) = r_0(1+4\epsilon')\ , \quad \eta_{00} = \frac{<\pi^0\pi^0|\mathcal{T}|K_L^0>}{<\pi^0\pi^0|\mathcal{T}|K_S^0>} = \epsilon_0 - 2\epsilon'\ . \tag{43}$$

In the discussion of ϵ_0 presented here, the real part has its origin in the CP violating phase between m_{12} and $-\Gamma_{12}$, and the imaginary part in the CP violating phase between q/p and $\overline{A}/A$. Why are these two phases related to each other in Kaon phenomenology; why is $\mathrm{Im}(\epsilon_0)/\mathrm{Re}(\epsilon_0) = 2\Delta m/\Gamma_S$? This is a speciality of the K meson system which is completely absent for B mesons. It has to do with unitarity and with the fact that K^0 and $\overline{K}^0$ decays are dominated by one single final state, $f = \pi\pi, I = 0$. The derivation requires saturation of Γ_{11} and Γ_{12} by this decay mode,

$$\Gamma_{11} = \mathrm{const}\cdot A_0 A_0^*\ , \quad \Gamma_{12} = \mathrm{const}\cdot \overline{A}_0 A_0^*\ , \tag{44}$$

the explicit expression for q/p in Eq. 11, and the approximations in Eq. 17. We obtain

$$\Gamma_{12} = -\gamma \cdot m_{12} \cdot (1 - i\phi_m) \, , \quad \frac{\overline{A}_0}{A_0} = \frac{\Gamma_{12}}{|\Gamma_{12}|} \, ,$$

$$r_0 = \frac{q\overline{A}_0}{pA_0} = \frac{-2m_{12}^*\Gamma_{12} + i|\Gamma_{12}|^2}{2|m_{12}\Gamma_{12}| + i|\Gamma_{12}|^2} = 1 - \frac{2i\gamma\phi_m}{2\gamma + i\gamma^2} \, ,$$

$$\epsilon_0 = \frac{1 - r_0}{1 + r_0} = \frac{i\gamma\phi_m}{2\gamma + i\gamma^2} \, , \quad \frac{\text{Im}(\epsilon_0)}{\text{Re}(\epsilon_0)} = \frac{2}{\gamma} = \frac{2\Delta m}{\Gamma_S} \, . \quad (45)$$

In the $K^0\overline{K}^0$ system, CP violation in oscillations and CP violation in the dominant decay mode are strongly linked by unitarity. This is not the case for the $B^0\overline{B}^0$ system in which no single decay mode has such a dominance.

I would like to warmly thank H. Schröder and M. Beyer for their invitation to this School, for their kind hospitality, and the well chosen scientific programme of the School. I also thank W. Kluge, M. Paulini, and R. Waldi for stimulating discussions on the subject presented in this article.

References

1. A. Apostolakis et al (CPLEAR), Phys. Lett. B 458 (1999) 545
2. E. P. Wigner and V. F. Weisskopf,
   Z. Physik 63 (1930) 54 and 65 (1930) 18
3. M. Gell-Mann and A. Pais, Phys. Rev. 97 (1955) 1387
4. R. H. Good et al, Phys. Rev. 124 (1961) 1223
5. J. H. Christensen et al, Phys. Rev. Lett. 13 (1964) 138
6. e. g. Q. Ho-Kim and X.-Y. Pham,
   "Elementary Particles and Their Interactions", Springer-Verlag 1998
   or G. C. Branco, L. Lavoura, and J. P. Silva,
   "CP Violation", Clarendon Press Oxford 1999
7. Review of Particle Physics, Particle Data Group,
   Eur. Phys. J. C 3 (1998) 1

Time Reversal Invariance in Nuclear Physics: From Neutrons to Stochastic Systems

Christopher R. Gould[1] and Edward David Davis[1,2]

[1] North Carolina State University, Raleigh NC 27695, USA
   and Triangle Universities Nuclear Laboratory, Durham NC 27708 USA*
[2] Physics Department, Kuwait University, P.O. Box 5969, Safat, Kuwait

Abstract. To test models of CP violation it is important to look for T-violating effects in as many different systems as possible. In this experimental search, nuclear tests with neutrons have played an important role. We review the basic issues underlying measurements of the neutron's T-odd electric dipole moment, and discuss a promising new line of investigation involving transmission of polarized neutrons through spin polarized and aligned nuclear targets, particularly at epithermal neutron energies. We give a self-contained derivation of the generalization of the optical theorem to include polarization degrees of freedom, and derive the expressions required to analyze polarized neutron transmission in targets of arbitrary polarization state.

1 Introduction

Nuclear tests of time reversal invariance have a long and rich history. In these lectures we focus on two categories of test with neutrons: searches for electric dipole moments (edm), and transmission tests with nuclear spin polarized and aligned targets. Neutron edm searches have been a focus of research for nearly fifty years. The techniques and implications for the experiments are well described in a number of excellent texts [1,2]. Neutron transmission tests, first proposed in the late 1980's, are much more recent. The motivations and methods have been discussed in the literature, and in conference proceedings [3,4].

This paper is divided into two distinct parts. In the first, we give a simple overview of edm searches and transmission tests, emphasizing the underlying physical ideas and the new experiments to be carried out over the next decade. This part will be accessible to anyone who has a background in undergraduate modern physics. In the second part of the paper, we give a more detailed description of the formalism associated with the transmission tests, an area that has been less well covered in a pedagogically coherent fashion. This part requires a strong background in nuclear reaction theory and angular momentum coupling and is appropriate for upper-level graduate students.

A third category of time reversal tests with neutrons involves beta decay correlation studies. New experiments are planned here, taking advantage of

---
* This work supported in part by US DOE under Grant No.DE-FG02-97ER41042.

many of the developments in ultra cold neutron (UCN) physics that are driving new edm experiments. We do not treat this topic, referring the reader instead to recent discussions in the literature.

2 Overview of Electric Dipole Moment and Transmission Tests of Time Reversal Invariance

In this section, the presentation is kept deliberately simple in the hope that it will whet the curious reader's appetite for the more formal treatments of the topics covered elsewhere in the literature (and indeed in the second half of this paper).

2.1 Electric Dipole Moments

Many molecules have edm's so the connection of a non-zero edm with time reversal invariance is not immediately obvious. The key point is that in a non-degenerate quantum system with non-zero spin (an electron, a neutron, an atom but not a polar molecule), the only vector characterizing the state is the spin s. The electric dipole operator $d = \Sigma_i e_i r_i$ also being a vector, must therefore, by the Wigner–Eckart theorem, be proportional to s. However, s is odd under the time reversal operator T [5], whereas d is even. In an electric field E the Hamiltonian of the system $H = -d \cdot E \sim s \cdot E$ is therefore odd under T. In fact H is also odd under P. But for our purposes the main point is that the presence of an edm, first-order in E in a non-degenerate system, implies that T is violated. No such edm has yet been observed.

Particles and atoms get edm's in a hierarchical fashion:

- nucleons – from quark edm's or from T-odd interactions
- nuclei – from nucleon edm's or from T-odd nucleon-nucleon (N-N) interactions
- atoms – from nuclear edm's, from electron edm's, or from T-odd e-N interactions

A first guess at the size of the neutron edm might be (nuclear size) (size of P-violation)(size of T-violation in kaon system) which is a number of order $10^{-13} \times 10^{-7} \times 10^{-3} = 10^{-23}$ e cm. Current limits $d_n < 0.63 \times 10^{-25}$ e cm are about a factor of a hundred better and correspond (if a neutron is expanded to the size of the earth) to a ten micron bump at the North Pole. In the standard model $d_n \sim 10^{-32}$ e cm and no measurement is foreseen that could achieve this accuracy. However, many models of supersymmetry predict edm's of order 10^{-28} e cm and another factor of hundred improvement in measurements would be extremely interesting.

To understand the difficulty of measuring a neutron edm of this size, we can put in some numbers. The Hamiltonian for a neutron with spin s in a magnetic field $\boldsymbol{B}$, electric field $\boldsymbol{E}$ is:

$$H = -(\mu_n \boldsymbol{s} \cdot \boldsymbol{B} + d_n \boldsymbol{s} \cdot \boldsymbol{E})/s \qquad (1)$$

where $s = 1/2$, the magnetic moment of the neutron is $\boldsymbol{\mu} = 2\mu_n \boldsymbol{s}$, $\mu_n = -1.91\mu_N$ and the electric dipole moment d_n is defined by $\boldsymbol{d} = 2d_n \boldsymbol{s}$. We can look for an energy change $\Delta W = W_+ - W_-$ when $\boldsymbol{E}$ is reversed compared to $\boldsymbol{B}$:

$$W_\pm = 2\mu_n B \pm 2d_n E \equiv \hbar\omega_B \pm \hbar\omega_E, \qquad (2)$$

or (more realistically) we can look for a change in the precession frequencies $\omega_\pm$, which from

$$d\boldsymbol{s}/dt = (\mu_n \boldsymbol{s} \times \boldsymbol{B} + d_n \boldsymbol{s} \times \boldsymbol{E})/s \qquad (3)$$

are given by:

$$\omega_\pm = \omega_B \pm \omega_E. \qquad (4)$$

With $E \sim 10$ kV/cm, and $d_n \sim 10^{-26}$ ecm we see an energy shift $\Delta W = 8 \times 10^{-22}$ eV and a frequency shift $\Delta\omega = 7 \times 10^{-7}$ Hz when $\boldsymbol{E}$ is reversed compared to $\boldsymbol{B}$. This may seem plausible to measure except by comparison, $\mu_n = 6 \times 10^{-12}$ eV/G is huge and $\omega_B = \omega_E$ for B of order only 10^{-10} G, a field comparable to that of a small bar magnet 100 m from an unshielded experiment. Clearly, extraordinary magnetic shielding is required to keep B stable enough to see the effect of reversing E during a measurement.

The measurements are carried out by an ingenious interference technique in which the precession due to $\boldsymbol{E}$ and $\boldsymbol{B}$ is compared against an rf oscillator tuned to match closely the Larmor frequency ω_B of the magnetic field B. The method was first developed by Ramsey and relied on two rf coils, the first to prepare the polarized neutrons, and the second to probe the amount of precession the neutrons undergo as they pass through the region with non-zero $\boldsymbol{E}$ and $\boldsymbol{B}$ fields. The neutrons gain a phase $\phi = (\omega_B + \omega_E - \omega_{RF})t$ passing through the system (t is the transit time) and the intensity at the exit varies as $I = I_0 \sin^2(\phi/2)$. Experiments look for a shift in this fringe pattern as $\boldsymbol{E}$ is periodically reversed over a many months series of cycles.

A useful bound on the ultimate precision of any edm measurement can be obtained as follows. Send two neutrons into fields $\pm E$ and measure the energy difference $\Delta W = 4d_n E$. The uncertainty in each $W_\pm$ measurement is by the uncertainty principle given by $\hbar/t$ where t is the measuring time. The uncertainty δd_n is therefore given by $\sqrt{2}\hbar/t \sim 4\delta d_n E$, or, with N total neutrons instead of two:

$$\delta d_n \sim \hbar/(2Et\sqrt{N}) \qquad (5)$$

The highest precision comes from large E (10 kV/cm), a long storage time t (100 s for UCN's), and a long counting time to maximize N (10^8 in ILL experiments). Interestingly these experiments can get surprisingly close to the quantum mechanical limits. However, the precision is ultimately determined by systematic effects.

The dominant systematic for beam experiments (neutrons or atoms) is the $\boldsymbol{v} \times \boldsymbol{E}$ effect. This term gives rise to a motional magnetic field B_m field which reverses with E and is indistinguishable from the effect of a non-zero edm. The effect enters in first order if the external E and B fields are not exactly parallel. But even if they are parallel, there is still a second order contribution. The solution for neutron experiments was to move from cold neutrons (300 m/s) to UCN's (6 m/s). All modern neutron edm experiments have been carried out with UCN's in storage bottles.

The remaining dominant systematic is accurately knowing the B field in the region where the neutrons are. ^{199}Hg comagnetometers have been used, but the most interesting new possibility [6] relies on doping small amounts of polarized into superfluid helium, and using the strong spin dependence of the n-^{3}He reaction to a) monitor the spin precession of the neutrons, and b) very precisely monitor the magnetic field in the region in which neutrons are contained. A proposal at Los Alamos aims to improve the accuracy of present neutron edm measurements by a factor of 250 over the next five years, potentially reaching sensitivities of 4×10^{-28} e cm.

Atomic edm's provide another route to measurements of nucleon and electron edm's. As first discussed in detail by Schiff, when an atom is placed in an external electric field, the electrons rearrange to give zero field at each constituent (proton, nucleon, electron). However, the cancellation is not complete due to relativistic effects and the finite size of the nucleus. Very sensitive measurements are possible in diamagnetic atoms – ^{199}Hg in particular – and the huge electric field enhancements inside paramagnetic atoms and polar molecules can potentially provide exceptional sensitivity. A useful figure of merit is the ratio of effective field to applied field: $R = E_{\text{effective}}/E_{\text{applied}}$.

For the neutron R=1. The maximum field that can typically be applied in a vacuum is 10 kV/cm. However, larger fields are proposed in liquid helium experiments where the nominal dielectric strength is of order 300 kV/cm.

For diamagnetic atoms (^{3}He, ^{199}Hg, ^{225}Ra), Schiff shielding operates and $R \sim 10 \times Z^2 (r_{\text{nuc}}/r_{\text{atom}})^2 \sim 10^{-3}$ for Hg. Despite this suppression, the measurements can be so precise that in many cases the bounds on particular T-violating processes are as good or better than those from the neutron edm. The case of Ra is of particular interest because an additional huge nuclear collective enhancement is expected due to close lying parity doublets in the nuclear level scheme. Large amounts of ^{225}Ra will be available at future radioactive ion beam facilities such as RIA in the US, Ganil in France or GSI in Germany.

For paramagnetic atoms (Cs, Tl), R scales as Z^3 and for heavy atoms, $R \gg 1$, ~ 600 for Tl example. The most precise bounds on electron edm's come from paramagnetic atom experiments.

For polar molecules (TlF, PbO), R can approach 10^7. The theoretical interpretation becomes much more complicated, but the huge enhancements are attractive. The best bounds on proton edm's come from these systems.

2.2 Neutron Transmission Tests of Symmetry Violation

In contrast to the fifty year history of neutron edm measurements, experiments on P-and T-violating effects in neutron transmission are much more recent.

P-odd effects in neutron transmission are due to the weak potential $U_W \sim \gamma_W \mathbf{s} \cdot \mathbf{p}$. Here $\mathbf{p} = \hbar \mathbf{k}$ is the momentum of the neutron. The potential is P-odd and T-even, and causes a precession of the spin around the direction of propagation, in analogy to the familiar Larmor precession due to the magnetic moment interaction $U_B \sim -\gamma_B \mathbf{s} \cdot \mathbf{B}$. The precession per unit length scales with G_F, the strength of the weak interaction, and absent nuclear enhancements, is given by $d\phi/dz \sim G_F Z \rho/\hbar c$, where Z is the atomic number of the target material and ρ is the number density. For tin this amounts to about 10^{-8} rad/cm. It was a pleasant surprise when the first experiments, using cold neutrons at ILL, saw an effect 400 times greater. The enhancement was understood to be due to the presence of a sub-threshold compound nuclear resonance in ^{118}Sn. Soon after, an experiment at Dubna saw a close to ten percent transmission asymmetry in the 0.7 eV resonance in ^{139}La. The relative strength of the weak and strong interactions is 10^{-7}, so this corresponds to a remarkable million fold enhancement in the size of the effect [7]. Similar enhancements were soon predicted for T-violating interactions using spin-polarized targets, but the experiments have proved to be much more challenging. To date only one set of measurements has been carried out, at TUNL using 6-MeV neutron beams and a spin aligned ^{165}Ho target [8].

Formally, neutron transmission experiments search for symmetry violating terms in the neutron-nucleus forward scattering amplitude $f_{el}(0)$. Heuristic considerations suggest that, for a polarized or aligned target, the dependence of $f_{el}(0)$ on the target spin $\mathbf{I}$ and the neutron spin $\mathbf{s}$ should be given by an expression of the form ([9] and references therein)

$$f_{el}(0) = f_0 + f_M \mathbf{s} \cdot \mathbf{I} + f_P \mathbf{s} \cdot \mathbf{p} + f_T \mathbf{s} \cdot (\mathbf{p} \times \mathbf{I}), \quad (6)$$

where the complex coefficients f_0, f_M, f_P and f_T are, in general, polynomials in $\mathbf{I} \cdot \mathbf{p}$. The symmetry terms of particular interest are:

- $\mathbf{s} \cdot \mathbf{p}$ – the P-odd T-even helicity term, with a polarized beam and an unpolarized target.
- $\mathbf{s} \cdot (\mathbf{I} \times \mathbf{p})$ – the P-odd, T-odd triple correlation (TC) with a polarized beam and a vector-polarized target with spin $\mathbf{I}$.
- $\mathbf{s} \cdot (\mathbf{I} \times \mathbf{p})(\mathbf{I} \cdot \mathbf{p})$ – the P-even, T-odd fivefold correlation (FC), with a polarized beam and a spin aligned target (tensor polarization).

In general the scattering amplitudes have real and imaginary parts, which respectively cause precession and attenuation, and there are three directions involved: $\mathbf{p}$, $\mathbf{I}$, and $\mathbf{p} \times \mathbf{I}$. All terms can be acting simultaneously, including symmetry conserving terms like the strong interaction spin-spin term $\mathbf{s} \cdot \mathbf{I}$.

The resulting propagation of the beam will be complicated. It is conveniently handled using the language of neutron optics, valid even for high energy neutron beams as we are considering strictly forward scattering. The amplitude $f_{el}(0)$ and the refractive index n are 2 by 2 matrices in the spin space of the neutron, related by:

$$n = 1 + (2\pi/k^2)\rho f_{el}(0). \tag{7}$$

Spinor transport (precession and attenuation) through a thickness z of target material is described by $\chi_f = M\chi_i$ with

$$M = \exp[i(n-1)kz] \tag{8}$$

The time reversal tests TC and FC probe the symmetry of the S-matrix: $S^J_{LS,L'S'} = S^J_{L'S',LS}$ in the channel spin representation for example. The partial amplitudes $f_{kK\lambda}$ [related to the $f_{el}(0)$ by (58)] are linear combinations of the anti-symmetric part of S when $k+K+\lambda$ is odd. For TC experiments, mixing of s- and p-wave amplitudes is required and the expectation is that the measurement will be carried out where parity violation is large. For FC experiments, mixing of s- and d-waves, or two separate p-waves is required.

An example of ^{232}Th parity violation data from TRIPLE collaboration work at Los Alamos is discussed in Mitchell et al [10]. Multiple cases of parity violation are seen where strong s-wave resonances mix into nearby weak p-wave resonances with the same spin. The helicity dependent p-wave resonance cross sections are given by $\sigma^\pm = \sigma_0(1 \pm P)$. Due to the complexity of the compound nuclear wave functions, the quantity P varies stochastically. In a two level mixing approximation: $P = (2M/\Delta E)\sqrt{\Gamma^s_n/\Gamma^p_n}$ where M, the matrix element of the weak interaction Hamiltonian between the compound nuclear states, is expected to be a Gaussian distributed random variable. The energy separation ΔE and the neutron widths $\Gamma^{s,p}_n$ of the s and p states are in general known, so a root mean square value M_{RMS} can be obtained if multiple cases are observed. In principle, M_{RMS} reflects the strength of the meson coupling constants for the strangeness conserving non-leptonic part of the weak interaction: $H^{NL}_{wk}(\Delta S = 0)$. These constants can only be studied in nuclear processes. The two most important are H^0_ρ the strength of the isoscalar ρ coupling, and H^1_π, the strength of the isovector pion coupling. These constants are expected to be of order 10^{-6} but are not well determined experimentally, particularly H^1_π for which anapole moment measurements and ^{18}F circular polarization studies give conflicting results. Considerable theoretical work has been carried out with a view to relating M_{RMS} to these coupling constants. The problem is exceptionally challenging because it involves linking the quark sector, the meson sector, and the nuclear sector in a complicated many-body system.

In contrast to parity violation where the underlying symmetry violating interaction is known, there is no accepted theory of time reversal violation. P-odd TRV can arise within the standard model but turns out to be second order weak, and therefore unobservably small. This is true for the neutron edm within the standard model also. In fact, a search for the TC term is analogous to a search for a neutron edm since a polarized target $\boldsymbol{I}$ represents an effective magnetic field $\boldsymbol{B}$, and the resulting motional electric field $\boldsymbol{E} \sim (\boldsymbol{B} \times \boldsymbol{p})$ acts on the neutron to probe the presence of the same $\boldsymbol{s} \cdot \boldsymbol{E}$ interaction. Pion exchange is expected to dominate P-odd TRV, and TC experiments are parametrized in terms of a T violating coupling constant g_π^T which from the neutron edm bound is expected to be $< 2 \times 10^{-11}$.

P-even TRV does not occur in first order in the standard model, and its absence is a general feature of gauge theories with elementary quarks. The pion does not contribute to P-even TRV, and the longest range contribution is due to ρ exchange. Experiments are parametrized by the ratio of the T-violating to T-conserving ρ couplings: g_ρ^T. The best bounds to date are from quite different experiments:

- FC studies of 6-MeV polarized neutron scattering from nuclear spin aligned ^{165}Ho at TUNL: $g_\rho^T < 6 \times 10^{-2}$
- The ^{199}Hg edm bound: $g_\rho^T < 1 \times 10^{-2}$
- Charge symmetry breaking studies of 200-MeV $\vec{n} - \vec{p}$ scattering at TRIUMF: $g_\rho^T < 0.7 \times 10^{-2}$
- γ-decay in ^{57}Fe at CALTECH: $g_\rho^T < 6 \times 10^{-2}$

The FC measurements at TUNL used a holmium target in the form of a cylindrical single crystal with the crystal symmetry axis pointing radially out. At 100 mK the nuclei are almost completely aligned due to the internal hyperfine coupling field. The cylinder can be rotated, thus varying θ the angle between the holmium crystal axis $\boldsymbol{I}$ and the beam direction $\boldsymbol{p}$. The deformation effect signal associated with the large ($\beta_2 \sim 0.3$) prolate deformation of ^{165}Ho shows up as a $P_2(\cos\theta)$ modulation of the transmission asymmetry, confirming the nuclear alignment. The FC effect varies as $\sin 2\theta$. By also flipping the neutron spin rapidly, a clean double modulation signal is obtained, allowing a 10^{-6} bound to be set on the FC signal amplitude. The reduced sensitivity to g_ρ^T is due in part to the $1/A$ effect where only the last valence nucleon truly contributes to the signal. The experiments with aligned ^{2}H planned at COSY will potentially have greater sensitivity.

Resonance tests have the potential to improve the bounds for both P-odd and P-even T-violation. Difficult experimental problems have to be overcome first however.

For the TC, the problems lie in a) accurately accounting for the effect of so-called sequential interaction terms, and b) identifying a sufficient number of cases in polarizable targets where parity violation is large. The effect of the TC term is equivalent to a "strong" rotation due to the spin-spin interaction term (pseudomagnetism) $\boldsymbol{s} \cdot \boldsymbol{I}$ followed by "weak" attenuation due to the helicity interaction term $\boldsymbol{s} \cdot \boldsymbol{p}$. The action of these sequential terms is quadratic in the target thickness, in contrast to the linear dependence of the TC term, and they can also be detected by a polarizer-analyzer combination.

Nevertheless, the expectation is that the search for a TC term will be carried out where parity violation is ten percent or more, and exceptional alignment of the apparatus will be needed to carry out a competitive measurement [9].

Even then it will not in general be sufficient to measure just a single case. As for parity violation, the observables are stochastic and a null bound in one resonance may just be due to the unlucky fact that the matrix element is zero for that case. The observable of interest is a ratio of the TC signal to the helicity signal: in terms of the partial cross sections $\sigma_{kK\lambda}$ defined in (44), $R_{TP} \equiv TC/PV = \sigma_{111}/\sigma_{101}$. This ratio is a function of two stochastic matrix elements, and one unknown channel spin mixing ratio characterizing the neutron partial width amplitudes. If the measurements yield only null bounds, we have shown [11] a minimum of four cases will be needed to set a bound on the T-violating coupling. We estimate that measurements at the 10^{-6} level on resonances with ten percent parity violation will be needed to be competitive with the edm bounds. Of course a single non-zero measurement will be a very exciting result regardless of the issue of null bounds. Experiments are planned at KEK and in Dubna using frozen spin polarized lanthanum targets.

For the FC, the problem lies in identifying suitable pairs of resonances with the same J^π. The two possibilities are s/d or p/p mixing. Fortunately, there is a strong interaction signature – the deformation effect – which makes the s/d measurement especially attractive in ^{165}Ho. The deformation effect arises from the term $(\boldsymbol{I} \cdot \boldsymbol{p})^2$ and is due to interference between the s- and d-wave amplitudes. In contrast to the (unmeasurably small) d-wave cross section, which scales with the d-wave width, the deformation effect cross section scales with the d-wave amplitude. It is an unambiguous signal of the presence of a d-wave component [11]. The s-wave resonance spacing is only 4 eV, and hundreds of resonances are known up to one keV or more. A nearby strong s-wave resonance with the same J^π is practically guaranteed by this high level density. As a result, multiple cases can be studied. The prospects for improving the bounds in the FC correlation are excellent. An additional advantage of the FC experiment is that the sequential interaction problem is absent because the two terms that contribute are both P-odd: "weak" rotation due to the helicity term $\boldsymbol{s} \cdot \boldsymbol{p}$ followed by "weak" attenuation due to the deformation effect helicity interaction term $(\boldsymbol{s} \cdot \boldsymbol{I})(\boldsymbol{I} \cdot \boldsymbol{p})$. As a result, the measurement can be carried out without an analyzer. The observable of interest is a ratio of the FC signal to the deformation effect signal: $R_{FD} \equiv FC/DE = \sigma_{122}/\sigma_{022}$. We estimate that twelve ten-percent measurements of the FC effect will match the csb bound on g_ρ^T. Additional cases, would lead to the tightest bound of any experiment.

3 Total Cross Section and Forward Elastic Scattering Amplitude for Arbitrary States of Polarization

It is an embarrassing fact that despite the substantial reliance of analyses of transmission tests on the decomposition of the forward elastic scattering amplitude in (6), there appears to be no complete formal derivation of the result in the literature. Instead, there are several presentations of a related decomposition of the total cross section (see [12] and the references therein). When specialized to the case of polarization states of the beam and target with exact axial symmetry, this latter decomposition yields results which are consistent with (6). The purpose of this part of the paper is twofold. First, to give a complete and careful derivation of the decomposition of the total cross section stated in [12]. In view of the significance which has been attached to this result and the length of its derivation, we feel that an independent check is justified. We have attempted to pick a path through the sometimes daunting angular momentum calculations which makes each major step explicit. The second goal is to obtain for the first time a similar decomposition for the full forward elastic scattering amplitude. Our result is given in full generality in (58). It, like the decomposition of the total cross section, rests only on the assumption that the dynamics of the projectile-target system are conservative and invariant under spatial translations and rotations. The specialization of (58) to specific choices of projectile spin s and target spin I is a little tedious but, for the sake of comparison with (6), has been carried through for the case $s = 1/2 = I$.

3.1 Generalization of the Optical Theorem to Include Spin Degrees of Freedom

In the remaining sections we develop the formalism for polarized neutron transmission from first principles. We have included a number of exercises which extend the results and fill in details of the derivations.

For a projectile (of spin s) and target (of spin I) with initial (mixed) spin states described initially by the (statistical) spin space density matrix $\rho_i(s, I)$ of rank $(2s + 1)(2I + 1)$, the optical theorem generalizes to [13]

$$\sigma_{\text{total}} = \frac{4\pi}{k_i} \text{Im} \left\{ \text{Tr} \left[\rho_i(s, I) f_{\text{el}}(0) \right] \right\}, \tag{9}$$

where k_i is the magnitude of the initial relative wave vector (in the center of mass frame), $f_{\text{el}}(0)$ is the forward elastic scattering amplitude generalized to accommodate changes in spin degrees of freedom (its relation to reduced transition matrix elements is given below) and the trace is evaluated in the $(2s + 1)(2I + 1)$-dimensional product spin space of the projectile and target. The derivation of (9) is straightforward (once a suitable notation has been decided upon), but is worth repeating here because it brings out the fact that the result is an *exact* consequence of the unitarity of the S-matrix.

Unitarity of the S-matrix implies that reduced T-matrix elements on the energy-momemtum shell satisfy the relation (see, for example, Sect. 15.6 of [14])

$$\mathrm{i}[T_{ba} - T^*_{ab}] = 2\pi \sum_{\text{All } c} \delta(E_a - E_c)\delta(\boldsymbol{P}_a - \boldsymbol{P}_c) T^*_{cb} T_{ca}, \tag{10}$$

where the indices $a, b, c, \ldots$ denote schematically the various reaction channels. (A reaction channel is specified by the nature of the fragmentation of the reactants and by the state vector for the system of fragments when they are at infinite relative separations.) On the other hand, the total cross-section $\sigma_{a,\text{total}}$ for transitions from a certain initial (two-body) channel a to all other channels c is related to reduced T-matrix elements by (see, for example, Sect. 15.2 of [14])

$$\sigma_{a,\text{total}} = \frac{(2\pi)^4}{\hbar v_a} \sum_{c \neq a} \delta(E_a - E_c)\delta(\boldsymbol{P}_a - \boldsymbol{P}_c) |T_{ca}|^2, \tag{11}$$

where v_a is the magnitude of the relative velocity in channel a, E_a is the total energy in this channel and $\boldsymbol{P}_a$ the total (or center-of-mass) momentum. [We assume that the cartesian components of the initial relative wavevector $\boldsymbol{k}_a$ are among the quantum numbers specifying channel a. In addition, the plane wave states $|\boldsymbol{k}\rangle$ are assumed to be normalized so that $\langle \boldsymbol{k}' | \boldsymbol{k} \rangle = \delta(\boldsymbol{k}' - \boldsymbol{k})$.] The generalized summations in (10) and (11) include integration over all directions of motion of the (infinitely separated) fragments in a channel. The contribution excluded in the summation in (11) [but retained in (10)] corresponds to elastic scattering *exactly* in the forward direction. Ignoring this contribution of measure zero, we may substitute for the sum in (11) using the specialization of (10) to $a = b$ to obtain

$$\sigma_{a,\text{total}} = -4\pi \frac{(2\pi)^2}{\hbar v_a} \operatorname{Im}\{T_{aa}\}. \tag{12}$$

Introducing the forward elastic scattering amplitude (for channel a)

$$f_{a,\text{el}}(0) \equiv -(2\pi)^2 k_a \frac{\mathrm{d}k_a}{\mathrm{d}E_a} T_{aa} \tag{13}$$

into (12) and invoking the (relativistic) relation $\hbar v_a = \mathrm{d}E_a/\mathrm{d}k_a$, we recover the standard form of the (generalized) optical theorem for *pure* states (see, for example, Sect. 15.6 of [14]).

To extend the optical theorem to mixed spin states of the initial projectile-target system, it is helpful to adopt a somewhat more explicit notation for the relevant initial two-body reaction channels, namely $a = (\alpha_a, \boldsymbol{k}_a, m_a, M_a)$, where m_a is the z-component of the spin of the projectile, M_a is the z-component of the spin of the target, $\boldsymbol{k}_a$ is their initial relative wavevector and α_a denotes any further information required to specify channel the a.

(In what follows, we assume that $\boldsymbol{k}_a$ and α_a are fixed uniquely in the initial projectile-target system.)

We consider first the special case in which the initial spin space density matrix $\rho_i(s, I)$ of the system is diagonal. Then, in view of the probablistic interpretation of a diagonal entry $[\rho_i(s, I)]_{m_a M_a}$ (as the fraction of projectile-target pairs for which the z-component of the projectile spin is m_a and the z-component of the target spin is M_a) and of the cross-section $\sigma_{a,\text{total}}$ (as the rate of transitions from channel a to all other channels), the total cross section is given by the incoherent sum

$$\sigma_{\alpha_a \boldsymbol{k}_a,\text{total}} = \sum_{m_a M_a} [\rho_i(s, I)]_{m_a M_a} \sigma_{(\alpha_a, \boldsymbol{k}_a, m_a, M_a),\text{total}} \tag{14}$$

$$= -4\pi \frac{(2\pi)^2}{\hbar v_a} \text{Im} \left\{ \sum_{m_a, M_a} [\rho_i(s, I)]_{m_a M_a} T_{(\alpha_a, \boldsymbol{k}_a, m_a, M_a)(\alpha_a, \boldsymbol{k}_a, m_a, M_a)} \right\},$$

where we have substituted for $\sigma_{(\alpha_a, \boldsymbol{k}_a, m_a, M_a),\text{total}}$ using (12) and exploited the reality of the density matrix eigenvalues $[\rho_i(s, I)]_{m_a M_a}$ to postpone the calculation of the imaginary part until the summation over m_a and M_a has been performed.

If the initial spin space density matrix $\rho_i(s, I)$ is *not* diagonal, we can always find a new (orthonormal) spin space basis $\{|\mu_a\rangle\}$ such that it *is* diagonal (because $\rho_i(s, I)$ is hermitian). Under the unitary transformation

$$|\mu_a\rangle = \sum_{m_a M_a} U_{\mu_a, m_a M_a} |m_a M_a\rangle$$

relating the new basis to the old orthonormal basis $\{|m_a M_a\rangle\}$, the T-matrix elements $T_{ca} = T_{c(\alpha_a, \boldsymbol{k}_a, m_a, M_a)}$ in (11) transform to

$$T_{c(\alpha_a, \boldsymbol{k}_a, \mu_a)} = \sum_{m_a M_a} U_{\mu_a, m_a M_a} T_{c(\alpha_a, \boldsymbol{k}_a, m_a, M_a)}.$$

In this new basis, we can repeat the analysis of the previous paragraph (with the index μ_a replacing the indices m_a and M_a) to obtain the general result

$$\sigma_{\alpha_a \boldsymbol{k}_a,\text{total}} = -4\pi \frac{(2\pi)^2}{\hbar v_a} \text{Im} \left\{ \sum_{\mu_a} [\rho_i(s, I)]_{\mu_a} T_{(\alpha_a, \boldsymbol{k}_a, \mu_a)(\alpha_a, \boldsymbol{k}_a, \mu_a)} \right\}. \tag{15}$$

Reference to the particular choice of spin space basis in (15) can be eliminated by noting that, because it is the eigenbasis of the reduced density matrix $\rho_i(s, I)$, the sum over μ_a is a trace (over spin space), and as such its value is unchanged under unitary changes of spin space basis. In the original product basis $\{|m_a M_a\rangle\}$, this trace is given by the sum

$$\sum_{m_a M_a, m'_a M'_a} [\rho_i(s, I)]_{m_a M_a, m'_a M'_a} T_{(\alpha_a, \boldsymbol{k}_a, m'_a M'_a)(\alpha_a, \boldsymbol{k}_a, m_a M_a)}.$$

Thus, defining the matrix-valued forward elastic scattering amplitude $f_{\alpha_a \boldsymbol{k}_a,\mathrm{el}}(0)$ with elements (in the product basis for the projectile-target spin space)

$$[f_{\alpha_a \boldsymbol{k}_a,\mathrm{el}}(0)]_{m'_a M'_a, m_a M_a} \equiv -(2\pi)^2 k_a \frac{\mathrm{d}k_a}{\mathrm{d}E_a} T_{(\alpha_a,\boldsymbol{k}_a,m'_a M'_a)(\alpha_a,\boldsymbol{k}_a,m_a M_a)}, \quad (16)$$

we may write

$$\sigma_{\alpha_a \boldsymbol{k}_a,\mathrm{total}} = \frac{4\pi}{k_a} \mathrm{Im}\left\{\mathrm{Tr}\left[\rho_i(s,I) f_{\alpha_a \boldsymbol{k}_a,\mathrm{el}}(0)\right]\right\}, \quad (17)$$

which is a more notationally precise version of (9).

In working with (17) [or (9)], we are, of course, not restricted to using the product basis $\{|m_a M_a\rangle\}$ for the projectile-target spin space. In fact, in deriving the decomposition of the total cross-section presented in [12], we find it convenient to invoke the spin-coupled basis $\{|S_a \nu_a\rangle\}$ (defined explicitly in Sect. 3.2). Matrix elements of the statistical density matrix in this basis are derived in Sect. 3.2 and the partial wave decomposition of T-matrix elements $T_{(\alpha_a,\boldsymbol{k}_a,S'_a,\nu'_a)(\alpha_a,\boldsymbol{k}_a,S_a,\nu_a)}$ is discussed in Sect. 3.3.

Exercise: [Choice of plane wave normalization.]
What form do (11) to (13) take if the plane wave normalization

$$\langle \boldsymbol{k}' | \boldsymbol{k} \rangle = (2\pi)^3 \delta(\boldsymbol{k}' - \boldsymbol{k})$$

is adopted? How does one reconcile these changes with the fact that cross-sections are independent of the choice of plane wave normalization?

3.2 Statistical Density Matrix for the Projectile Target Spin Space

In working with the statistical density operator $\widehat{\rho}(s,I)$ describing the mixed spin states of the system comprising a projectile (of spin s) and a target (of spin I), it is convenient to adopt a representation in terms of the statistical tensors $t_k(s)$ and $t_K(I)$ of the projectile and target, respectively. In (19) and (21) below, we present expressions of this kind for two obvious choices of basis for the $(2s+1)(2I+1)$-dimensional spin space of the projectile-target system, namely the direct product basis and the spin-coupled basis. It is (21) which we invoke subsequently (in Sect. 3.4); (19) is needed only in our derivation of (21). The relation between the statistical spin space density operator of a *single* species of particle and statistical tensors is reviewed in the Appendix.

In our notation, the *direct product basis* for the spin space of the projectile-target system has members

$$|m_s M_I\rangle \equiv |s m_s\rangle |I M_I\rangle,$$

where $m_s = -s, -s+1, \ldots, s$ and $M_I = -I, -I+1, \ldots, I$, and the *spin-coupled basis* has members

$$|S\nu\rangle \equiv \sum_{m_s M_I} \langle sm_s I M_I | S\nu \rangle | m_s M_I \rangle, \tag{18}$$

where $S = |I-s|, |I-s|+1, \ldots, I+s$ and $\nu = -S, -S+1, \ldots, S$. Above, $|sm_s\rangle$ and $|IM_I\rangle$ denote simultaneous eigenkets of the sets of spin operators $\{\hat{s}^2, \hat{s}_z\}$ and $\{\hat{I}^2, \hat{I}_z\}$, respectively, and the $\langle sm_s I M_I | S\nu \rangle$'s are Clebsch-Gordan (or vector addition) coefficients. (In [15], for example, the notation used the Clebsch-Gordan coefficient $\langle j_1 m_1 j_2 m_2 | jm \rangle$ is $C^{jm}_{j_1 m_1 j_2 m_2}$. Other commonly used notations are given on p. 268 of [15].)

In the direct product basis (invoked in Sect. 3.1), $\hat{\rho}(s, I)$ has matrix elements which factor as

$$[\rho(s, I)]_{m'_s M'_I, m_s M_I} = [\rho(s)]_{m'_s m_s} [\rho(I)]_{M'_I M_I}$$

because the beam and target are prepared separately (and so the statistical properties of their spins are independent). Using (72) to expand $[\rho(s)]_{m'_s m_s}$ and $[\rho(I)]_{M'_I M_I}$ in terms of the statistical tensors $t_k(s)$ ($k = 0, 1, \ldots, 2s$) and $t_K(I)$ ($K = 0, 1, \ldots, 2I$), respectively, and decomposing components $t_{kq}(s) t_{KQ}(I)$ of the direct product of a $t_k(s)$ with a $t_K(I)$ in terms of components of irreducible tensor products of the $t_k(s)$ with the $t_K(I)$, we find that

$$[\rho(s, I)]_{m'_s M'_I, m_s M_I} = \frac{1}{(2s+1)(2I+1)} \cdot \tag{19}$$

$$\sum_{\substack{kK\lambda \\ qQ\mu}} \hat{k}\hat{K} \langle sm'_s kq | sm_s \rangle \langle IM'_I KQ | IM_I \rangle \langle kqKQ | \lambda\mu \rangle \{t_k(s) \otimes t_K(I)\}_{\lambda\mu},$$

where, following the notation of [15] (Sect. 3.1.7), $\{t_k(s) \otimes t_K(I)\}_{\lambda\mu}$ denotes the μth component of the rank λ irreducible tensor product of the statistical tensors $t_k(s)$ and $t_K(I)$, i.e.

$$\{t_k(s) \otimes t_K(I)\}_{\lambda\mu} \equiv \sum_{qQ} \langle kqKQ | \lambda\mu \rangle t_{kq}(s) t_{KQ}(I),$$

and $\hat{k} \equiv (2k+1)^{1/2}$.

¿From (18) (and the real-valuedness of the Clebsch-Gordan coefficients), matrix elements of $\hat{\rho}(s, I)$ in the spin-coupled basis $[\rho(s, I)]_{S'\nu', S\nu}$ are related to those in the product basis by the similarity transformation

$$[\rho(s, I)]_{S'\nu', S\nu} = \sum_{\substack{m_s M_I \\ m'_s M'_I}} \langle sm'_s I M'_I | S'\nu' \rangle \langle sm_s I M_I | S\nu \rangle [\rho(s, I)]_{m'_s M'_I, m_s M_I}.$$

$$\tag{20}$$

Substituting for the matrix elements $[\rho(s,I)]_{m'_s M'_I, m_s M_I}$ in (20) using (19), the summations over m_s, m'_s, M_I and M'_I can be eliminated (using, for example, (8.7.4(26)) in [15]) and then the summations over q and Q (using the unitarity of the Clebsch–Gordon coefficients – see, for example, (8.1.1.(8)) of [15]) to obtain

$$[\rho(s,I)]_{S'\nu',S\nu} = \frac{\widehat{S}'}{\widehat{s}\widehat{I}} \sum_{\substack{kK\lambda \\ \mu}} \widehat{k}\,\widehat{K}\widehat{\lambda}\, \langle S'\nu'\lambda\mu|S\nu\rangle \begin{Bmatrix} S & s & I \\ S' & s & I \\ \lambda & k & K \end{Bmatrix} \{t_k(s) \otimes t_K(I)\}_{\lambda\mu}, \tag{21}$$

where the array in the large curly braces is a Wigner $9j$ symbol (see, for example, Chap. 10 of [15]).

Exercise: [Properties of (21).]
Confirm that (21) implies that $\rho(s,I)$ is a hermitian matrix of unit trace [provided $t_{00}(s) = 1 = t_{00}(I)$] and reduces to (72) when either s or I is zero.

3.3 Partial Wave Expansions on the Energy Momentum Shell

Our interest is in the generalized partial-wave expansion of the reduced T-matrix elements $T_{(\alpha_a, \boldsymbol{k}_a, S'_a \nu'_a)(\alpha_a, \boldsymbol{k}_a, S_a \nu_a)}$ (refered to the end of Sect. 3.1). We assume that we are working in the center-of-mass frame of the projectile-target system so that it is possible to characterize the states of definite total linear momentum $\widehat{\boldsymbol{P}}$ needed to define (on-shell) reduced T-matrix elements as states of good total angular momentum $\widehat{\boldsymbol{J}}$ as well (in general, $[\widehat{\boldsymbol{P}}, \widehat{\boldsymbol{J}}] \neq 0$, but in the center-of-mass frame $\widehat{\boldsymbol{P}} = 0$). Our primary result is (29). The derivations of (26) and (34) below are standard and are included for completeness and to establish our notation.

In this context, it is appropriate to transform from the characterization of plane wave states in terms of the cartesian components $\{k_x, k_y, k_z\}$ of the relative wavevector $\boldsymbol{k}$ (used implicitly in Sect. 3.1) to a characterization using the center-of-mass energy E_k [determined by the magnitude k of $\boldsymbol{k}$ through a dispersion relation $E_k = E(k)$] and the unit vector $\boldsymbol{n}$ in the direction of $\boldsymbol{k}$. We take the state vector

$$|E_k \boldsymbol{n}\rangle = k\sqrt{\frac{dk}{dE}}|\{k_x, k_y, k_z\}\rangle, \tag{22}$$

where the constant of proportionality is chosen so that

$$\langle E_{k'} \boldsymbol{n}' | E_k \boldsymbol{n}\rangle = \delta(E_{k'} - E_k)\delta(\boldsymbol{n}' - \boldsymbol{n})$$

since we assume that $\langle k'_x, k'_y, k'_z | k_x, k_y, k_z \rangle = \delta(\boldsymbol{k}' - \boldsymbol{k})$.
Using (22), the elastic reduced T-matrix elements

$$T_{(\alpha_a, \boldsymbol{k}_a, S'_a \nu'_a)(\alpha_a, \boldsymbol{k}_a, S_a \nu_a)} = \frac{1}{k_a^2}\frac{dE_a}{dk_a}\langle \alpha_a E_a\, \boldsymbol{n}_a S'_a \nu'_a | T | \alpha_a E_a \boldsymbol{n}_a S_a \nu_a \rangle. \tag{23}$$

The partial wave expansion relates these matrix elements to matrix elements for another choice of basis for the subspace of fixed α_a and E_a (corresponding to another choice of complete commuting set of observables), namely the simultaneous eigenstates $|\alpha_a E_a L_a S_a J_a M_a\rangle$ of the total angular momentum squared $\hat{J}^2$ of the projectile-target system in the center-of-mass frame [eigenvalue $J_a(J_a+1)\hbar^2$], its z-component $\hat{J}_z$ (eigenvalue $M_a\hbar$), the relative orbital angular momentum squared $\hat{L}^2$ [eigenvalue $L_a(L_a+1)\hbar^2$] and the channel spin squared $\hat{S}^2$ [eigenvalue $S_a(S_a+1)\hbar^2$]. The unitary transformation between these two bases has matrix elements

$$\langle \alpha_a E_a \boldsymbol{n}_a \, S_a \, \nu_a | \alpha_a E_a L_a S_a' J_a M_a \rangle \quad (24)$$
$$= \sum_{\substack{LM \\ S\nu}} \langle \alpha_a E_a \boldsymbol{n}_a S_a \nu_a | \alpha_a E_a LMS\nu \rangle \langle \alpha_a E_a LMS\nu | \alpha_a E_a L_a S_a' J_a M_a \rangle$$
$$= \delta_{S_a S_a'} \langle L_a(M_a - \nu_a) S_a \nu_a | J_a M_a \rangle Y_{L_a(M_a-\nu_a)}(\boldsymbol{n}_a),$$

where, in the intermediate summmation above, we have invoked the product basis $|\alpha_a E_a LmS\nu\rangle$ comprising states of definite relative orbital angular momentum $\hat{\boldsymbol{L}}$ and channel spin $\hat{\boldsymbol{S}}$ and $Y_{LM}(\boldsymbol{n}) = \langle \boldsymbol{n}| LM\rangle$ is a spherical harmonic evaluated at the polar and azimuthal angles corresponding to the direction of the unit vector $\boldsymbol{n}$. [It is a simple matter to check that the matrix elements given in (24) do indeed constitute matrix elements of a unitary matrix.] The advantage in working with states of definite total angular momentum $|\alpha_a E_a L_a S_a J_a M_a\rangle$ is that the rotational invariance of the full Hamiltonian implies that the corresponding T-matrix elements must be diagonal in J_a and M_a and independent of M_a (see, for example, Sect. 16.7 in [14]), i.e.

$$\langle \alpha_a E_a L_a' S_a' J_a' M_a' | T | \alpha_a E_a L_a S_a J_a M_a \rangle = \delta_{J_a' J_a} \delta_{M_a' M_a} T^{\alpha_a J_a}_{L_a' S_a', L_a S_a}(E_a). \quad (25)$$

Thus, introducing $\mathcal{U}(L_a S_a' J_a M_a, \boldsymbol{n}_a S_a \nu_a) \equiv \langle \alpha_a E_a L_a S_a' J_a M_a | \alpha_a E_a \boldsymbol{n}_a S_a \nu_a \rangle$,

$$\langle \alpha_a E_a \boldsymbol{n}_a S_a' \nu_a' | T | \alpha_a E_a \boldsymbol{n}_a S_a \nu_a \rangle \quad (26)$$
$$= \sum_{\substack{J_a M_a \\ L_a' L_a}} \mathcal{U}^*(L_a' S_a' J_a M_a, \boldsymbol{n}_a S_a' \nu_a') \mathcal{U}(L_a S_a J_a M_a, \boldsymbol{n}_a S_a \nu_a) T^{\alpha_a J_a}_{L_a' S_a', L_a S_a}(E_a),$$

which, together with (24), amounts to the specialization of the standard partial wave expansion of the T-matrix for elastic two-body reactions (see, for example, (18.101) in [14]) to forward angle scattering.

The geometrical character of the T-matrix elements in (26) can be clarified by expanding the product of spherical harmonics implicit in the sum $\sum_{M_a} \mathcal{U}^*(L_a' S_a' J_a M_a, \boldsymbol{n}_a S_a' \nu_a') \mathcal{U}(L_a S_a J_a M_a, \boldsymbol{n}_a S_a \nu_a)$ in terms of components of the irreducible tensor products $\{Y_{L_a'}(\boldsymbol{n}_a) \otimes Y_{L_a}(\boldsymbol{n}_a)\}_L$. Invoking the decomposition

$$Y_{l'm'}(\theta', \phi') Y_{lm}^*(\theta, \phi) = (-1)^m \sum_{LM} \langle l'm'l(-m)| LM \rangle \{Y_{l'}(\theta', \phi') \otimes Y_l(\theta, \phi)\}_{LM}$$

$$= \frac{(-1)^l}{\widehat{l'}} \sum_{LM} \widehat{L} \langle lmLM|l'm'\rangle \{Y_{l'}(\theta',\phi') \otimes Y_l(\theta,\phi)\}_{LM}, \qquad (27)$$

it is possible [using, for example, (8.7.3(13)) in [15] followed by the substitution of $\langle LMS'_a\nu'_a|S_a\nu_a\rangle$ by $(-1)^{L+S'_a-S_a}\langle S'_a\nu'_a LM|S_a\nu_a\rangle$] to transcribe the summation $\sum_{M_a} \mathcal{U}^*(L'_a S'_a J_a M_a, \boldsymbol{n}'_a S'_a \nu'_a) \mathcal{U}(L_a S_a J_a M_a, \boldsymbol{n}_a S_a \nu_a)$ as follows:

$$\sum_{M_a} \mathcal{U}^*(L'_a S'_a J_a M_a, \boldsymbol{n}'_a S'_a \nu'_a) \mathcal{U}(L_a S_a J_a M_a, \boldsymbol{n}_a S_a \nu_a) = (-1)^{J_a+S_a} \frac{(2J_a+1)}{\widehat{S_a}}$$

$$\sum_{LM} \widehat{L} \langle S'_a \nu'_a LM|S_a\nu_a\rangle \begin{Bmatrix} L'_a & L_a & L \\ S_a & S'_a & J_a \end{Bmatrix} \{Y_{L'_a}(\boldsymbol{n}'_a) \otimes Y_{L_a}(\boldsymbol{n}_a)\}_{LM}, \qquad (28)$$

where the array in the curly braces denotes a Wigner $6j$ symbol (see, for example, Chap. 9 of [15]). The generalization to distinct directions $\boldsymbol{n}_a$ and $\boldsymbol{n}'_a$ in (28) facilitates a more complete check of its consistency with the constraints implied by the unitarity of the $\mathcal{U}(L_a S_a J_a M_a, \boldsymbol{n}_a S_a \nu_a)$'s [see part (d) of Exercise below]. Setting $\boldsymbol{n}'_a = \boldsymbol{n}_a$ does not change the structure of the result in (28) in any essential way except that $\{Y_{L'_a}(\boldsymbol{n}_a) \otimes Y_{L_a}(\boldsymbol{n}_a)\}_{LM}$ may be replaced by its relation to the spherical harmonic $Y_{LM}(\boldsymbol{n}_a)$ (see, for example, (5.6.2(14)) in [15]). Accordingly,

$$\langle \alpha_a E_a \boldsymbol{n}_a S'_a \nu'_a | T | \alpha_a E_a \boldsymbol{n}_a S_a \nu_a \rangle = \qquad (29)$$

$$\frac{1}{4\pi \widehat{S_a}} \sum_{\substack{J_a L'_a L_a \\ LM}} \Big[(-1)^{J_a+S_a} (2J_a+1) \widehat{L'_a}\widehat{L_a}\widehat{L} \langle L'_a 0 L_a 0 | L0 \rangle \begin{Bmatrix} L'_a & L_a & L \\ S_a & S'_a & J_a \end{Bmatrix}$$

$$\langle S'_a \nu'_a LM | S_a \nu_a \rangle C_{LM}(\boldsymbol{n}_a) T^{\alpha_a J_a}_{L'_a S'_a, L_a S_a}(E_a) \Big],$$

where we have introduced the rescaled spherical harmonic

$$C_{lm}(\theta,\phi) \equiv \sqrt{\frac{4\pi}{2l+1}} Y_{lm}(\theta,\phi). \qquad (30)$$

[We choose to work with the C_{lm}'s because they appear naturally in statistical tensors for the axially symmetric polarization states considered below – see (77).]

The connection of the T-matrix elements $T^{\alpha_a J_a}_{L'_a S'_a, L_a S_a}$ with the S-matrix in the center-of-mass frame of the projectile-target system can be established by starting from the relation for two-body reactions (implied, for example, by (14.174) of [14])

$$\langle \alpha_b, \boldsymbol{k}_b, S_b\nu_b | S | \alpha_a, \boldsymbol{k}_a, S_a\nu_a \rangle = \qquad (31)$$
$$\delta_{\alpha_b\alpha_a}\delta(\boldsymbol{k}_b-\boldsymbol{k}_a)\delta_{S_bS_a}\delta_{\nu_b\nu_a} - 2\pi i\delta(E_b-E_a)T_{(\alpha_b,\boldsymbol{k}_b,S_b\nu_b)(\alpha_a,\boldsymbol{k}_a,S_a\nu_a)}.$$

With our choice of the states $|E_k\boldsymbol{n}\rangle$, the (off-shell) S-matrix elements

$$\langle \alpha_a, \boldsymbol{k}'_a, S'_a\nu'_a | S | \alpha_a, \boldsymbol{k}_a, S_a\nu_a \rangle = \qquad (32)$$

$$\frac{1}{k_a^2}\frac{dE_a}{dk_a}\langle \alpha_a E_a \boldsymbol{n}'_a S'_a \nu'_a | S | \alpha_a E_a \boldsymbol{n}_a S_a \nu_a \rangle \delta(E'_a - E_a).$$

Together, (23), (31) and (32) imply that the elastic S-matrix elements

$$\langle \alpha_a E_a \boldsymbol{n}'_a S'_a \nu'_a | S | \alpha_a E_a \boldsymbol{n}_a S_a \nu_a \rangle =$$
$$\delta(\boldsymbol{n}'_a - \boldsymbol{n}_a) \delta_{S'_a S_a} \delta_{\nu'_a \nu_a} - 2\pi\mathrm{i} \langle \alpha_a E_a \boldsymbol{n}'_a S'_a \nu'_a | T | \alpha_a E_a \boldsymbol{n}_a S_a \nu_a \rangle.$$

If we repeat the partial wave analysis outlined above and write the S-matrix elements with respect to states of total angular momentum $|\alpha_a E_a L_a S_a J_a M_a\rangle$ (which are again diagonal in the total angular momentum and its z-component and independent of the z-component) as

$$\langle \alpha_a E_a L'_a S'_a J'_a M'_a | S | \alpha_a E_a L_a S_a J_a M_a \rangle = \delta_{J'_a J_a} \delta_{M'_a M_a} S^{\alpha_a J_a}_{L'_a S'_a, L_a S_a}(E_a), \quad (33)$$

then we find that

$$S^{\alpha_a J_a}_{L'_a S'_a, L_a S_a}(E_a) = \delta_{L'_a L_a} \delta_{S'_a S_a} - 2\pi\mathrm{i} T^{\alpha_a J_a}_{L'_a S'_a, L_a S_a}(E_a). \quad (34)$$

An analogous relation holds between the T-matrix elements $T^{\alpha_a J_a}_{l'_a j'_a, l_a j_a}(E_a)$ considered below in the Exercise and the S-matrix elements $S^{\alpha_a J_a}_{l'_a j'_a, l_a j_a}(E_a)$. [In this exercise, we use lowercase letters for the eigenvalues of $\hat{L}^2$, i.e. we write an eigenvalue as $l_a(l_a + 1)\hbar^2$ not $L_a(L_a + 1)\hbar^2$.]

Exercise: [Partial wave expansion for a spin-orbit coupled basis]
An alternative partial wave expansion arises if one starts from the basis $\{|\alpha_a E_a \boldsymbol{n}_a m_a \widetilde{m}_a\rangle\}$ in which the spins of the projectile and target have definite z-components ($m_a \hbar$ and $\widetilde{m}_a \hbar$, respectively) and transforms to the basis $\{|\alpha_a E_a l_a j_a J_a M_a\rangle\}$ in which the angular momentum $\hat{\boldsymbol{j}} = \hat{\boldsymbol{L}} + \hat{\boldsymbol{s}}$ has a definite magnitude squared $[j_a(j_a + 1)\hbar^2]$.
a) Show that the coefficients in the expansion

$$|\alpha_a E_a \boldsymbol{n}_a m_a \widetilde{m}_a\rangle = \sum_{\substack{J_a M_a \\ l_a j_a}} U_{l_a j_a J_a M_a, \boldsymbol{n}_a m_a \widetilde{m}_a} |\alpha_a E_a l_a j_a J_a M_a\rangle$$

are given by

$$U_{l_a j_a J_a M_a, \boldsymbol{n}_a m_a \widetilde{m}_a} = \langle l_a(M_a - \widetilde{m}_a - m_a) s m_a | j_a(M_a - \widetilde{m}_a) \rangle$$
$$\langle j_a(M_a - \widetilde{m}_a) I \widetilde{m}_a | J_a M_a \rangle Y^*_{l_a(M_a - \widetilde{m}_a - m_a)}(\boldsymbol{n}_a)$$

and use this result to prove that these coefficients satisfy the unitarity relations

$$\sum_{\substack{JM \\ lj}} U_{ljJM, \boldsymbol{n}_1 m_1 \widetilde{m}_1} U^*_{ljJM, \boldsymbol{n}_2 m_2 \widetilde{m}_2} = \delta_{m_1 m_2} \delta_{\widetilde{m}_1 \widetilde{m}_2} \delta(\boldsymbol{n}_1 - \boldsymbol{n}_2) \quad (35)$$

$$\sum_{m\widetilde{m}} \int U_{l_1 j_1 J_1 M_1, \boldsymbol{n} m \widetilde{m}} U^*_{l_2 j_2 J_2 M_2, \boldsymbol{n} m \widetilde{m}} \, d\boldsymbol{n} = \delta_{l_1 l_2} \delta_{j_1 j_2} \delta_{J_1 J_2} \delta_{M_1 M_2}. \quad (36)$$

b) Observe that, in terms of the unitary matrix elements $U_{ljJM,\boldsymbol{n}m\widetilde{m}}$, we may write the partial wave decomposition of $\langle \alpha_a E_a \boldsymbol{n}_a m'_a \widetilde{m}'_a | T | \alpha_a E_a \boldsymbol{n}_a m_a \widetilde{m}_a \rangle$ as

$$\langle \alpha_a E_a \boldsymbol{n}_a m'_a \widetilde{m}'_a | T | \alpha_a E_a \boldsymbol{n}_a m_a \widetilde{m}_a \rangle = \tag{37}$$

$$\sum_{J_a} \sum_{l'_a j'_a l_a j_a} \sum_{M_a} U^*_{l'_a j'_a J_a M_a, \boldsymbol{n}_a m'_a \widetilde{m}'_a} U_{l_a j_a J_a M_a, \boldsymbol{n}_a m_a \widetilde{m}_a} T^{\alpha_a J_a}_{l'_a j'_a, l_a j_a}(E_a).$$

Show that, after expanding the product of the two spherical harmonics present in

$$\sum_{M_a} U^*_{l'_a j'_a J_a M_a, \boldsymbol{n}'_a m'_a \widetilde{m}'_a} U_{l_a j_a J_a M_a, \boldsymbol{n}_a m_a \widetilde{m}_a}$$

in terms of irreducible tensor products [as in (27)], the sum (which now involves a product of five Clebsch-Gordan coefficients) can be transformed to read

$$\sum_{M_a} U^*_{l'_a j'_a J_a M_a, \boldsymbol{n}'_a m'_a \widetilde{m}'_a} U_{l_a j_a J_a M_a, \boldsymbol{n}_a m_a \widetilde{m}_a} = \tag{38}$$

$$(-1)^{l_a}(2J_a+1)\frac{\widehat{j'_a}\widehat{j_a}}{\widehat{s}\widehat{I}} \sum_{\substack{kKL \\ qQM}} \Big[(-1)^{k+K} \widehat{k}^2 \widehat{K}^2 \langle sm'_a kq | sm_a \rangle \langle I\widetilde{m}'_a KQ | I\widetilde{m}_a \rangle$$

$$\langle kqKQ|LM\rangle W(J_a j_a IK; I j'_a) \begin{Bmatrix} l_a & s & j_a \\ L & k & K \\ l'_a & s & j'_a \end{Bmatrix} \{ Y_{l'_a}(\boldsymbol{n}'_a) \otimes Y_{l_a}(\boldsymbol{n}_a) \}_{LM} \Big],$$

where we have introduced the Racah coefficient $W(J_a j_a I K; I j'_a)$ (defined, for example, in (9.1.2.(11)) of [15]).

c) If the projectile is spinless ($s=0$), the overlap matrix element $U_{l_a j_a J_a M_a, \boldsymbol{n}_a m_a \widetilde{m}_a}$ in (38) coincides with one of the overlap matrix elements introduced in connection with (26), namely $\mathcal{U}(l_a I J_a M_a, \boldsymbol{n}_a I \widetilde{m}_a)$. Show that, consistent with this observation, (38) reduces to (28) with $S_a = I = S'_a$, $\nu_a = \widetilde{m}_a$, $\nu'_a = \widetilde{m}'_a$, $L_a = l_a$ and $L'_a = l'_a$.

d) Show that we can recover from (38) the unitarity relation (35) and

$$\sum_{m\widetilde{m}} \int U_{l_1 j_1 JM,\boldsymbol{n}m\widetilde{m}} U^*_{l_2 j_2 JM,\boldsymbol{n}m\widetilde{m}} \, d\boldsymbol{n} = (2J+1)\delta_{l_1 l_2}\delta_{j_1 j_2}$$

implied by (36). [Corresponding results expected for the $\mathcal{U}(LSJM, nS'\nu)$'s follow from (28).]

3.4 Decomposition of the Total Cross Section

We are now in a position to derive the decomposition of the total cross section stated in [12]. To this end, we use the spin-coupled basis $\{|S_a\nu_a\rangle\}$ (defined in Sect. 3.2) to evaluate the trace in (17). The overall structure which emerges is given in (42) below (tantamount to (2) in [12]). The form of this result is, in fact, independent of the choice of channel states but the relation of the associated cross section coefficients $A^{J_a}_{kK\lambda}(E_a)$ to S-matrix elements is not. An expression for the $A^{J_a}_{kK\lambda}(E_a)$'s appropriate to a spin-coupled channel basis [namely, (43) together with (40)] is a by-product of our derivation of (42). Thereafter, we obtain the $A^{J_a}_{kK\lambda}(E_a)$'s when spin-orbit coupled angular momenta are used to define channels [(46) and (48)]. Modulo a phase factor discussed below in connection with (49), our results are consistent with those of [12].

Expressing the forward elastic scattering amplitude in terms of the T-matrix [using the analogue of (16) for the spin-coupled basis] and then invoking (23), the trace in (17) reads

$$\text{Tr}\,[\rho_i(s,I)f_{\alpha_a \mathbf{k}_a,\text{el}}(0)] =$$
$$-\frac{(2\pi)^2}{k_a}\sum_{\substack{S_a \nu_a \\ S'_a \nu'_a}}[\rho_i(s,I)]_{S'_a \nu'_a, S_a \nu_a}\langle \alpha_a E_a\, \mathbf{n}_a S_a \nu_a | T | \alpha_a E_a \mathbf{n}_a S'_a \nu'_a \rangle.$$

Substituting for the elements of the density matrix $[\rho_i(s,I)]$ using (21) and the T-matrix elements using (29), the double summation over ν_a and ν'_a reduces to one of the form

$$\sum_{\nu'_a \nu_a} \langle S'_a \nu'_a \lambda\mu | S_a \nu_a \rangle \langle S_a \nu_a L M | S'_a \nu'_a \rangle = (-1)^{S'_a - S_a + \mu}\frac{\widehat{S'_a S_a}}{(2\lambda+1)}\delta_{\lambda L}\delta_{(-\mu)M}$$

so that the irreducible tensors $\{t_k(s)\otimes t_K(I)\}_\lambda$ [from (21)] and $C_L(\mathbf{n}_a)$ [from (29)] appear in the rotationally invariant combination

$$\sum_\mu (-1)^\mu \{t_k(s)\otimes t_K(I)\}_{\lambda\mu} C_{\lambda(-\mu)}(\mathbf{n}_a) = (-1)^\lambda \hat{\lambda}\{\{t_k(s)\otimes t_K(I)\}_\lambda \otimes C_\lambda(\mathbf{n}_a)\}_{00}.$$

Thus, we find (after an interchange of the dummy variables of summation S_a and S'_a) that the trace in (17) naturally decomposes as

$$\text{Tr}\,[\rho_i(s,I)f_{\alpha_a \mathbf{k}_a,\text{el}}(0)] = -\frac{\pi}{k_a}\sum_{kK\lambda}(-i)^{k+K+\lambda}\mathcal{S}_{kK\lambda}\sum_{J_a}(2J_a+1) \quad (39)$$
$$\sum_{L'_a S'_a L_a S_a} F^{(sI)}_{kK\lambda}(L'_a S'_a L_a S_a J_a) T^{\alpha_a J_a}_{L'_a S'_a, L_a S_a}(E_a),$$

where the real coefficients

$$F^{(sI)}_{kK\lambda}(L'S'LSJ) \equiv \quad (40)$$
$$(-1)^{J+S'+\lambda}\frac{\widehat{kK\lambda}}{\widehat{sI}}\widehat{S'}\widehat{S}\widehat{L'}\widehat{L}\,\langle L'0L0|\lambda 0\rangle \begin{Bmatrix} L' & L & \lambda \\ S & S' & J \end{Bmatrix}\begin{Bmatrix} S' & s & I \\ S & s & I \\ \lambda & k & K \end{Bmatrix}$$

and we have introduced the *polarization geometry* correlation factors

$$\mathcal{S}_{kK\lambda} \equiv i^{k+K+\lambda}\{\{t_k(s)\otimes t_K(I)\}_\lambda \otimes C_\lambda(\boldsymbol{n}_a)\}_{00} \tag{41}$$

$$= \frac{i^{k+K-\lambda}}{\hat{\lambda}} \sum_{\mu,q,Q} \langle kqKQ|\lambda\mu\rangle t_{kq}(s) t_{KQ}(I) C^*_{\lambda\mu}(\boldsymbol{n}_a).$$

The inclusion of the phase factor $i^{k+K+\lambda}$ in the definition of the $\mathcal{S}_{kK\lambda}$'s guarantees that they are also real-valued [which may be proved using (71) for the statistical tensors and the analogous result $C^*_{lm}(\theta,\phi) = (-1)^m C_{l(-m)}(\theta,\phi)$]. The normalization of the $\mathcal{S}_{kK\lambda}$'s is such that $\mathcal{S}_{000} = 1$.

Substituting (40) into (17) and invoking (34), we conclude that the total cross section

$$\sigma_{\alpha_a \boldsymbol{k}_a,\text{total}} = \frac{2\pi}{k_a^2} \sum_{kK\lambda} \mathcal{S}_{kK\lambda} \sum_{J_a} (2J_a+1) A^{J_a}_{kK\lambda}(E_a), \tag{42}$$

where the cross section coefficients

$$A^{J_a}_{kK\lambda}(E_a) \equiv \sum_{L'_a S'_a L_a S_a} F^{(sI)}_{kK\lambda}(L'_a S'_a L_a S_a J_a) \tag{43}$$

$$\text{Re}\left\{(-i)^{k+K+\lambda}[\delta_{L'_a L_a}\delta_{S'_a S_a} - S^{\alpha_a J_a}_{L'_a S'_a, L_a S_a}(E_a)]\right\}.$$

Equation (42) constitutes the desired decomposition of the total cross section. In [12], this decomposition is written in terms of the combinations

$$\sigma_{kK\lambda} = \frac{2\pi}{k_a^2} \sum_{J_a} (2J_a+1) A^{J_a}_{kK\lambda}(E_a), \tag{44}$$

which are termed *partial* cross sections.

To obtain the expression for the cross section coefficients $A^{J_a}_{kK\lambda}(E_a)$ in terms of S-matrix elements $S^{\alpha_a J_a}_{l'_a j'_a, l_a j_a}(E_a)$ for the spin-orbit coupled channel states $\{|\alpha_a E_a l_a j_a J_a M_a\rangle\}$ (we adopt the notation of the Exercise), we invoke the unitary transformation (in a subspace of fixed α_a, E_a, J_a and M_a) between this basis and the basis of spin-coupled states $\{|\alpha_a E_a L_a S_a J_a M_a\rangle\}$. The transformation has real matrix elements (see, for example, (9.1.1.(5)) in [15] and Exercise below)

$$\langle \alpha_a E_a l_a j_a J_a M_a | \alpha_a E_a L_a S_a J_a M_a\rangle = \tag{45}$$

$$(-1)^{s+I+J_a+l_a}\hat{j}_a\hat{S}_a \begin{Bmatrix} I & s & S_a \\ l_a & J_a & j_a \end{Bmatrix} \delta_{l_a L_a}.$$

Substituting for $2\pi i T^{\alpha_a J_a}_{L'_a S'_a, L_a S_a}(E_a) = \delta_{L'_a L_a}\delta_{S'_a S_a} - S^{\alpha_a J_a}_{L'_a S'_a, L_a S_a}(E_a)$ in (43) in terms of $2\pi i T^{\alpha_a J_a}_{l'_a j'_a, l_a j_a}(E_a) = \delta_{l'_a l_a}\delta_{j'_a j_a} - S^{\alpha_a J_a}_{l'_a j'_a, l_a j_a}(E_a)$, we find that

$$A^{J_a}_{kK\lambda}(E_a) = \sum_{l'_a j'_a l_a j_a} G^{(sI)}_{kK\lambda}(l'_a j'_a l_a j_a J_a) \tag{46}$$

$$\text{Re}\left\{(-i)^{k+K+\lambda}[\delta_{l'_a l_a}\delta_{j'_a j_a} - S^{\alpha_a J_a}_{l'_a j'_a, l_a j_a}(E_a)]\right\}$$

in which

$$G^{(sI)}_{kK\lambda}(l'_a j'_a l_a j_a J_a) \equiv \qquad (47)$$

$$(-1)^{l'_a+l_a} \widehat{j'_a}\widehat{j_a} \sum_{S'_a S_a} \widehat{S'_a}\widehat{S_a} \begin{Bmatrix} I & s & S'_a \\ l'_a & J_a & j'_a \end{Bmatrix} \begin{Bmatrix} I & s & S_a \\ l_a & J_a & j_a \end{Bmatrix} F^{(sI)}_{kK\lambda}(l'_a S'_a l_a S_a J).$$

The double summation over S'_a and S_a in (47) can be performed (using (40) and, for example, (12.2.4.(36)) in [15]) to yield

$$G^{(sI)}_{kK\lambda}(l'j'lj J) = (-1)^{l+j+I+J+K} \frac{\widehat{k}\widehat{K}\widehat{\lambda}}{\widehat{s}\widehat{I}} \widehat{\widetilde{j'}}\widehat{\widetilde{j}}\widehat{\widetilde{l'}}\widehat{\widetilde{l}} \langle l'0l0|\lambda 0\rangle \begin{Bmatrix} I & I & K \\ j' & j & J \end{Bmatrix} \begin{Bmatrix} j & s & l \\ j' & s & l' \\ K & k & \lambda \end{Bmatrix}$$

$$= (2\lambda+1) \frac{\widehat{k}\widehat{K}}{\widehat{s}\widehat{I}} \widehat{\widetilde{j'}}\widehat{\widetilde{j}}\widehat{\widetilde{l}} \langle l0\lambda 0|l'0\rangle W(JjIK;Ij') \begin{Bmatrix} l & s & j \\ \lambda & k & K \\ l' & s & j' \end{Bmatrix}, \qquad (48)$$

where the replacements used to obtain the result in the second line from that in the first parallel those used to obtain (49) below from (40).

Equations (46) and (48) are in complete agreement with the corresponding result for $\sigma_{kK\lambda}$ [(3)] reported in [12]. However, our result for $A^{J_a}_{kK\lambda}(E_a)$ in channel-spin coupled representation (43) *differs* in one small respect that implicit in (4) of [12]. To make the comparison, we invoke an equivalent form of $F^{(sI)}_{kK\lambda}(L'S'LSJ)$ containing the Racah coefficient $W(JLS'\lambda;SL')$, namely

$$F^{(sI)}_{kK\lambda}(L'S'LSJ) = \qquad (49)$$

$$(2\lambda+1)\frac{\widehat{k}\widehat{K}}{\widehat{s}\widehat{I}} \widehat{S'}\widehat{S}\widehat{L} \langle L0\lambda 0|L'0\rangle W(JLS'\lambda;SL') \begin{Bmatrix} S' & s & I \\ S & s & I \\ \lambda & k & K \end{Bmatrix}$$

obtained from (40) with the replacements

$$\begin{Bmatrix} L' & L & \lambda \\ S & S' & J \end{Bmatrix} = (-1)^{J+S'+\lambda+L} W(JLS'\lambda;SL')$$

and $\widehat{L'}\langle L'0L0|\lambda 0\rangle = (-1)^L \widehat{\lambda} \langle L0\lambda 0|L'0\rangle$. We note that the result for $\sigma_{kK\lambda}$ implied by (43), (44) and (49) does not contain the factor of $(-1)^{S-S'}$ present in (4) of [12]. The discrepancy can be traced back to the phase factor in (40) which we claim should be $(-1)^{J+S'+\lambda}$ and not $(-1)^{J+S+\lambda}$.

The decomposition of the total cross section in (42) serves to distinguish geometrical ingredients peculiar to a given experimental configuration (the $\mathcal{S}_{kK\lambda}$'s) from the dynamical behaviour of the projectile-target system as encapsulated by the corresponding S-matrix (the $A^{J_a}_{kK\lambda}$'s) and holds under very general conditions. The only assumptions we have made about the projectile-target system are that its dynamics are conservative and invariant under spatial translations and rotations. If, in addition, it is assumed that the dynamics are time-reversal invariant (i.e. the Hamiltonian $\hat{H}$ of the system in the Schrödinger representation commutes with the anti-unitary operator $\hat{T}$ effecting the time-reversal transformation $\mathcal{T}$), then, the cross-section coefficients $A_{kK\lambda}^{J_a}(E_a)$ for which $k+K+\lambda$ is odd necessarily vanish. This observation forms the foundation for null transmission tests of time-reversal. Unlike its derivation, the experimental implications of (42) (or equivalent forms of this result) for tests of time-reversal reversal have been discussed at length in various papers (notably [16]).

If the further restriction is made (as in [12]) that the polarization states of the beam and target possess axial symmetry (see Appendix), then the polarization geometry correlation factors reduce to

$$\mathcal{S}_{kK\lambda} = i^{k+K+\lambda}\widetilde{t}_{k0}(s)\widetilde{t}_{K0}(I)\{\{C_k(\boldsymbol{s}) \otimes C_K(\boldsymbol{I})\}_\lambda \otimes C_\lambda(\boldsymbol{n}_a)\}_{00},$$

where $\boldsymbol{s}$ and $\boldsymbol{I}$ denote unit vectors along the symmetry axes of the beam and target polarizations, respectively. The assumption of axial symmetry is inessential, but it helps to bring out the geometrical content of the correlation factors $\mathcal{S}_{kK\lambda}$ (see, for example, the analysis in [12]). It is, in fact, the simplified expression for the forward elastic scattering amplitude which emerges under the assumption of axial symmetry which has formed the point of departure for the most penetrating analyses published to date of the experimental difficulties confronting null transmission tests of time-reversal. It would perhaps be worthwhile to consider the impact on these analyses of relaxing the assumption of axial symmetry.

Exercise: [Transformation between different coupling schemes]
The (implicit) choices of angular momentum states used in the channel states $|\alpha_a E_a l_a j_a J_a M_a\rangle$ *and* $|\alpha_a E_a L_a S_a J_a M_a\rangle$ *are*

$$|ljJM\rangle = \sum_{m_l m_s m_I} \langle jm_j Im_I|JM\rangle\langle lm_l sm_s|jm_j\rangle|lm_l\rangle|sm_s\rangle|Im_I\rangle$$

and

$$|LSJM\rangle = \sum_{m_L m_s m_I} \langle Lm_L S\nu|JM\rangle\langle sm_s Im_I|S\nu\rangle|Lm_L\rangle|sm_s\rangle|Im_I\rangle,$$

respectively. Confirm that the forms of $\mathcal{U}(LSJM, \boldsymbol{n}S'\nu)$ *and* $U_{ljJM,\boldsymbol{n}m\widetilde{m}}$ *adopted in Sect. 3.3 are consistent with these choices and deduce (45) above.*

3.5 Decomposition of the Forward Elastic Scattering Amplitude

To discuss the coherent propagation of a beam through a target, one requires the forward elastic scattering amplitude $\boldsymbol{f}_{\alpha_a \boldsymbol{k}_a, \text{el}}(0)$ in the spin space of the projectile alone. This amplitude is obtained from the forward elastic scattering amplitude in the full spin space of the projectile-target system $f_{\alpha_a \boldsymbol{k}_a,\text{el}}(0)$ [defined in (16)] after averaging over the mixed spin states of the target nuclei [described by the density matrix $\rho(I)$]:

$$[\boldsymbol{f}_{\alpha_a \boldsymbol{k}_a,\text{el}}(0)]_{m'_a,m_a} = \sum_{M_a,M'_a} [\rho(I)]_{M_a,M'_a} [f_{\alpha_a \boldsymbol{k}_a,\text{el}}(0)]_{m'_a M'_a, m_a M_a}. \quad (50)$$

We seek a decomposition of $\boldsymbol{f}_{\alpha_a \boldsymbol{k}_a,\text{el}}(0)$ paralleling that of the total cross-section $\sigma_{\alpha_a \boldsymbol{k}_a,\text{total}}$ in (42). In fact, as we demonstrate below, it proves fairly straightforward to obtain this decomposition using our earlier result (40) for the trace $\text{Tr}\,[\rho_i(s,I) f_{\alpha_a \boldsymbol{k}_a,\text{el}}(0)]$. To this end, we note that it is useful to rewrite (40) in a form which brings out its structure, namely

$$\text{Tr}\,[\rho_i(s,I) f_{\alpha_a \boldsymbol{k}_a,\text{el}}(0)] = \frac{1}{2k_a} \sum_{kK\lambda} S_{kK\lambda} \sum_{J_a} (2J_a+1) \mathcal{A}^{J_a}_{kK\lambda}(E_a), \quad (51)$$

where, for the spin-coupled basis used in our derivation of (40), the amplitude coefficients

$$\mathcal{A}^{J_a}_{kK\lambda}(E_a) \equiv -2\pi(-i)^{k+K+\lambda} \sum_{L'_a S'_a L_a S_a} F^{(sI)}_{kK\lambda}(L'_a S'_a L_a S_a J_a) T^{\alpha_a J_a}_{L'_a S'_a, L_a S_a}(E_a).$$

[Observe that the cross section coefficient $A^{J_a}_{kK\lambda}$ introduced in (43) is the imaginary part of the amplitude coefficient $\mathcal{A}^{J_a}_{kK\lambda}$.] As with (42), the form of the decomposition in (51) is independent of the choice of channel states.

The scattering amplitude $\boldsymbol{f}_{\alpha_a \boldsymbol{k}_a,\text{el}}(0)$ is a $(2s+1) \times (2s+1)$ square matrix and so admits the expansion

$$\boldsymbol{f}_{\alpha_a \boldsymbol{k}_a,\text{el}}(0) = \frac{1}{2s+1} \sum_{k=0}^{2s} \sum_{q=-\lambda}^{\lambda} f^*_{kq}(s) \boldsymbol{T}_{kq}(s) \quad (52)$$

in terms of the matrices $\boldsymbol{T}_{kq}(s)$ of the polarization operators $\widehat{T}_{kq}(s)$ [see Appendix, specifically (67) and the related discussion]. Using the orthonormality relation (66), the expansion coefficients

$$f^*_{kq}(s) = \text{tr}\left[\boldsymbol{T}^\dagger_{kq}(s) \boldsymbol{f}_{\alpha_a \boldsymbol{k}_a,\text{el}}(0)\right], \quad (53)$$

where, for clarity below [in (55)], we have introduced the notation tr to denote the trace over the spin space of the projectile alone (as opposed to the trace over the full spin space of the projectile-target system, which we continue to denote by Tr). Since the scattering amplitude $\boldsymbol{f}_{\alpha_a \boldsymbol{k}_a,\text{el}}(0)$ is invariant under spatial rotations, the expansion coefficients $f^*_{kq}(s)$ for fixed k are [like the $\boldsymbol{T}^\dagger_{kq}(s)$'s] *contravariant* components of a spherical tensor of rank k, implying that their complex conjugates $f_{kq}(s)$ are *covariant* components of this tensor (see, for example, Sect. 3.1.2 in [15] on the distinction between covariant and contravariant components of a spherical tensor). In writing the expansion coefficients in (52) as complex conjugates, we have chosen to conform to the convention adopted for the expansion of the (scalar) density operator in terms of polarization operators [see (68) in Appendix or (54) below.]

The $f_{kq}^*(s)$'s are simply related to the trace in (51). In terms of $\boldsymbol{f}_{\alpha_a \boldsymbol{k}_a,\text{el}}(0)$ and the density matrix of the beam

$$\rho(s) = \frac{1}{2s+1} \sum_{kq} t_{kq}^*(s) \boldsymbol{T}_{kq}(s), \tag{54}$$

we have

$$\text{Tr}\left[\rho_i(s,I) f_{\alpha_a \boldsymbol{k}_a,\text{el}}(0)\right] = \text{tr}\left[\rho(s)\boldsymbol{f}_{\alpha_a \boldsymbol{k}_a,\text{el}}(0)\right] = \frac{1}{2s+1} \sum_{kq} t_{kq}(s) f_{kq}^*(s), \tag{55}$$

where, to obtain the second line, we have invoked the hermicity of $\rho(s)$ [to replace $\rho(s)$ by its hermitian adjoint $\rho(s)^\dagger$] and then used (53) and the hermitian adjoint of (54). By judicious recoupling of the product of spherical tensors in the polarization geometry correlation factor $\mathcal{S}_{kK\lambda}$ (using, for example, (3.1.8.(35)) and (3.3.1.(2)) in [15]), it can be cast into the form

$$\mathcal{S}_{kK\lambda} = i^{k+K+\lambda} \left\{ t_k(s) \otimes \{t_K(I) \otimes C_\lambda(\boldsymbol{n}_a)\}_k \right\}_{00},$$

which facilitates the comparison of (51) and (55). In view of the independence of the $t_{kq}(s)$'s, we can immediately infer from this comparison that

$$f_{kq}^*(s) = \frac{(-1)^k}{\hat{k}} \frac{2s+1}{2k_a} \sum_{K\lambda} \left[\sum_{J_a}(2J_a+1)\mathcal{A}_{kK\lambda}^{J_a}(E_a)\right] \mathcal{S}_{kq}^*(K\lambda), \tag{56}$$

where $\mathcal{S}_k(K\lambda)$ is a rank k spherical tensor with (covariant) components

$$\mathcal{S}_{kq}(K\lambda) \equiv i^{k+K+\lambda}\left\{t_K(I) \otimes C_\lambda\right\}_{kq}.$$

Equation (56) makes explicit the expected tensorial character of the $f_{kq}^*(s)$'s. The choice of phase factors in $\mathcal{S}_{kq}(K\lambda)$ guarantees that, in common with many other spherical tensors (notably, statistical tensors and spherical harmonics),

$$\mathcal{S}_{kq}^*(K\lambda) = (-1)^q \mathcal{S}_{k(-q)}(K\lambda). \tag{57}$$

Observe that $f_{kq}^*(s) = (-1)^q f_{k(-q)}(s)$ *only* if all the amplitude coefficients $\mathcal{A}_{kK\lambda}^{J_a}$ are all real-valued.

Substituting for the $f_{kq}^*(s)$'s in (52) using (56) and (57), we obtain the decomposition

$$\boldsymbol{f}_{\alpha_a \boldsymbol{k}_a,\text{el}}(0) = \sum_{kK\lambda} f_{kK\lambda} \boldsymbol{s}_{kK\lambda}, \tag{58}$$

where the partial scattering amplitudes

$$f_{kK\lambda} \equiv \frac{1}{2k_a} \sum_{J_a} (2J_a + 1) \mathcal{A}_{kK\lambda}^{J_a}(E_a) \tag{59}$$

and the polarization geometry correlation matrices

$$\boldsymbol{s}_{kK\lambda} \equiv \frac{(-1)^k}{\widehat{k}} \sum_q (-1)^q \mathcal{S}_{k(-q)}(K\lambda) \boldsymbol{T}_{kq}(s)$$
$$= i^{k+K+\lambda} \{\{t_K(I) \otimes C_\lambda(\boldsymbol{n}_a)\}_k \otimes \boldsymbol{T}_k(s)\}_{00}. \tag{60}$$

The parallels between (58) and (42) are obvious. Time-reversal invariance implies that the partial amplitudes $f_{kK\lambda}$ for which $k + K + \lambda$ is odd vanish. The correlation matrices $\boldsymbol{s}_{kK\lambda}$ are hermitian but not, in general, real-valued [unlike the correlation factors $\mathcal{S}_{kK\lambda}$ in (42)].

For the neutron transmission work discussed earlier, it is the specialization of (58) to spin $s = 1/2$ projectiles which is of interest. In this case, the forward elastic scattering amplitude $\boldsymbol{f}_{\alpha_a k_a, \mathrm{el}}(0)$ is ultimately a linear combination of the unit 2×2 matrix $\mathbf{1}$ and the Pauli matrices: under the Madison convention (see Appendix), $\boldsymbol{T}_{00}(1/2) = \mathbf{1}$ and

$$\boldsymbol{T}_{1q}(1/2) = \sigma_q,$$

where the σ_q's are the (covariant) spherical components of the vector $\boldsymbol{\sigma}$ of Pauli matrices (in terms of the more familiar Cartesian components σ_x, σ_y and σ_z, $\sigma_{\pm 1} = (\sigma_x \pm i\sigma_y)/(\mp\sqrt{2})$ and $\sigma_0 = \sigma_z$). In Table 1, we list the non-zero polarization geometry correlation matrices $\boldsymbol{s}_{kK\lambda}$ for $K \leq 1$ (sufficient to deal with target nuclei of spin $I = 1/2$). In this table, $\widehat{n}_a$ denotes the unit vector in the beam direction and the vector $\boldsymbol{P}_I$ is proportional to the polarization vector $\boldsymbol{P}$ of the target:

$$\boldsymbol{P}_I \equiv \sqrt{\frac{I}{I+1}} \boldsymbol{P} = \frac{1}{\sqrt{I(I+1)}} \mathrm{Tr}\left[\widehat{\rho}(I)\widehat{\boldsymbol{I}}\right],$$

where $\widehat{\boldsymbol{I}}$ is the spin operator (in units of $\hbar$) for target nuclei and $\widehat{\rho}(I)$ is the statistical spin space density operator of these nuclei. If the target nuclei have spin $I = 1/2$, our results imply that

$$\boldsymbol{f}_{\alpha_a k_a, \mathrm{el}}(0) = a_0 \mathbf{1} + a_M \boldsymbol{P}_{1/2} \cdot \boldsymbol{\sigma} + a_P \widehat{n}_a \cdot \boldsymbol{\sigma} + a_T (\widehat{n}_a \times \boldsymbol{P}_{1/2}) \cdot \boldsymbol{\sigma}, \tag{61}$$

where, in terms of the partial scattering amplitudes $f_{kK\lambda}$,

$$\begin{aligned} a_0 &= f_{000} + f_{011}\widehat{n}_a \cdot \boldsymbol{P}_{1/2}, & a_M &= f_{110} - \tfrac{1}{\sqrt{10}} f_{112}, \\ a_P &= \tfrac{1}{\sqrt{3}} f_{101} + \tfrac{3}{\sqrt{10}} f_{112} \widehat{n}_a \cdot \boldsymbol{P}_{1/2}, & a_T &= \tfrac{1}{\sqrt{2}} f_{111}. \end{aligned} \tag{62}$$

Clearly, (61) and (62) are consistent with (6) (once allowance has been made for the differences in notation). However, the generalization to targets of

Table 1. Some polarization geometry correlation matrices $s_{kK\lambda}$ for $K \leq 1$.

$(kK\lambda)$	$s_{kK\lambda}$
(00λ)	$\delta_{\lambda 0} 1$
(01λ)	$\delta_{\lambda 1}(\hat{n}_a \cdot \boldsymbol{P}_I) 1$
(10λ)	$\delta_{\lambda 0} \frac{1}{\sqrt{3}}(\hat{n}_a \cdot \boldsymbol{\sigma})$
(110)	$\boldsymbol{P}_I \cdot \boldsymbol{\sigma}$
(111)	$\frac{1}{\sqrt{2}}(\hat{n}_a \times \boldsymbol{P}_I) \cdot \boldsymbol{\sigma}$
(112)	$\frac{3}{\sqrt{10}}\left[(\hat{n}_a \cdot \boldsymbol{P}_I)(\hat{n}_a \cdot \boldsymbol{\sigma}) - \frac{1}{3}(\hat{n}_a \cdot \boldsymbol{P}_I)1\right]$

higher spin introduces dependence on higher (tensorial) moments of $\hat{\rho}(I)$ ($\boldsymbol{P}_I$ is, in effect, the first moment). To our knowledge, explicit results of this kind for $\boldsymbol{f}_{\alpha_a \boldsymbol{k}_a,\mathrm{el}}(0)$ have yet to be presented in the literature.

Exercise: [Pedestrian derivation of (58)]
a) Show that, in terms of the T-matrix elements

$$\langle \alpha_a E_a \boldsymbol{n}_a m'_a \widetilde{m}'_a | T | \alpha_a E_a \boldsymbol{n}_a m_a \widetilde{m}_a \rangle$$

considered in the Exercise at the end of Sect. 3.3,

$$\left[\boldsymbol{f}_{\alpha_a \boldsymbol{k}_a,\mathrm{el}}(0)\right]_{m'_a, m_a} = \qquad (63)$$
$$-\frac{(2\pi)^2}{k_a} \sum_{\widetilde{m}_a, \widetilde{m}'_a} [\rho(I)]_{\widetilde{m}_a, \widetilde{m}'_a} \langle \alpha_a E_a \boldsymbol{n}_a m'_a \widetilde{m}'_a | T | \alpha_a E_a \boldsymbol{n}_a m_a \widetilde{m}_a \rangle.$$

b) Repeat the derivation of (58) using the partial wave expansion in (37) and (38) for the T-matrix elements $\langle \alpha_a E_a \boldsymbol{n}_a m'_a \widetilde{m}'_a | T | \alpha_a E_a \boldsymbol{n}_a m_a \widetilde{m}_a \rangle$.

Appendix

Representation of a Statistical Density Operator in Spin Space

In the $(2s+1)$-dimensional spin space of a particle of spin s, the action of the statistical density operator $\hat{\rho}(s)$ describing mixed spin states of this particle is represented by a $(2s+1) \times (2s+1)$ square matrix $\rho(s)$. The set of *all* $(2s+1) \times (2s+1)$ square matrices forms a linear vector space with respect to matrix addition. Different parameterizations of $\rho(s)$ [and hence $\hat{\rho}(s)$] correspond to different choices of basis for this $(2s+1)^2$-dimensional space of matrices. The choice convenient for our purposes is provided by the set of polarization operators $\hat{T}_{\lambda\mu}(s)$ defined below. The coefficients in the corresponding expansion of $\rho(s)$ involve statistical tensors $t_\lambda(s)$. This representation of $\rho(s)$ [see (72) below] and the inferences we can draw about the statistical tensors are the primary object of this appendix, but to keep the presentation self-contained we begin with a review of the properties of polarization operators.

By definition, polarization operators $\widehat{T}_{\lambda\mu}(s)$ ($\mu = -\lambda, -\lambda+1, \ldots, \lambda-1, \lambda$) are the components of an irreducible spherical tensor of rank λ acting in the spin space of a particle of spin s (see, for example, Sect. 2.4 of [15]). The admissable values of the rank are $\lambda = 0, 1, \ldots, 2s$. [It is only for these values of λ that the Clebsch-Gordon coefficient in (64) below and, hence, the matrix representing $\widehat{T}_{\lambda\mu}(s)$ is non-zero.]

Appealing to the Wigner-Eckart theorem (see, for example, Sect. 13.1.1 in [15]), matrix elements of $\widehat{T}_{\lambda\mu}(s)$ in the spherical basis $\{|sm_s\rangle\}$ for this spin space must be related to the Clebsch-Gordon coefficient $\langle sm_s\lambda\mu | sm'_s\rangle$ by

$$\langle sm'_s | \widehat{T}_{\lambda\mu}(s) | sm_s \rangle = \langle sm_s\lambda\mu | sm'_s \rangle \alpha_{s\lambda}, \tag{64}$$

where, as the notation suggests, the factor $\alpha_{s\lambda}$ may depend on either or both of s and λ, but does *not* depend on μ, m_s and m'_s. Under the Madison convention [17] (which we adopt), the Condon-Shortley phase convention for Clebsch-Gordon coefficients is assumed and the choice for $\alpha_{s\lambda}$ made is

$$\alpha_{s\lambda} = \sqrt{2\lambda+1}.$$

(By contrast, the choice of $\alpha_{s\lambda}$ in [15] is $\alpha_{s\lambda} = \sqrt{2\lambda+1}/\sqrt{2s+1}$.) Then, (64) together with the symmetries of Clebsch-Gordon coefficients implies that the hermitian conjugate

$$\widehat{T}^\dagger_{\lambda\mu}(s) = (-1)^\mu \widehat{T}_{\lambda(-\mu)}(s) \tag{65}$$

Invoking the unitarity of Clebsch-Gordon coefficients, (64) may also be used to show that

$$\frac{1}{2s+1}\operatorname{Tr}\left[\widehat{T}^\dagger_{\lambda\mu}(s)\widehat{T}_{\lambda'\mu'}(s)\right] = \delta_{\lambda\lambda'}\delta_{\mu\mu'}, \tag{66}$$

which may be interpreted as an orthonormality relation for the polarization operators.

The full set of polarization operators $\widehat{T}_{\lambda\mu}(s)$ are represented by

$$\sum_{\lambda=0}^{2s} 2\lambda+1 = (2s+1)^2$$

linearly independent $(2s+1) \times (2s+1)$ square matrices in the spin space of a particle of spin s. [Linear independence may be proved with (66).] Hence, any arbitrary operator $\widehat{A}$ acting in this space [and so represented by some $(2s+1) \times (2s+1)$ square matrix] may be expanded as

$$\widehat{A} = \sum_{\lambda=0}^{2s}\sum_{\mu=-\lambda}^{\lambda} a_{\lambda\mu}\widehat{T}_{\lambda\mu}(s), \tag{67}$$

where, using (66), the expansion coefficient

$$a_{\lambda\mu} = \frac{1}{2s+1} \text{Tr}\left[\widehat{T}^\dagger_{\lambda\mu}(s)\widehat{A}\right].$$

In particular, a statistical spin space density operator $\widehat{\rho}(s)$ for the particle may be decomposed as

$$\widehat{\rho}(s) = \frac{1}{2s+1} \sum_{\lambda=0}^{2s} \sum_{\mu=-\lambda}^{\lambda} t^*_{\lambda\mu}(s)\widehat{T}_{\lambda\mu}(s), \tag{68}$$

where (according to the Madison convention) the parameters

$$t_{\lambda\mu}(s) \equiv \text{Tr}\left[\widehat{T}_{\lambda\mu}(s)\widehat{\rho}(s)\right]. \tag{69}$$

In (68), we have made use of the hermiticity of $\widehat{\rho}(s)$ to relate $\text{Tr}\left[\widehat{T}^\dagger_{\lambda\mu}(s)\widehat{\rho}(s)\right]$ to $t^*_{\lambda\mu}(s)$:

$$\text{Tr}\left[\widehat{T}^\dagger_{\lambda\mu}(s)\widehat{\rho}(s)\right] = \text{Tr}\left[\left(\widehat{\rho}(s)\widehat{T}_{\lambda\mu}(s)\right)^\dagger\right] = \text{Tr}\left[\left(\widehat{\rho}(s)\widehat{T}_{\lambda\mu}(s)\right)^*\right] = t^*_{\lambda\mu}(s). \tag{70}$$

Together, (65), (69) and (70) imply that

$$t^*_{\lambda\mu}(s) = (-1)^\mu t_{\lambda(-\mu)}(s). \tag{71}$$

In writing down matrix elements of $\widehat{\rho}(s)$ in the spherical basis $\{|sm_s\rangle\}$ using (64) and (68), it is convenient to invoke (71). Thus,

$$[\rho(s)]_{m'_s m_s} \equiv \langle sm'_s | \widehat{\rho}(s) | sm_s \rangle = \frac{1}{2s+1} \sum_{\lambda,\mu} \sqrt{2\lambda+1} \langle sm'_s \lambda\mu | sm_s \rangle t_{\lambda\mu}(s), \tag{72}$$

where the identity

$$\langle sm_s \lambda\mu | sm'_s \rangle = (-1)^\mu \langle sm'_s \lambda(-\mu) | sm_s \rangle$$

and a change of dummy variable of summation (from μ to $-\mu$) have been employed.

Since the statistical density operator $\widehat{\rho}(s)$ is a scalar, we can infer from (69) that the $t_{\lambda\mu}(s)$'s (for fixed λ) are like the $\widehat{T}_{\lambda\mu}(s)$'s components of an irreducible spherical tensor of rank λ, namely the statistical tensor $t_\lambda(s)$. Consequently, under a rotation of the co-ordinate system specified by the Euler angles α, β and γ, statistical tensor components transform as (see, for example, (3.1.3(11)) in [15])

$$t'_{\lambda\mu'}(s) = \sum_\mu D^\lambda_{\mu\mu'}(\alpha,\beta,\gamma) t_{\lambda\mu}(s), \tag{73}$$

where $t'_{\lambda\mu'}(s)$ and $t_{\lambda\mu}(s)$ are the components in the (new) rotated and the initial co-ordinate systems, respectively, and $D^\lambda_{\mu\mu'}$ is a Wigner D-function (discussed, for example, in Chap. 4 of [15]). We adopt the definition of Euler angles given in Sect. 1.4.1 of [15] (which is the same as the definition used in [18] and several standard texts on quantum mechanics). According to this definition, the rotation of the initial xyz co-ordinate system into the new XYZ co-ordinate system (with the same origin O) is the result of the following sequence of three rotations: (i) a rotation through the angle α about the z-axis (the y-axis goes over into the line of nodes); (ii) a rotation through the angle β about the line of nodes (such that the z-axis goes over into the Z-axis); (iii) a rotation through the angle γ about the Z-axis.

Exercise: [Interpretation of polarization operators as an orthonormal basis]
a) Show that the scalar function

$$F(A, B) \equiv \frac{1}{N} Tr\left[A^\dagger B\right]$$

of two square $N \times N$ matrices A and B has the (three) properties required of a scalar product in the space of all square $N \times N$ matrices (and is therefore one candidate for scalar product in this space).
b) Prove (66).

Implications of Axial Symmetry for Statistical Tensors

Using (73), we can establish the consequences of *axial* symmetry in the orientation of the spins of an ensemble described by the statistical spin space density operator $\widehat{\rho}(s)$. When the z-axis is aligned with the axis of symmetry, we obtain the simple result given in (76). For any other choice of z-axis, it is (77) which applies.

Let $\widetilde{t}_{\lambda\mu}(s)$ denote the corresponding statistical tensor components for a co-ordinate system with its z-axis directed along the symmetry axis. In a second co-ordinate system which differs from this co-ordinate system by a rotation about the z-axis through an angle ϕ, the new statistical tensor components $\widetilde{t}'_{\lambda\mu'}(s)$ must, in view of the symmetry about the z-axis, be identical with the old components, i.e.

$$\widetilde{t}'_{\lambda\mu'} = \widetilde{t}_{\lambda\mu'} \tag{74}$$

According to (73) (and (4.16(2)) in [15]),

$$\widetilde{t}'_{\lambda\mu'} = \sum_\mu D^\lambda_{\mu\mu'}(\phi, 0, 0)\widetilde{t}_{\lambda\mu}(s) = e^{-i\mu'\phi}\widetilde{t}_{\lambda\mu'}. \tag{75}$$

Clearly, (74) and (75) are consistent for arbitary ϕ only if

$$\widetilde{t}_{\lambda\mu}(s) = \widetilde{t}_{\lambda 0}(s)\delta_{\mu 0}. \tag{76}$$

The specialization of (71) to $\mu = 0$ implies that the $\widetilde{t}_{\lambda 0}(s)$'s are real-valued. Of these parameters, only $2s$ are non-trivial since $\widetilde{t}_{00}(s) = \text{Tr}[\widehat{\rho}(s)] = 1$.

Let us now consider a Cartesian co-ordinate system xyz where the axis of symmetry still passes through the origin but lies along the unit vector

$$\boldsymbol{e} = \sin\theta_s \cos\phi_s \boldsymbol{i} + \sin\theta_s \sin\phi_s \boldsymbol{j} + \cos\theta_s \boldsymbol{k}$$

instead of the z-axis ($\boldsymbol{i}$, $\boldsymbol{j}$ and $\boldsymbol{k}$ denote unit vectors along the x, y and z axes, respectively). Let $x_s y_s z_s$ denote the Cartesian co-ordinate system with the same origin but oriented so that the z_s-axis points along $\boldsymbol{e}$ and the y_s-axis along $\boldsymbol{e} \times \boldsymbol{k}$. A rotation with Euler angles $\alpha = 0$, $\beta = \theta_s$ and $\gamma = \pi - \phi_s$ transforms the $x_s y_s z_s$ system into the xyz system. Thus, using (73) and (76) (which applies in the $x_s y_s z_s$ system), statistical tensors in the xyz system have components

$$t_{\lambda\mu}(s) = D^\lambda_{0\mu}(0, \theta_s, \pi - \phi_s)\widetilde{t}_{\lambda 0}(s) = \sqrt{\frac{4\pi}{2\lambda+1}} Y_{\lambda\mu}(\theta_s, \phi_s)\widetilde{t}_{\lambda 0}(s), \qquad (77)$$

where we have invoked the relation of the matrix elements $D^\lambda_{0\mu}$ to spherical harmonics $Y_{\lambda\mu}$ (see, for example, (4.17(1)) in [15]) and the identities $Y_{\lambda(-\mu)}(\theta, \pi - \phi) = (-1)^\mu Y_{\lambda\mu}(\theta, \phi - \pi) = Y_{\lambda\mu}(\theta, \phi)$.

Exercise: [Properties of (77).]
Show that (77) is consistent with (71) and (76) and the requirement that $t_{00}(s) = 1$.

Acknowledgements

We would like to thank Tyler Pulis for her efforts in reformatting our material.

References

1. R. Golub, D.J. Richardson, S.K. Lamoreaux: *Ultracold Neutrons* (Adam-Hilger, Bristol 1991)
2. S. Lamoreaux, I. Khriplovich: *CP Violation Without Strangeness - Electric Dipole Moments of Particles, Atoms and Molecules* (Springer, New York 1997)
3. N.R. Roberson, C.R. Gould, J.D. Bowman: *Tests of Time Reversal Invariance in Neutron Physics* (World Scientific, Singapore 1987)
4. C.R. Gould, J.D. Bowman, Yu.P. Popov: *Time Reversal Invariance and Parity violation in Neutron Reactions* (World Scientific, Singapore 1994)
5. see E. Henley (this book) for a detailed discussion of the time reversal and parity operators
6. R. Golub, S.K. Lamoreaux: Phys. Reports **237**, 1 (1994)
7. For a pedagogical discussion of weak neutron optics, see C.R. Gould: Am. J. Physics **65**, 1213 (1997)

8. P. Huffman et al.: Phys. Rev. C **55**, 2684 (1997)
9. S.K. Lamoreaux, R. Golub : Phys. Rev. D **50**, 5632 (1994)
10. G.E. Mitchell, J.D. Bowman, H.A. Weidenmuller: Rev. Mod. Phys. **71**, 445 (1999)
11. E.D. Davis, C.R. Gould: Phys. Lett. B **447**, 209 (1999)
12. V. Hnizdo: Phys. Rev. **50**, 2639 (1994).
13. R.J.N. Phillips: Nucl. Phys. **43**, 413 (1963)
14. C.J. Joachain: *Quantum Collision Theory*, 3rd Ed. (North-Holland, Amsterdam 1983)
15. D.A. Varshalovich, A.N. Moskalev, V.K. Khersonskii: *Quantum Theory of Angular Momentum* (World Scientific, Singapore 1988)
16. C.R. Gould, D.G. Haase, N.R. Roberson, H. Postma, J.D. Bowman: Int. J. Mod. Phys. **A5**, 2181 (1990)
17. M. Simonius: 'Polarization Nuclear Physics'. In: *Lecture Notes in Physics*, vol. 30 ed. by D. Fick (Springer-Verlag, Berlin 1974) p. 38
18. A. R. Edmonds: *Angular Momentum in Quantum Mechanics* (Princeton University Press, New York 1957)

CP Violation and Baryogenesis

Werner Bernreuther

Institut für Theoretische Physik, RWTH Aachen, 52056 Aachen, Germany

Abstract. In these lecture notes an introduction is given to some ideas and attempts to understand the origin of the matter-antimatter asymmetry of the universe. After the discussion of some basic issues of cosmology and particle theory the scenarios of electroweak baryogenesis, GUT baryogenesis, and leptogenesis are outlined.

1 Introduction

CP violation has been observed so far in the neutral K meson system, both in $|\Delta S| = 2$ and $|\Delta S| = 1$ processes, and recently also in neutral B meson decays. These phenomena are very probably caused by the Kobayashi-Maskawa (KM) mechanism, that is to say by a non-zero phase $\delta_{\rm KM}$ in the coupling matrix of the charged weak quark currents to W bosons. CP violation found so far in these meson systems does not catch the eye: either the value of the CP observable or/and the branching ratio of the associated mesonic decay mode is small. However, the interactions that give rise to these subtle effects may have also been jointly responsible for an enormous phenomenon, namely for the apparent matter-antimatter asymmetry of the universe. In this context it has been a long-standing question whether or not CP violation in $K^0/\bar{K}^0$ mixing, i.e. the parameter ϵ_K, is related to the baryon asymmetry of the universe (BAU) $\eta = (n_b - n_{\bar{b}})/n_\gamma \sim 10^{-10}$. In particular, is the experimental result Re $\epsilon_K > 0$ related to the fact that our universe is filled with matter rather than antimatter? Because the CP effects observed so far in K and B meson decays are consistently explained by the KM mechanism, one may paraphrase these questions in more specific terms by asking whether the standard model of particle physics (SM) combined with the standard model of cosmology (SCM) can explain the value of η? This has been answered in recent years and, surprisingly, the answer does not refer to the role the KM phase $\delta_{\rm KM}$ may play in these explanatory attempts. Theoretical progress in understanding the SM electroweak phase transition in the early universe in conjunction with the experimental lower bound on the mass of the SM Higgs boson, $m_H^{\rm SM} > 114$ GeV, leads to the conclusion: no! In these lecture notes an introduction is given to concepts and results which are necessary to understand how this conclusion is reached. Furthermore I shall discuss a few viable (so far) and rather plausible baryogenesis scenarios beyond the SM.

The plan of these notes is as follows: Section 2 contains some basics of the SCM which are used in the following chapters. Equilibrium distributions and rough criteria for the departure from local thermal equilibrium are recalled. In section 3 a heuristic discussion of the BAU η is given. Then the Sakharov conditions for generating a baryon asymmetry within the SCM are discussed and illustrated. In section 4 we review how baryon number (B) violation occurs in the SM and how strong B-violating SM reaction rates are below and above the electroweak phase transition. Section 5 is devoted to electroweak baryogenesis scenarios. The electroweak phase transition is discussed, including results concerning its nature in the SM which reveal why the SM fails to explain the observed BAU. Nevertheless, electroweak baryogenesis is still a viable scenario in extensions of the SM, for instance in 2-Higgs doublet and supersymmetric (SUSY) extensions. We shall outline this in the context of one of the several non-SM electroweak baryogenesis mechanisms which were developed. In section 6 we discuss the perhaps most plausible, in any case most popular, baryogenesis scenario above the electroweak phase transition, namely the out-of-equilibrium decay of (a) superheavy particle(s). After having recalled a textbook example of baryogenesis in grand unified theories (GUTs), we turn to a viable and attractive scenario that has found much attention in recent years, which is baryogenesis through leptogenesis caused by the decays of heavy Majorana neutrinos. A summary and outlook is given in section 7. Some formulae concerning the transformation properties of the baryon number operator and the properties of Majorana neutrino fields are contained in appendices A and B, respectively.

Throughout these lectures the natural units of particle physics are used in which $\hbar = c = k_B = 1$, where k_B is the Boltzmann constant. In these units we have, for instance, that 1 GeV $\simeq 10^{13}$K and $1(\text{GeV})^{-1} \simeq 6 \times 10^{-25} s$. Moreover, it is useful to recall that the present extension of the visible universe is characterized by the Hubble distance $H_0^{-1} \sim 10$ Gpc, where 1 pc $\simeq$ 3.2 light years.

These lectures were intended as an introduction to the subject for graduate students. The reader who wants to delve more deeply into these topics should consult the textbook [1], the reviews [2–7] and, of course, the original literature.

2 Some Basics of Cosmology

2.1 The Standard Model of Cosmology

The current understanding of the large-scale evolution of our universe is based on a number of observations. These include the expansion of the universe and the approximate isotropic and homogeneous matter and energy distribution on large scales. The Einstein field equations of general relativity imply that the metric of space-time shares these symmetry properties of the sources of gravitation on large scales. It is represented by the Robertson-Walker (RW) metric which corresponds to the line element

$$ds^2 = dt^2 - R^2(t)\left\{\frac{dr^2}{1-kr^2} + r^2 d\theta^2 + r^2 \sin^2\theta\, d\theta d\phi^2\right\}, \quad (1)$$

where (t, r, θ, ϕ) are the dimensionless comoving coordinates and $k = 0, 1, -1$ for a space of vanishing, positive, or negative spatial curvature. Cosmological data are consistent with $k = 0$ [9]. The dynamical variable $R(t)$ is the cosmic scale factor and has dimension of length. The matter/energy distribution on large scales may be modeled by the stress-energy tensor of a perfect fluid, $T^\mu_\nu = \text{diag}(\rho, -p, -p, -p)$, where $\rho(t)$ is the total energy density of the matter and radiation in the universe and $p(t)$ is the isotropic pressure.

The dynamical equations which determine the time-evolution of the scale factor follow from Einstein's equations. Inserting the metric tensor which is encoded in (1) and the above form of $T_{\mu\nu}$ into these equations one obtains the Friedmann equation

$$H^2 \equiv \left(\frac{\dot R}{R}\right)^2 = \frac{8\pi G_N}{3}\rho - \frac{k}{R} + \frac{\Lambda}{3}. \quad (2)$$

Here $H(t) \equiv \dot R(t)/R(t)$ is the Hubble parameter which measures the expansion rate of the universe at time t, and Λ denotes the cosmological "constant" at time t. According to the inflationary universe scenario the Λ term played a crucial role at a very early epoch when vacuum energy was the dominant form of energy in the universe, leading to an exponential increase of the scale factor. Recent observations indicate that today the largest component of the energy density of the universe is some dark energy which can also be described by a non-zero cosmological constant [9]. The baryogenesis scenarios that we shall discuss in these lecture notes are associated with a period in the evolution of the early universe where, supposedly, a Λ term in the evolution equation (2) for H can be neglected.

The covariant conservation of the stress tensor $T_{\mu\nu}$ yields another important equation, namely

$$d(\rho R^3) = -p\, d(R^3). \quad (3)$$

This can be read as the first law of thermodynamics: the total change of energy is equal to the work done on the universe, $dU = dA = -pdV$. Moreover, it turns out (see section 2.2) that the various forms of matter/energy which determine the state of the universe during a certain epoch can be described, to a good approximation, by the equation of state

$$p = w\rho, \quad (4)$$

where, for instance, $w = 1/3, 0, -1$ if the energy of the universe is dominated by relativistic particles (i.e., radiation), non-relativistic particles, and vacuum energy, respectively.

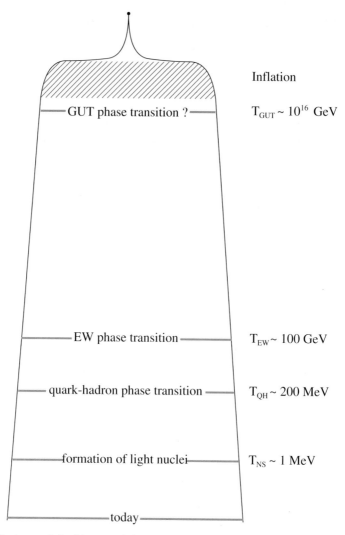

Fig. 1. Cartoon of the history of the universe. The slice of a cake, stretched at the top, illustrates the expansion of the universe as it cooled off. Inflation may have ended well below $T_{\rm GUT}$ [6].

Integrating (3) with (4) one obtains that the energy density evolves as $\rho \propto R^{-3(1+w)}$. In the radiation-dominated era, $\rho \propto R^{-4}$. Inserting this scaling law into the Friedmann equation, one finds that in this epoch the expansion rate behaves as

$$H(t) \propto t^{-1}. \tag{5}$$

Fig. 1 illustrates the history of the early universe, as reconstructed by the SCM and by the SM of particle physics. The baryogenesis scenarios which will be discussed in sections 5 and 6 apply to some instant in the – tiny – time interval after inflation and before or at the time of the electroweak phase transition. In this era, where the SM particles were massless, the energy of the universe was – according to what is presently known – essentially due to relativistic particles.

2.2 Equilibrium Thermodynamics

As was just mentioned the baryogenesis scenarios which we shall discuss in sections 5 and 6 apply to the era between the end of inflation and the electroweak phase transition. During this period the universe expanded and cooled off to temperatures $T \gtrsim T_{EW} \sim 100$ GeV. For most of the time during this stage the reaction rates of the majority of particles were much faster than the expansion rate of the cosmos. The early universe, which we view as a (dense) plasma of particles, was then to a good approximation in thermal equilibrium. In several situations it is reasonable to treat this gas as dilute and weakly interacting[1]. Let's therefore recall the equilibrium distributions of an ideal gas. Because particles in the early universe were created and destroyed, it is natural to describe the gas by means of the grand canonical ensemble. Consider an ensemble of a relativistic particle species A. Its phase space distribution or occupancy function is given by

$$f_A(\mathbf{p}) = \frac{1}{e^{(E_A - \mu_A)/T_A} \mp 1}, \tag{6}$$

where T_A is the temperature, μ_A is the chemical potential of the species which is associated with a conserved charge of the ensemble, and the minus (plus) sign refers to bosons (fermions). If different species are in chemical equilibrium then their chemical potentials are related. For instance, suppose the particle reaction $A + B \leftrightarrow C$ takes place rapidly. Then the relation $\mu_A + \mu_B = \mu_C$ holds. Take the standard example $e^+ + e^- \leftrightarrow n\gamma$. Because $\mu_\gamma = 0$ we have $\mu_{e^+} = -\mu_{e^-}$.

From (6) one obtains the number density n_A, the energy density ρ_A, the isotropic pressure p_A, and the entropy density s_A. Defining $d\tilde{p} \equiv d^3p/(2\pi)^3$ we have

$$n_A = g_A \int d\tilde{p}\, f_A(\mathbf{p}), \tag{7}$$

$$\rho_A = g_A \int d\tilde{p}\, E_A(\mathbf{p}) f_A(\mathbf{p}), \tag{8}$$

$$p_A = g_A \int d\tilde{p}\, \frac{\mathbf{p}^2}{3 E_A} f_A(\mathbf{p}), \tag{9}$$

$$s_A = \frac{\rho_A + p_A}{T}. \tag{10}$$

---
[1] This is of course not true in general. The early universe contained, in particular, particles that carried unscreened non-abelian gauge charges. Such a plasma behaves in many ways differently than an ideal gas.

Here $E_A = \sqrt{\mathbf{p}^2 + m_A^2}$, where m_A is the mass of A, and g_A denotes the internal degrees of freedom of A; for instance, $g_e = 2$ for the electron and $g_\nu = 1$ for a massless neutrino.

In the following we need these expressions in the ultra-relativistic ($T_A \gg m_A$) and nonrelativistic ($T_A \ll m_A$) limits. Integrating eqs. (7) - (9) one obtains the well-known textbook formulae for n_A, ρ_A, and p_A. For relativistic particles A (and $T_A \gg \mu_A$)

$$n_A = a_A g_A T_A^3, \tag{11}$$

$$\rho_A = b_A g_A T_A^4, \tag{12}$$

$$p_A \simeq \rho_A/3, \tag{13}$$

while for nonrelativistic particles the number density becomes exponentially suppressed for decreasing temperature:

$$n_A = g_A \left(\frac{m_A T_A}{2\pi}\right)^{3/2} e^{-(m_A - \mu_A)/T_A}, \tag{14}$$

$$\rho_A = n_A m_A, \tag{15}$$

$$p_A \simeq n_A T_A \ll \rho_A. \tag{16}$$

In eqs. (11), (12) a_A and b_A are numbers depending on whether A is a boson or fermion. Eqs. (13), (16) are the equations of state that we used already above.

When considering the total energy density and pressure of all particle species it is useful to express these quantities in terms of the photon temperature T. The corresponding formulae are obtained in a straightforward fashion by summing the respective contributions, taking into account that some species A may have a thermal distribution with a temperature $T_A \neq T$. When the universe was in thermal equilibrium its entropy remained constant. Its entropy density is given by

$$s = \frac{S}{V} = \frac{\rho + p}{T} = \frac{2\pi^2}{45} g_{*s} T^3, \tag{17}$$

where the last equality comes from the fact that s is dominated by the contributions from relativistic particles. During the epoch we are interested in, the factor g_{*s} was equal to the total number of relativistic degrees of freedom g_* [1]. (For $T \gg m_{top}$ we have $g_* \simeq 106$ in the SM.) The entropy being constant implies $s \propto R^{-3}$, hence $g_{*s} T^3 R^3 = $ const. From this we obtain that in the radiation dominated epoch the temperature of the universe decreased as

$$T \propto R^{-1}. \tag{18}$$

From these relations we can draw another important conclusion. Consider the number N_A of some particle species A. Because $N_A \equiv R^3 n_A \propto n_A/s$ this ratio also remained constant, in the absence of "A number" violation and/or entropy production, during the expansion of the universe. Therefore in the context of baryogenesis the relevant quantity is the baryon-to-entropy ratio $n_B/s \equiv (n_b - n_{\bar{b}})/s$, where n_b and $n_{\bar{b}}$ denotes the number density of baryons and antibaryons, respectively. The BAU $\eta \equiv n_B/n_\gamma$ is given in terms of this ratio by $\eta = 1.8 g_{*s} n_B/s$. The relativistic degrees of freedom g_{*s} decreased during the expansion of the early universe. This number and, hence, η remained constant only after the time of $e^+ e^-$ annihilation. From then on

$$\eta \simeq 7 \frac{n_B}{s}. \tag{19}$$

2.3 Departures from Thermal Equilibrium

Departures from thermal equilibrium (DTE) were, of course, crucial for the development of the universe to that state that we perceive today. Examples for DTEs include the decoupling of neutrinos, the decoupling of the photon background radiation, and primordial nucleosynthesis. More speculative examples are inflation, first order phase transitions in the early universe (see below), the decoupling of weakly interacting massive particles, and the topic of these lectures, baryogenesis. In any case the DTEs have led to the (light) elements, to a net baryon number of the visible universe, and to the neutrino and the microwave background.

A rough criterion for whether or not a particle species A is in local thermal equilibrium is obtained by comparing reaction rate Γ_A with the expansion rate H. Let $\sigma(A + \text{target} \to X)$ be the total cross section of the reaction(s) of A that is (are) crucial for keeping A in thermal equilibrium. Then Γ_A is given by

$$\Gamma_A = \sigma(A + \text{target} \to X) n_{\text{target}} |\mathbf{v}|, \tag{20}$$

where n_{target} is the target density and $\mathbf{v}$ is the relative velocity. Keep in mind that $[\Gamma_A] = \text{s}^{-1}$. If

$$\Gamma_A \gtrsim H, \tag{21}$$

then the reactions involving A occur rapidly enough for A to maintain thermal equilibrium. If

$$\Gamma_A < H, \tag{22}$$

then the ensemble of particles A will fall out of equilibrium. The Hubble parameter $H(t)$ which is relevant for the baryogenesis scenarios to be discussed below is the expansion rate during the radiation dominated epoch. It follows from eqs. (2) and (12) that in this era

$$H = \sqrt{\frac{8\pi G_N}{3} \rho} = 1.66 \sqrt{g_*} \frac{T^2}{m_{\text{Pl}}}, \tag{23}$$

where $m_{\text{Pl}} = 1.22 \times 10^{19}\,\text{GeV}$ denotes the Planck mass.

Eqs. (21) and (22) constitute a useful rule of thumb that is often quite accurate. It is sufficient for the purpose of these lectures. A proper treatment involves the determination of the time evolution of the particle's phase space distribution f_A which is governed by the Boltzmann equation (cf. for instance [1]). Comparing the number density $n_A(t)$, obtained from solving this equation, with the equilibrium distribution n_A^{eq} (which was discussed above for (non)relativistic particles) one sees whether or not A has decoupled from the thermal bath. Rather than going into details let us sketch in Fig. 2 the behaviour of the ratio

$$Y_A \equiv \frac{n_A}{s} \qquad (24)$$

as a function of the decreasing temperature when an ensemble of massive particles A decouples from the thermal bath. In thermal equilibrium Y_A is constant for $T \gg m_A$. At later times, when $T \lesssim m_A$, $Y_A \propto (m_A/T)^{3/2} \times \exp(-m_A/T)$ if the reaction rate still obeys (21). Thus, if A would have remained in thermal equilibrium until today its abundance would be completely negligible. However, if Γ_A becomes smaller than H, the interactions of A "freeze out", and the actual abundance of A deviates from its equilibrium value at temperature T. The larger the $A\bar{A}$ annihilation cross section the smaller the decoupling temperature and the actual abundance Y_A. The further fate of the decoupled species depends on whether or not A is stable. If a (quasi)stable species A – a weakly interacting massive particle – froze out at a temperature T not much smaller than m_A then its abundance today can be significant.

3 The Baryon Asymmetry of the Universe

3.1 Heuristic Considerations

Now to the main topic, the matter-antimatter asymmetry of our observable universe. So far, no primordial antimatter has been observed in the cosmos. Cosmic rays contain a few antiprotons, $n_{\bar{p}}/n_p \sim 10^{-4}$, but that number is consistent with secondary production by protons hitting interstellar matter, for instance, $p + p \to 3p + \bar{p}$. Also, in the vicinity of the earth no antinuclei such as $\bar{D}$, $\overline{He}$ were found [11,12]. In fact if large, separated domains of matter and antimatter in the universe exist, for instance galaxies and antigalaxies, then one would expect annihilation at the boundaries, leading to a diffuse, enhanced γ ray background. However, no anomaly was observed in such spectra. A phenomenological analysis led the authors of ref. [13] to the conclusion that on scales larger than 100 Mpc to 1 Gpc the universe consists only of matter. While this does not preclude a universe with net baryon number equal to zero, no mechanism is known that separates matter from antimatter on such large scales.

Thus for the visible universe

$$n_b - n_{\bar{b}} \simeq n_b \quad \Rightarrow \quad \eta \simeq \frac{n_b}{n_\gamma}. \qquad (25)$$

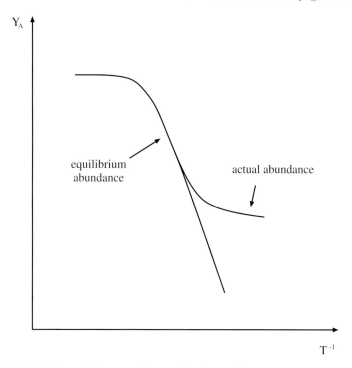

Fig. 2. The behaviour of $Y_A \equiv n_A/s$ as a function of decreasing temperature for a massive, (non)relativistic particle species A falling out of thermal equilibrium.

How is η determined? The most direct estimate is obtained by counting the number of baryons in the universe and comparing the resulting n_b with the number density of the $T = 2.7$ K microwave photon background (CMB), $n_\gamma = 2\zeta(3)T^3/\pi^2 \simeq 420/\text{cm}^3$. In fact this not very precise method yields a number for η that is not too far off from the one that comes from the still most accurate determination to date, the theory of primordial nucleosynthesis – a theory that is one of the triumphs of the SCM. There the present abundances of light nuclei, p, D, ^{3}He, ^{4}He, etc. are predicted in terms of the input parameter η. Comparison with the observed abundances yields [10]

$$\eta \simeq (1.2 - 5.7) \times 10^{-10}. \qquad (26)$$

It is gratifying that the recent determination of η from the CMB angular power spectrum measured by the Boomerang and MAXIMA collaborations [14] is consistent with (26).

Can the order of magnitude of the BAU η be understood within the SCM, without further input? The answer is no! The following textbook exercise shows nicely the point; namely, in order to understand (26) the universe must have been baryon-asymmetric already at early times. The usual, plausible starting point of the SCM is that the big bang produces equal numbers of quarks and antiquarks that end up in equal numbers of nucleons and antinucleons if there were no baryon number violating interactions. Let's compute the nucleon and antinucleon densities. At temperatures below the nucleon mass m_N we would have, as long as the (anti)nucleons are in thermal equilibrium,

$$\frac{n_b}{n_\gamma} = \frac{n_{\bar{b}}}{n_\gamma} \simeq \left(\frac{m_N}{T}\right)^{3/2} \exp\left(-m_N/T\right). \qquad (27)$$

The freeze-out of (anti)nucleons occurs when the $N\bar{N}$ annihilation rate $\Gamma_{\text{ann}} = n_b < \sigma_{\text{ann}}|\mathbf{v}| >$ becomes smaller than the expansion rate. Using $\sigma_{\text{ann}} \sim 1/m_\pi^2$ and using eq. (23) we find that this happens at $T \simeq 20$ MeV. Then we have from (27) that at the time of freeze-out $n_b/n_\gamma = n_{\bar{b}}/n_\gamma \simeq 10^{-18}$, which is 8 orders of magnitude below the observed value! In order to prevent $N\bar{N}$ annihilation some unknown mechanism must have operated at $T \gtrsim 40$ MeV, the temperature when $n_b/n_\gamma = n_{\bar{b}}/n_\gamma \simeq 10^{-10}$, and separated nucleons from antinucleons. However, the causally connected region at that time contained only about 10^{-7} solar masses! Hence this separation mechanism were completely useless for generating our universe made of baryons. Therefore the conclusion to be drawn from these considerations is that the universe possessed already at early times ($T \gtrsim 40$ MeV) an asymmetry between the number of baryons and antibaryons.

How does this asymmetry arise? There might have been some (tiny) excess of baryonic charge already at the beginning of the big bang – even though that does not seem to be an attractive idea. In any case, in the context of inflation such an initial condition becomes futile: at the end of the inflationary period any trace of such a condition had been wiped out.

3.2 The Sakharov Conditions

In the early days of the big bang model η was accepted as one of the fundamental parameters of the model. In 1967, three years after CP violation was discovered by the observation of the decays of $K_L \to 2\pi$, Sakharov pointed out in his seminal paper [15] that a baryon asymmetry can actually arise dynamically during the evolution of the universe from an initial state with baryon number equal to zero if the following three conditions hold:
- baryon number (B) violation,
- C and CP violation,
- departure from thermal equilibrium (i.e., an "arrow of time").

Many models of particle physics have these ingredients, in combination with the SCM. The theoretical challenge has been to find out which of them support (plausible) scenarios that yield the correct order of magnitude of the BAU. Before turning to some of these models, let us briefly discuss the Sakharov conditions. The first one seems obvious – see, however, the remark below. The second requirement is easily understood, noticing that the baryon number operator $\hat{B}$ is odd both under C and CP (see Appendix A). Therefore a non-zero baryon number, i.e., a non-zero expectation value $<\hat{B}>$

requires that the Hamiltonian H of the world violates C and CP. A formal argument for condition three is as follows: First, recall that a system which is in thermal equilibrium is stationary and is described by a density operator $\rho = \exp(-H/T)$. Using $\hat{B}(t) = e^{iHt}\hat{B}(0)e^{-iHt}$ we have

$$<\hat{B}(t)>_T = \mathrm{tr}(e^{-H/T}e^{iHt}\hat{B}(0)e^{-iHt}) = \mathrm{tr}(e^{-iHt}e^{-H/T}e^{iHt}\hat{B}(0)) = <\hat{B}(0)>_T,$$

If the Hamiltonian H is $\Theta \equiv CPT$ invariant, $\Theta^{-1}H\Theta = H$, we get for the quantum mechanical equilibrium average of $\hat{B} \equiv \hat{B}(0)$:

$$<\hat{B}>_T = \mathrm{tr}(e^{-H/T}\hat{B}) = \mathrm{tr}(\Theta^{-1}\Theta e^{-H/T}\hat{B})$$
$$= \mathrm{tr}(e^{-H/T}\Theta\hat{B}\Theta^{-1}) = -<\hat{B}>_T, \qquad (28)$$

where we used that $\hat{B}$ is odd under CPT (see Appendix A). Thus $<\hat{B}>_T = 0$ in thermal equilibrium.

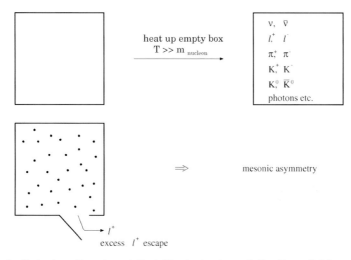

Fig. 3. A *Gedanken-Experiment* that illustrates two of the three Sakharov conditions.

How the average baryon number is kept equal to zero in thermal equilibrium is a bit tricky, as the following example shows [2]. Consider an ensemble of a heavy particle species X that has 2 baryon-number violating decay modes $X \to qq$ and $X \to \ell\bar{q}$ into quarks and leptons. (Take $q = d$ and $\ell = e$.) Further, assume that there is C and CP violation in these decays such that an asymmetry in the partial decay rates of X and its antiparticle $\bar{X}$ is induced:

$$\Gamma(X \to qq) = (1+\epsilon)\Gamma_0, \qquad \Gamma(\bar{X} \to \bar{q}\bar{q}) = (1-\epsilon)\Gamma_0, \qquad (29)$$

and there will also be an asymmetry for the other channel. CPT invariance is supposed to hold. Then the total decays rates of X and $\bar{X}$ are equal. In the decays of $X, \bar{X}$ a non-zero baryon number ΔB is generated. The ensemble is supposed to be in thermal equilibrium. One might be inclined to appeal to the principle of detailed balance which would tell us that the inverse decay $qq \to X$ is more likely than $\bar{q}\bar{q} \to \bar{X}$, and the temporary excess $\Delta B \neq 0$ would be erased this way. However, this principle is based on T invariance – but CPT invariance implies that this symmetry is broken because of CP violation. In fact applying a CPT transformation to the above decays, CPT invariance tells us that the inverse decays push ΔB into the same direction as (29):

$$\Gamma(qq \to X) = (1-\epsilon)\Gamma_0, \qquad \Gamma(\bar{q}\bar{q} \to \bar{X}) = (1+\epsilon)\Gamma_0. \qquad (30)$$

The elimination of the baryon number ΔB is achieved by the B-violating reactions $qq \to \ell \bar{q}$, $\bar{q}\bar{q} \to \bar{\ell}q$, and the CPT-transformed reactions, where the $X, \bar{X}$ resonance contributions are to be taken out of the scattering amplitudes. It is the unitarity of the S matrix which does the job of keeping $<\hat{B}>_T = 0$ in thermal equilibrium.

The following *Gedanken-Experiment*, sketched in Fig. 3, illustrates two of the three Sakharov conditions [16]. Let's simulate the big bang by taking an empty box and heat it up to a temperature, say, above the nucleon mass. Pairs of particles and antiparticles are produced that start interacting with each other, instable particles decay, etc. The K^0 and $\bar{K}^0$ evolve in time as coherent superpositions of K_L and K_S, and these states have CP-violating decays, for instance the observed non-leptonic modes $K_L \to \pi\pi$, and there is the observed CP-violating charge asymmetry in the semileptonic decays $K_L \to \pi^\mp \ell^\pm \nu$ [8]. When analyzing the semileptonic decays of K^0 and $\bar{K}^0$ one finds that slightly more $\pi^-\ell^+\nu_\ell$ are produced than $\pi^+\ell^-\bar{\nu}_\ell$, by about one part in 10^3. Hence, although initially there were equal numbers of K^0 and $\bar{K}^0$, their decays produce more π^- than π^+. Yet as long as the system is in thermal equilibrium, CP violation in the reactions including $\pi^+\ell^- \leftrightarrow \pi^+\pi^-\nu_\ell$ and $\pi^-\ell^+ \leftrightarrow \pi^+\pi^-\bar{\nu}_\ell$ will wash out the temporary excess of π^-. However, if a thermal instability is created by opening the box for a while, the excess ℓ^+ from neutral kaon decay have a chance to escape. Then the inverse reactions involving ℓ^+ are blocked to some degree, and a mesonic asymmetry $(N_{\pi^-} - N_{\pi^+}) > 0$ is generated. Of course, we haven't yet produced the real thing, as no B-violating interactions came into play.

In general, the Sakharov conditions are sufficient but not necessary for generating a non-zero baryon number. Each of them can be circumvented in principle [2]. For instance, if H is not CPT invariant, the argumentation of eq. (28) fails. However, such ideas have so far not led to a satisfactory explanation of (26). For the baryogenesis scenarios that will be discussed in sections 5,6 the Sakharov conditions are necessary ones.

4 CP and B Violation in the Standard Model

The standard model of particle physics combined with the SCM has, it seems, all the ingredients for generating a baryon asymmetry. First we recall the salient features of the SM at temperatures $T \simeq 0$ which apply to present-day physics. The observed particle spectrum tells us that the electroweak gauge symmetry $SU(2)_L \times U(1)_Y$, for which there is solid empirical evidence, cannot be a symmetry of the ground state. In the SM this spontaneous symmetry breaking is accomplished by a $SU(2)_L$ doublet of scalar fields $\Phi(x)$, the Higgs field, that is assumed to have a non-zero ground state expectation value $<0|\Phi|0> = 246$ GeV (see below). This classical field selects a direction in the internal $SU(2)_L \times U(1)_Y$ space and hence breaks the electroweak symmetry, leaving intact the gauge symmetry of electromagnetism. The W and Z bosons, quarks, and leptons acquire their masses by coupling to this field (which may be viewed as a Lorentz-invariant ether).

C and CP are violated by the charged weak quark interactions

$$\mathcal{L}_{\rm cc} = -\frac{g_w}{\sqrt{2}} \bar{U}_L \gamma^\mu V_{\rm CKM} D_L W_\mu^+ + h.c. \qquad (31)$$

Here $U_L = (u_L, c_L, t_L)^T$, $D_L = (d_L, s_L, b_L)^T$, denote the left-handed quark fields ($q_L = (1-\gamma_5)q/2$), W_μ^+ is the W boson field, g_w is the weak gauge coupling, and $V_{\rm CKM}$ is the Cabibbo-Kobayashi-Maskawa mixing matrix. CP is violated if the KM phase angle $\delta_{\rm KM} \neq 0, \pm\pi$. By this "mechanism" the CP effects observed so far in the K and B meson systems (cf., e.g., [17–19]) can be explained.

There is also baryon number violation in the SM, but this is a subtle, non-perturbative effect which is completely negligible for particle reactions in the laboratories at present-day collision energies, but very significant for the physics of the early universe. Let us outline how this effect arises. From experience we know that baryon and lepton number, which are conventionally assigned to quarks and leptons as given in the table, are good quantum numbers in particle reactions in the laboratory.

	q	$\bar{q}$	ℓ	$\bar{\ell}$
B	1/3	-1/3	0	0
L	0	0	1	-1

In the SM this is explained by the circumstance that the SM Lagrangian $\mathcal{L}_{\rm SM}(x)$, with its strong-interaction (QCD) and electroweak parts, has a global $U(1)_B$ and $U(1)_L$ symmetry: $\mathcal{L}_{\rm SM}$ is invariant under the following two sets of global phase transformations of the quark and lepton fields[2] $q = u, ..., t$;

---

[2] Possible right-handed Dirac-neutrino degrees of freedom are of no concern to us here. Majorana neutrinos that lead to violation of lepton number – see Appendix B – would be evidence for physics beyond the SM.

$\ell = e, ..., \nu_\tau$:

$$q(x) \to e^{i\omega/3}q(x), \quad \ell(x) \to \ell(x), \tag{32}$$
$$\ell(x) \to e^{i\lambda}\ell(x), \quad q(x) \to q(x). \tag{33}$$

Applying Noether's theorem we obtain the associated symmetry currents J_μ^B and J_μ^L, which are conserved at the Born level:

$$\partial^\mu J_\mu^B = \partial^\mu \sum_q \frac{1}{3}\bar{q}\gamma_\mu q = 0, \tag{34}$$

$$\partial^\mu J_\mu^L = \partial^\mu \sum_\ell \bar{\ell}\gamma_\mu \ell = 0. \tag{35}$$

(The currents are to be normal-ordered.) Thus the associated charge operators

$$\hat{B} = \int d^3x\, J_0^B(x), \tag{36}$$

$$\hat{L} = \int d^3x\, J_0^L(x) \tag{37}$$

are time-independent. At the level of quantum fluctuations beyond the Born approximation these symmetries are, however, explicitly broken because eqs. (34), (35) no longer hold. This is seen as follows. Decompose the vector current

$$\bar{f}\gamma_\mu f = \bar{f}_L\gamma_\mu f_L + \bar{f}_R\gamma_\mu f_R, \tag{38}$$

where $f = q, \ell$, into its left- and right-handed pieces. Because of the clash between gauge and chiral symmetry at the quantum level the gauge-invariant chiral currents are not conserved: in the quantum theory the current-divergencies suffer from the Adler-Bell-Jackiw anomaly [20,21]. For a gauge theory based on a gauge group G, which is a simple Lie group of dimension d_G, the anomaly equations for the L- and R-chiral currents $\bar{f}_L\gamma_\mu f_L$ and $\bar{f}_R\gamma_\mu f_R$ read

$$\partial^\mu \bar{f}_L\gamma_\mu f_L = -c_L\frac{g^2}{32\pi^2}F_{\mu\nu}^a\tilde{F}^{a\mu\nu}, \tag{39}$$

$$\partial^\mu \bar{f}_R\gamma_\mu f_R = +c_R\frac{g^2}{32\pi^2}F_{\mu\nu}^a\tilde{F}^{a\mu\nu}, \tag{40}$$

where $F^{a\mu\nu}$ is the (non)abelian field strength tensor ($a = 1, ..., d_G$) and $\tilde{F}^{a\mu\nu} = \epsilon^{\mu\nu\alpha\beta}F_{\alpha\beta}^a/2$ is the dual tensor,[3] g denotes the gauge coupling, and the constants c_L, c_R depend on the representation which the f_L and f_R form. Let us apply (38) - (40) to the above baryon and lepton number currents of the SM where the gauge group is $SU(3)_c \times SU(2)_L \times U(1)_Y$. Because gluons couple to right-handed and left-handed quark currents with the same strength,

---

[3] We use the convention $\epsilon_{0123} = +1$.

we have $c_L^{QCD} = c_R^{QCD}$. Therefore J_μ^B has no QCD anomaly. However, the weak gauge bosons W_μ^a, $a = 1, 2, 3$, couple only to left-handed quarks and leptons, while the weak hypercharge boson couples to f_L and f_R with different strength. Hence $c_R^W = 0$ and $c_L^Y \neq c_R^Y$. Putting everything together one obtains

$$\partial^\mu J_\mu^B = \partial^\mu J_\mu^L = \frac{n_F}{32\pi^2}(-g_w^2 W_{\mu\nu}^a \tilde{W}^{a\mu\nu} + g'^2 B_{\mu\nu}\tilde{B}^{\mu\nu}), \qquad (41)$$

where $W_{\mu\nu}^a$ and $B_{\mu\nu}$ denote the $SU(2)_L$ and $U(1)_Y$ field strength tensors, respectively, g' is the $U(1)_Y$ gauge coupling, and $n_F = 3$ is the number of generations.

Eq. (41) implies that $\partial^\mu(J_\mu^B - J_\mu^L) = 0$. Thus the difference of the baryonic and leptonic charge operators $\hat{B} - \hat{L}$ remains time-independent also at the quantum level and therefore the quantum number

$$B - L \text{ is conserved in the SM.}$$

How does $B + L$ number violation come about? We note that the right hand side of eq. (41) can also be written as the divergence of a current K^μ:

$$\text{r.h.s. of (41)} = n_F \partial_\mu K^\mu, \qquad (42)$$

where

$$K^\mu = -\frac{g_w^2}{32\pi^2} 2\epsilon^{\mu\nu\alpha\beta} W_\nu^a(\partial_\alpha W_\beta^a + \frac{g_w}{3}\epsilon^{abc}W_\alpha^b W_\beta^c) + \frac{g'^2}{32\pi^2}\epsilon^{\mu\nu\alpha\beta}B_\nu B_{\alpha\beta}. \qquad (43)$$

Let's integrate eq. (41), using (42), over space-time. Using Gauß's law we convert these integrals into integrals over a surface at infinity. Let's first do the surface integral for the right-hand side of (41). For hypercharge gauge fields B_μ with acceptable behaviour at infinity, that is, vanishing field strength $B_{\alpha\beta}$, the abelian part of K_μ makes no contribution to this integral. For the non-abelian gauge fields W_ν^a vanishing field strength implies that $2\epsilon_{\mu\nu\alpha\beta}\partial^\alpha W^{a\beta} = -g_w \epsilon_{\mu\nu\alpha\beta}\epsilon^{abc}W^{b\alpha}W^{c\beta}$ at infinity. Using this we obtain

$$\int d^4x\, \partial^\mu K_\mu = \frac{g_w^3}{96\pi^2}\int_{\partial V_4} dn^\mu\, \epsilon_{\mu\nu\alpha\beta}\epsilon^{abc}W^{a\nu}W^{b\alpha}W^{c\beta}. \qquad (44)$$

Now we choose the surface ∂V_4 to be a large cylinder with top and bottom surfaces at t_f and t_i, respectively, and let the volume of the cylinder tend to infinity. Because $\partial_\mu K^\mu$ is gauge-invariant, we may choose a special gauge. Choose the temporal gauge condition, $W_0^a = 0$. Then there is no contribution from the integral over the coat of the cylinder and we obtain

$$\int d^3x\, dt\, \partial_\mu K^\mu = N_{CS}(t_f) - N_{CS}(t_i) \equiv \Delta N_{CS}, \qquad (45)$$

where

$$N_{\rm CS}(t) = \frac{g_w^3}{96\pi^2} \int d^3x\, \epsilon_{ijk}\epsilon^{abc} W^{ai} W^{bj} W^{ck} \qquad (46)$$

is the Chern-Simons number. This integral assigns a topological "charge" to a classical gauge field. Actually $N_{\rm CS}$ is not gauge invariant but $\Delta N_{\rm CS}$ is. A nonabelian gauge theory like weak-interaction $SU(2)_L$ is topologically non-trivial, which is reflected by the fact that it has an infinite number of ground states whose vacuum gauge field configurations have different topological charges $\Delta N_{\rm CS} = 0, \pm 1, \pm 2, \ldots$ Imagine the set of gauge and Higgs fields and consider the energy functional $E[\text{field}]$ that forms a hypersurface over this infinite-dimensional space. The ground states with different topological charge are separated by a potential barrier. In Fig. 4 a one-dimensional slice through this hypersurface is drawn. The direction in field space has been chosen such that the classical path from one ground state to another goes over a pass of minimal height.

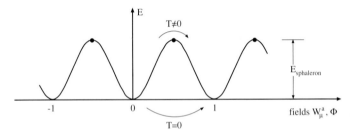

Fig. 4. The periodic vacuum structure of the standard electroweak theory. The direction in field space has been chosen as described in the text. The schematic diagram shows the energy of static gauge and Highs field configurations $W_\mu^a(\mathbf{x}), \Phi(\mathbf{x})$. The integers are the Chern-Simons number $N_{\rm CS}$ of the respective zero-energy field configuration.

Finally we perform the integral over the left-hand sides of (41) and get the result

$$\Delta\hat{B} = \Delta\hat{L} = n_F \Delta N_{\rm CS}, \qquad (47)$$

with $\Delta\hat{Q} \equiv \hat{Q}(t_f) - \hat{Q}(t_i)$, $Q = B, L$. Eq. (47) is to be interpreted as follows. As long as we consider small gauge field quantum fluctuations around the perturbative vacuum configuration $W_\mu^a = 0$ the right-hand side of (47) is zero, and B and L number remain conserved. This is the case in perturbation theory to arbitrary order where B- and L-violating processes have zero amplitudes. However, large gauge fields $W_\mu^a \sim 1/g_w$ with nonzero topological charge $\Delta N_{\rm CS} = \pm 1, \pm 2, \ldots$ exist. As discovered by 't Hooft [22] they can induce transitions at the quantum level between fermionic states $|i, t_i>$ and $|f, t_f>$ with baryon and lepton numbers that differ according to the

rule (47):
$$\Delta B = \Delta L = n_F \Delta N_{\mathrm{CS}}. \tag{48}$$

This selection rule tells us that B and L must change by at least 3 units.[4] A closer inspection of the global U(1) symmetries and associated currents shows that the selection rule can be refined: there is a change in quantum numbers by the same amount for every generation. Thus, e.g., $\Delta L_e = \Delta L_\mu = \Delta L_\tau = \Delta B/3 = \Delta N_{\mathrm{CS}}$.

The dominant B- and L-violating transitions are between states $|i, t_i>$ and $|f, t_f>$ where $|\Delta B| = |\Delta L|$ changes by 3 units. At temperature $T = 0$, transitions with $|\Delta B| = |\Delta L| = 3$ are induced by the (anti)instanton [23], a gauge field which connects two vacuum configurations whose topological charge differ by ± 1. When put into the temporal gauge $W_0^a = 0$ then the instanton field $W_i^a(\mathbf{x}, t)$ approaches, for instance, $W_i^a = 0$ at $t_i \to -\infty$ and a topologically non-trivial vacuum configuration with $N_{\mathrm{CS}} = 1$ at $t_f \to +\infty$, as indicated in Fig. 4. The corresponding amplitudes $< f, t_f | i, t_i >$ involve 9 left-handed quarks (right-handed $\bar{q}$) – where each generation participates with 3 different color states – and 3 left-handed leptons (right-handed $\bar{\ell}$), one of each generation. One of the possible amplitudes is depicted in Fig. 5. Hence we have, for instance, the antiinstanton induced reaction with $\Delta B = \Delta L = -3$:

$$u + d \to \bar{d} + 2\bar{s} + \bar{c} + 2\bar{b} + \bar{t} + \bar{\nu}_e + \bar{\nu}_\mu + \bar{\nu}_\tau. \tag{49}$$

What is the probability for such a transition to occur? It is clear from Fig. 4 that it corresponds to a tunneling process. Thus it must be exponentially suppressed. The classic computation of 't Hooft [22,24] implies, for energies $E_{\mathrm{c.m.}}(ud) \lesssim \mathcal{O}(1\,\mathrm{TeV})$, a cross-section

$$\sigma_{B+L} \propto e^{-4\pi/\alpha_w} \sim 10^{-164}, \tag{50}$$

where $\alpha_w = g_w^2/4\pi \simeq 1/30$.

When the standard model is coupled to a heat bath of temperature T, the situation changes. As was first shown in [25] (see also [26]), at very high temperatures $T \gtrsim T_{\mathrm{EW}} \sim 100$ GeV the B- and L-violating processes in the SM are fast enough to play a significant role in baryogenesis. In order to understand this we have again a look at Fig. 4. The ground states with different N_{CS} are separated by a potential barrier of minimal height

$$E_{\mathrm{sph}}(T) = \frac{4\pi}{g_w} v_T f(\frac{\lambda}{g_w}), \tag{51}$$

where $v_T \equiv < 0|\Phi|0>_T$ is the vacuum expectation value (VEV) of the SM Higgs doublet field $\Phi(x)$ at temperature T. At $T = 0$ we have $v_{T=0} = 246$ GeV. The parameter f varies between $1.6 < f < 2.7$ depending on the value of

---
[4] Notice that, even after taking these non-perturbative effects into account, the SM still predicts the proton to be stable.

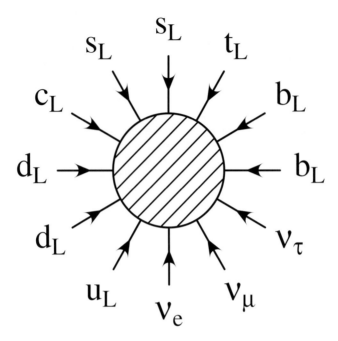

Fig. 5. An example of a $(B+L)$-violating standard model amplitude. The arrows indicate the flow of the fermionic quantum numbers.

the Higgs self-coupling λ, i.e., on the value of the SM Higgs mass. This yields $E_{\rm sph}(T=0) \simeq 8 - 13$ TeV. The subscript "sph" refers to the sphaleron, a gauge and Higgs field configuration of Chern-Simons number $1/2$ (+ integer) which is an (unstable) solution of the classical field equations of the SM gauge-Higgs sector [27,28]. These kind of field configurations (their locations are indicated by the dots in Fig. 4) lie on the respective minimum energy path from one ground state to another with different Chern-Simons number. Fig. 4 suggests that the rate of fermion-number non-conserving transitions will be proportional to the Boltzmann factor $\exp(-E_{\rm sph}(T)/T)$ as long as the energy of the thermal excitations is smaller than that of the barrier, while unsuppressed transitions will occur above that barrier.

At this point we recall that the electroweak (EW) $SU(2)_L \times U(1)_Y$ gauge symmetry was unbroken at high temperatures, that is, in the early universe. The critical temperature $T_{\rm EW}$ where – running backwards in time – the transition from the broken phase with Higgs VEV $v_T \neq 0$ to the symmetric phase with $v_T = 0$ occurs is, in the SM, about 100 GeV. (A discussion of this transition will be given in the next section.) Hence the B- and L-violating transition rates of the SM will no longer be exponentially suppressed above this temperature. Detailed investigations have led to the following results:

- In the phase where the EW gauge is broken, i.e., $T < T_{\rm EW} \sim 100$ GeV, the sphaleron-induced $\not{B} + \not{L}$ transition rate per volume V is given by (see, e.g., [4,29])

$$\frac{\Gamma^{\rm sph}_{\not{B}+\not{L}}}{V} = \kappa_1 \left(\frac{m_W}{\alpha_w T}\right)^3 m_W^4 \exp(-E_{\rm sph}(T)/T)\,, \qquad (52)$$

where $m_W(T) = g_w v_T/2$ is the temperature-dependent mass of the W boson and κ_1 is a dimensionless constant.

- The calculation of the transition rate in the unbroken phase is very difficult. On dimensional grounds we expect this rate per volume to be proportional to T^4. Recent investigations [30,31] yield for $T > T_{\rm EW} \sim 100$ GeV:

$$\frac{\Gamma^{\rm sph}_{\not{B}+\not{L}}}{V} = \kappa_2 \, \alpha_w^5 T^4\,, \qquad (53)$$

with $\kappa_2 \sim 21$.

By comparing $\Gamma^{\rm sph}_{\not{B}+\not{L}}$ above $T_{\rm EW}$ with the expansion rate H given in (23), we can assess whether the $(B+L)$-violating SM reactions, which conserve B-L, are fast enough to keep up with the expansion of the early universe in the radiation dominated epoch. From the requirement $\Gamma^{\rm sph}_{\not{B}+\not{L}} \gg H$ one obtains that these processes are in thermal equilibrium for temperatures

$$T_{\rm EW} \sim 100\,{\rm GeV} \, < \, T \, \lesssim \, 10^{12}\,{\rm GeV}\,. \qquad (54)$$

This result provides an important constraint on any baryogenesis mechanism which operates above $T_{\rm EW}$. If the B- and L-violating interactions involved in this mechanism conserve B-L, then any excess of baryon and lepton number generated above $T_{\rm EW}$ will be washed out by the B- and L-nonconserving SM sphaleron-induced reactions. Hence baryogenesis scenarios above $T_{\rm EW}$ must be based on particle physics models that violate also B-L. Examples will be discussed in section 6.

5 Electroweak Baryogenesis

We haven't discussed yet which phenomenon could possibly provide the third Sakharov ingredient, the departure from thermal equilibrium, if one attempts to explain the baryon asymmetry within the SM of particle physics. A little thought reveals that a baryogenesis scenario based on the SM requires that the thermal instability must come from the electroweak phase transition. First of all, the expansion rate of the universe at temperatures, say, $T \lesssim 10^{12}$ GeV is too slow for causing a departure from local thermal equilibrium: the reaction rates of most of the SM particles, which are typically of the order of $\Gamma \sim \alpha_w^2 T$ or larger, are much larger than the expansion rate (23), even for extremely high temperatures. Further, the SM charged weak quark current interactions lead to CP-violating effects only because, apart from a

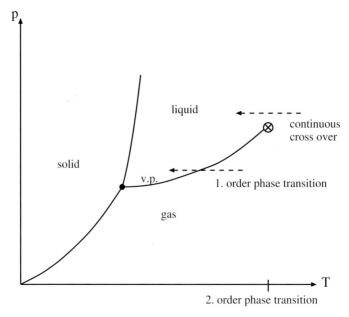

Fig. 6. The phase diagram of water.

non-trivial KM phase, the u- and d-type quarks have non-degenerate masses (see eq. (95) below). These masses are generated at the EW transition, while all SM particles are massless above T_{EW}. If $\Delta B \neq 0$ was created at the EW transition it would be – if the phase change was strongly first order – frozen in during the later evolution of the universe, as the B- and L-violating reactions below T_{EW} would be strongly suppressed (see eq. (52) and below). However, the investigations of refs. [34–36] have shown that the EW transition in the SM fails to provide the required thermal instability.

Before reviewing the results on the nature of the EW transition in the SM let us recall some basic concepts about phase transitions. Consider Fig. 6 where the pressure versus temperature phase diagram of water is sketched. We concentrate on the vapor ↔ liquid transition. The curve to the right of the triple point is the so-called vapor-pressure curve. For values of p, T along this line there is a coexistence of the liquid and gaseous phases. A change of the parameters across this curve leads to a first order phase transition which becomes weaker along the curve. The endpoint corresponds to a second order transition. Beyond that point there is a smooth cross-over from the gaseous to the liquid phase and vice versa. The nature of a phase transition can be characterized by an order parameter appropriate to the system. For the vapor-liquid transition the order parameter is the difference in the densities of water in the liquid and gaseous phase, $\tilde{\rho} = \rho_{\text{liquid}} - \rho_{\text{vapour}}$. In the case of a strong first order phase transition the order parameter has a strong discontinuity at the critical temperature T_c where the transition occurs: in the example at hand $\tilde{\rho}$ is very small in the vapor phase but it makes a sizeable jump at T_c because of the coexistence of both phases – see Fig. 7. That's what we need in a successful EW baryogenesis scenario! In case of a second order phase transition the order parameter changes also rapidly in the vicinity of T_c, but the change is continuous. In the cross-over region of the phase diagram the continuous change of $\tilde{\rho}$ as a function of T is less pronounced.

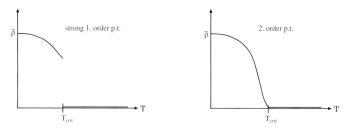

Fig. 7. The behaviour of the order parameter $\tilde{\rho}$ in the case of a strong first order and a second order transition.

So far to the statics of phase transitions. As to their dynamics, we know from experience how the first-order liquid-vapor transition evolves in time. Heating up water, vapor bubbles start to nucleate slightly below $T = T_c$ within the liquid. They expand and finally percolate above T_c. This is illustrated in Fig 8. Drawing the analogy to the early universe we should, of course, rather consider the cooling of vapor and its transition to a liquid through the formation of droplets.

A standard theoretical method to determine the nature of a phase transition in a classical system, like the vapor↔liquid or paramagnetic↔ferromagnetic transition is as follows. Let $\mathcal{H} = \mathcal{H}(s)$ be the classical Hamiltonian of the system, where $s(\mathbf{x})$ is a (multi-component) classical field. In the case of water $s(\mathbf{x})$ is the local density, while for a magnetic material $\mathbf{s}(\mathbf{x})$ denotes the three-component local magnetization. From the computation of the partition function Z we obtain the Helmholtz free energy $F = -T \ln Z$ from which the thermodynamic functions of interest can be derived. In particular we can compute the order parameter $s_{\mathrm{av}} = < \sum_{\mathbf{x}} s(\mathbf{x}) >_T$ and study its behaviour as a function of temperature.

The investigation of the static thermodynamic properties of gauge field theories proceeds along the same lines. In the case of the standard electroweak theory the role of the order parameter is played by the VEV of the $\mathrm{SU}(2)_L$ Higgs doublet field Φ. This becomes obvious when we recall the following. Experiments tell us that the $\mathrm{SU}(2)_L \times \mathrm{U}(1)_Y$ gauge symmetry is broken at $T = 0$. For the SM this means that the mass parameter in the Higgs potential must be tuned such that there is a non-zero Higgs VEV. On the other hand it was shown a long time ago [32] that at temperatures significantly larger

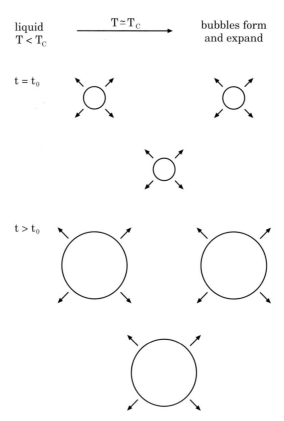

Fig. 8. Dynamics of a first-order liquid-vapor phase transition: Formation and expansion of vapor bubbles.

than, say, the W boson mass the Higgs VEV is zero and the $SU(2)_L \times U(1)_Y$ gauge symmetry is restored. (This will be shown below.) Hence during the evolution of the early universe the Higgs field must have condensed at some $T = T_c$. The order of this phase transition is deduced from the behaviour of the Higgs VEV (and other thermodynamic quantities) around T_c.

Let's couple the standard electroweak theory to a heat bath of temperature T. The free energy $F = -T \ln Z$ is obtained from the Euclidean functional integral

$$F(J,T) = -T \ln \left[\int_\beta \mathcal{D}[\text{fields}] \exp(-\int_\beta dx (\mathcal{L}_{\text{EW}} + J \cdot \Phi)) \right], \quad (55)$$

where $\mathcal{L}_{\text{EW}} = \mathcal{L}_{\text{EW}}(\Phi, W^a_\mu, B_\mu, q, \ell)$ denotes the Euclidian version of the electroweak SM Lagrangian, J is an auxiliary external field, $\beta = 1/T$,

$$\int_\beta dx = \int_0^\beta d\tau \int_V d^3x, \quad (56)$$

has been studied for the SM SU(2) gauge-Higgs model as a func-
 Higgs boson mass with analytical methods [34], and numerically
ensional [35] and 3-dimensional [36] lattice methods. These results
e qualitative features discussed above: the strength of the phase
changes from strongly first order ($m_H \lesssim 40$ GeV) to weakly first
e Higgs mass is increased, ending at $m_H \simeq 73$ GeV [37–39] where
transition is second order (cf. the liquid-vapor transition discussed
e corresponding critical temperature is $T_c \simeq 110$ GeV [40]. For
es of m_H there is a smooth cross-over between the symmetric and
 phase.
he result of the LEP2 experiments, $m_H^{SM} > 114$ GeV, leads to the
conclusion: if the SM Higgs mechanism provides the correct descrip-
ctroweak symmetry breaking then the EW phase transition in the
verse does not provide the thermal instability required for baryo-
he B-violating sphaleron processes are only adiabatically switched
 the transition from $T > T_c$ to $T < T_c$; they are still thermal for
hus the standard model of particle physics cannot explain the BAU
ective of the role that SM CP violation may play in this game.

Phase Transition in SM Extensions

, whether or not the SM Higgs field or some other mechanism pro-
 correct description of EW symmetry breaking remains to be clar-
 fact, this is the most important unsolved problem of present-day
 physics. Future collider experiments hope to resolve this issue. On
retical side, a number of extensions and alternatives to the SM Higgs
sm have been discussed for quite some time. One may distinguish
 models which postulate elementary Higgs fields (i.e., the associated
 particles have pointlike couplings up to some high energy scale
0 GeV) which trigger the breakdown of $SU(2)_L \times U(1)_Y$, and oth-
h assume that it is caused by the Bose condensation of (new) heavy
-antifermion pairs. The dynamics of the symmetry breaking sector
 models can change the order of the EW phase transition, as com-
ith the SM. Let's briefly discuss results for some models that belong
first class. The presently most popular extensions of the SM are su-
metric (SUSY) extensions, in particular the minimal supersymmetric
d model (MSSM), the Higgs sector of which contains two Higgs dou-
lthough the requirement of SUSY breaking to be soft does not allow
ependent quartic couplings in the Higgs potential $V(\Phi_1, \Phi_2)$, the num-
parameters of the scalar sector of this model is larger than that of
 and a first order transition can be arranged.[5] Investigations of V_{eff} at

odels with 2 Higgs doublets the EW phase transition typically proceeds in
ges, because the 2 neutral scalar fields condense, in general, at 2 different
eratures [42,43].

and the subscript β on the functional integral indicates that the bosonic (fermionic) fields satisfy (anti)periodic boundary conditions at $\tau = 0$ and $\tau = \beta$. From the free energy density $F(J,T)/V$ the effective potential $V_{\text{eff}}(\phi,T)$ is obtained by a Legendre transformation, where $\phi = \partial F/\partial J|_{J=0}$ is the expectation value of the Higgs doublet field, $\phi = <\Phi>_T$. (Actually in order to compare with numerical lattice calculations it is useful to employ a gauge-invariant order parameter.) Recall that the effective potential $V_{\text{eff}}(\phi,T)$ is the energy density of the system in that state $|a>_T$ in which the expectation value $<a|\Phi|a>_T$ takes the value ϕ. Hence by computing the stationary point(s), $\partial V_{\text{eff}}(\phi,T)/\partial \phi = 0$, the ground-state expectation value(s) $\phi = <0|\Phi|0>_T$ of Φ at a given temperature T are determined. If at some $T = T_c$ two minima are found then this signals two coexisting phases and a first order phase transition.

5.1 Why the SM Fails

Let us now discuss the effective potential of the SM. At $T = 0$ the tree-level effective potential is just the classical Higgs potential $V_{\text{tree}} = -\mu^2(\Phi^\dagger \Phi) + \lambda(\Phi^\dagger \Phi)^2$. Choosing the unitary gauge, $\Phi^{\text{unitary}} = (0, \phi/\sqrt{2})$ with $\phi \geq 0$ we have

$$V_{\text{tree}}(\phi) = -\frac{\mu^2}{2}\phi^2 + \frac{\lambda}{4}\phi^4, \qquad (57)$$

where $\lambda > 0$ and, by assumption, $\mu^2 > 0$ in order that the Higgs field is nonzero in the state of minimal energy: $\phi_0 \equiv <0|\Phi|0>_{T=0} \equiv v_{T=0}/\sqrt{2} = \sqrt{\mu^2/\lambda}$, and $v_{T=0}$ is fixed by, e.g., the experimental value of the W boson mass to $v_{T=0} = 246$ GeV. The mass of the SM Higgs boson is given by

$$m_H = v_{T=0}\sqrt{2\lambda} + \text{quantum corrections}. \qquad (58)$$

The experiments at LEP2 have established the lower bound $m_H > 114$ GeV [33]. Hence the SM Higgs self-coupling $\lambda > 0.33$.

At $T \neq 0$ the SM effective potential is computed at the quantum level as outlined above. Because the gauge coupling g' and the Yukawa couplings of quarks and leptons $f \neq t$ (t denotes the top quark) to the Higgs doublet Φ are small, the contributions of the hypercharge gauge boson and of $f \neq t$ may be neglected. This is usually done in the literature. Let us first discuss, for illustration, the effective potential computed to one-loop approximation for the now obsolete case of a very light Higgs boson. For high temperatures V_{eff} is given by

$$V_{\text{eff}}(\phi,T) = \frac{1}{2}a(T^2 - T_1^2)\phi^2 - \frac{1}{3}bT\phi^3 + \frac{1}{4}\lambda\phi^4, \qquad (59)$$

where

$$a = \frac{3}{16}g_w^2 + \left(\frac{1}{2} + \frac{m_t^2}{m_H^2}\right)\lambda, \; b = 9\frac{g_w^3}{32\pi}, \; T_1 = \frac{m_H}{2\sqrt{a}}, \qquad (60)$$

and m_t is the mass of the top quark. The term cubic in ϕ is due to fluctuations at $T \neq 0$. If the Higgs boson was light the quartic term would be small. Inspecting eq. (59) we recover the result quoted above that at high temperatures the Higgs field is zero in the ground state. When the temperature is lowered we find that at $T_c = T_1/\sqrt{1 - 2b^2/(9a\lambda)} > T_1$ a first order phase transition occurs: the effective potential V_{eff} has two energetically degenerate minima: one at $\phi = 0$ and the other at

$$v_{T_n} \equiv \phi_{\text{crit}} = \frac{2b}{3\lambda} T_c, \qquad (61)$$

separated by an energy barrier, see Fig. 9. At T_c the free energy of the symmetric and of the broken phase are equal; however, the universe remains for a while in the symmetric phase because of the energy barrier. As the universe expands and cools down further, bubbles filled with the Higgs condensate start to nucleate at some temperature below T_c. These bubbles become larger by releasing latent heat, percolate, and eventually fill the whole volume at $T = T_1$. Bubble nucleation and expansion are non-equilibrium phenomena which are difficult to compute.

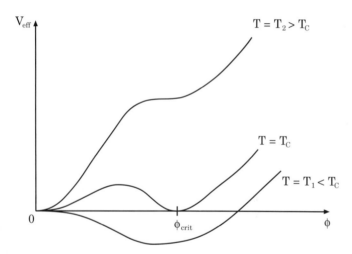

Fig. 9. Behaviour of V_{eff} in the case of a first order phase transition.

Fig. 10 shows the behaviour of $V_{\text{eff}}(\phi, T)$ in the case of a second order phase transition. In this case there are no energetically degenerate minima separated by a barrier at $T = T_c$, i.e., no bubble nucleation and expansion. The Higgs field gradually condenses uniformly at $T \lesssim T_c$ and grows to its present value as the system cools off.

The value of the critical temperature depends on the parameters of the respective model and is obtained by detailed computations (see the references

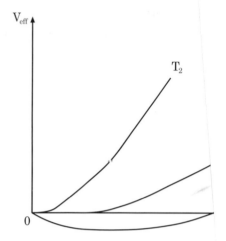

Fig. 10. Behaviour of V_{eff} in the case of a se[cond order phase transition.]

given below). Nevertheless, we may use the ab[ove] estimate and obtain $T_c \sim 70$ GeV for $m_H =$ value, see below.) With eqs. (5) and (23) we the[n find that the] transition took place at a time $t_{\text{EW}} \sim 5 \times 10^{-}$ implies that the causal domain, the diameter of in the radiation-dominated era, was then of the

Back to baryogenesis. It should be clear now phase transition is required. In this case the ti[me scale for] nucleation and expansion of Higgs bubbles is con[siderably larger than that] of the particle reactions. This causes a departur[e from equilibrium.] How is this to be quantified? Let's consider one which, after expansion and percolation, eventua[lly] bubble must get filled with more quarks than a[ntiquarks by 1 part in] 10^{10} and this ratio remains conserved. This me[ans that baryogenesis has] to take place outside of the bubble while the sp[haleron-induced B-] violating reactions must be strongly suppressed [within the bubble, such] that the sphaleron rate, which in the broken p[hase is given by] $\Gamma^{\text{sph}}_{B+L} \propto \exp(-4\pi f v_T/g_w T)$, is practically switche[d off. This means v_T] must jump at T_c, from $\phi = 0$ in the symmetric p[hase to a value in the] broken phase such that

$$\frac{v_{T_c}}{T_c} \gtrsim 1.$$

This is the condition for a first order transition to

In view of the experimental lower bound m_H^{SM} lae (59), (60) for V_{eff} which are valid only for a ve[ry light Higgs boson no] longer apply. Nevertheless, eq. (61) shows that the [] when the Higgs mass is increased. The strength o[f] at the critical temperature T_c where the transition occurs: in the example at hand $\tilde{\rho}$ is very small in the vapor phase but it makes a sizeable jump at T_c because of the coexistence of both phases – see Fig. 7. That's what we need in a successful EW baryogenesis scenario! In case of a second order phase transition the order parameter changes also rapidly in the vicinity of T_c, but the change is continuous. In the cross-over region of the phase diagram the continuous change of $\tilde{\rho}$ as a function of T is less pronounced.

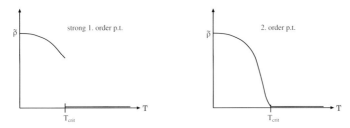

Fig. 7. The behaviour of the order parameter $\tilde{\rho}$ in the case of a strong first order and a second order transition.

So far to the statics of phase transitions. As to their dynamics, we know from experience how the first-order liquid-vapor transition evolves in time. Heating up water, vapor bubbles start to nucleate slightly below $T = T_c$ within the liquid. They expand and finally percolate above T_c. This is illustrated in Fig 8. Drawing the analogy to the early universe we should, of course, rather consider the cooling of vapor and its transition to a liquid through the formation of droplets.

A standard theoretical method to determine the nature of a phase transition in a classical system, like the vapor↔liquid or paramagnetic↔ferromagnetic transition is as follows. Let $\mathcal{H} = \mathcal{H}(s)$ be the classical Hamiltonian of the system, where $s(\mathbf{x})$ is a (multi-component) classical field. In the case of water $s(\mathbf{x})$ is the local density, while for a magnetic material $\mathbf{s}(\mathbf{x})$ denotes the three-component local magnetization. From the computation of the partition function Z we obtain the Helmholtz free energy $F = -T \ln Z$ from which the thermodynamic functions of interest can be derived. In particular we can compute the order parameter $s_{\mathrm{av}} = <\sum_{\mathbf{x}} s(\mathbf{x})>_T$ and study its behaviour as a function of temperature.

The investigation of the static thermodynamic properties of gauge field theories proceeds along the same lines. In the case of the standard electroweak theory the role of the order parameter is played by the VEV of the $SU(2)_L$ Higgs doublet field Φ. This becomes obvious when we recall the following. Experiments tell us that the $SU(2)_L \times U(1)_Y$ gauge symmetry is broken at $T = 0$. For the SM this means that the mass parameter in the Higgs potential must be tuned such that there is a non-zero Higgs VEV. On the other hand it was shown a long time ago [32] that at temperatures significantly larger

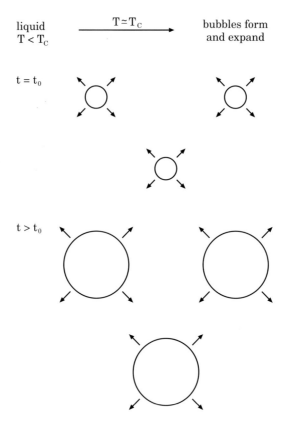

Fig. 8. Dynamics of a first-order liquid-vapor phase transition: Formation and expansion of vapor bubbles.

than, say, the W boson mass the Higgs VEV is zero and the $SU(2)_L \times U(1)_Y$ gauge symmetry is restored. (This will be shown below.) Hence during the evolution of the early universe the Higgs field must have condensed at some $T = T_c$. The order of this phase transition is deduced from the behaviour of the Higgs VEV (and other thermodynamic quantities) around T_c.

Let's couple the standard electroweak theory to a heat bath of temperature T. The free energy $F = -T \ln Z$ is obtained from the Euclidean functional integral

$$F(J,T) = -T \ln \left[\int_\beta \mathcal{D}[\text{fields}] \exp(-\int_\beta dx (\mathcal{L}_{\text{EW}} + J \cdot \varPhi)) \right], \quad (55)$$

where $\mathcal{L}_{\text{EW}} = \mathcal{L}_{\text{EW}}(\varPhi, W_\mu^a, B_\mu, q, \ell)$ denotes the Euclidian version of the electroweak SM Lagrangian, J is an auxiliary external field, $\beta = 1/T$,

$$\int_\beta dx = \int_0^\beta d\tau \int_V d^3x, \quad (56)$$

$T \neq 0$ show that there is a region in the MSSM parameter space which allows for a sufficiently strong first order EW phase transition (see, for instance, the reviews [40,41] and references therein). The condition for this is that the mass of the scalar partner $\tilde{t}_R$ of the right-handed top quark t_R must be sufficiently light and the mass of $\tilde{t}_L$ must be sufficiently heavy. An upper bound on the mass of the lightest neutral Higgs boson H_1 of the model obtains from the requirement that the mass of $\tilde{t}_L$ should not be unnaturally large. In summary, the MSSM predicts a sufficiently strong 1st order EW phase transition if

$$m_{H_1} \lesssim 105 - 115 \,\text{GeV}\,, \qquad m_{\tilde{t}_R} \lesssim 170 \,\text{GeV}\,. \tag{63}$$

In the next-to-minimal SUSY model which contains an additional gauge singlet Higgs field a strong first order transition can be arranged quite easily [68].

Non-supersymmetric SM extensions may be, in general, less motivated than SUSY models, but several of these models are, nevertheless, worth to be studied as they predict interesting phenomena. For illustrative purposes we mention here only the class of 2 Higgs doublet models (2HDM) where the field content of the SM is extended by an additional Higgs doublet, leading to a physical particle spectrum which includes 3 neutral and one charged Higgs particle. The general, renormalizable and $SU(2)_L \times U(1)_Y$ invariant Higgs potential $V(\Phi_1, \Phi_2)$ contains a large number of unknown parameters. Therefore, it is not surprising that in these models, too, the requirement of a strong 1st order EW transition can be arranged quite easily as studies of the finite-temperature effective potential show (see, for instance, [44]). No tight upper bound on the mass of the lightest Higgs boson obtains.

5.3 CP Violation in SM Extensions

Another aspect of SM extensions, namely non-standard CP violation, is also essential for baryogenesis scenarios. SM extensions as those mentioned above involve, in particular, an extended non-gauge sector; that is to say, a richer set of Yukawa and Higgs-boson self-interactions than in the SM. It is these interactions that break, in general, CP invariance. Thus, in SM extensions additional sources of CP violation besides the KM phase are usually present. We shall confine ourselves to 2 examples. (For a review, see for instance [45].)

Higgs sector CPV An interesting possibility is CP violation (CPV) by an extended Higgs sector which can occur already in the 2-Higgs doublet extensions of the SM. Consider the class of 2HDM which are constructed such that flavour-changing neutral (pseudo)scalar currents are absent at tree level. The appropriate[6] $SU(2)_L \times U(1)_Y$ invariant tree-level Higgs potential

---
[6] Neutral flavor conservation is enforced by imposing a discrete symmetry, say, $\Phi_2 \to -\Phi_2$, on $\mathcal{L}$ that may be softly broken by $V(\Phi_1, \Phi_2)$.

$V(\Phi_1, \Phi_2)$ of these models may be represented in the following way:

$$\begin{aligned}V_{\text{tree}}(\Phi_1, \Phi_2) &= \lambda_1(2\Phi_1^\dagger\Phi_1 - v_1^2)^2 + \lambda_2(2\Phi_2^\dagger\Phi_2 - v_2^2)^2 \\ &+ \lambda_3[(2\Phi_1^\dagger\Phi_1 - v_1^2) + (2\Phi_2^\dagger\Phi_2 - v_2^2)] \\ &+ \lambda_4[(\Phi_1^\dagger\Phi_1)(\Phi_2^\dagger\Phi_2) - (\Phi_1^\dagger\Phi_2)(\Phi_2^\dagger\Phi_1)] \\ &+ \lambda_5[2\text{Re}(\Phi_1^\dagger\Phi_2) - v_1 v_2 \cos\xi]^2 \\ &+ \lambda_6[2\text{Im}(\Phi_1^\dagger\Phi_2) - v_1 v_2 \sin\xi]^2,\end{aligned} \quad (64)$$

where λ_i, v_1, v_2 and ξ are real parameters and the parameterization of V_{tree} is chosen such that the Higgs fields have non-zero VEVs in the state of minimal energy.

Performing a CP transformation,

$$\Phi_{1,2}(\mathbf{x}, t) \xrightarrow{CP} e^{i\alpha_{1,2}} \Phi_{1,2}^\dagger(-\mathbf{x}, t), \quad (65)$$

we see that $H_V = \int d^3 x \, V_{\text{tree}}(\Phi_1, \Phi_2)$ is CP-noninvariant if $\xi \neq 0$. Notice that it is unnatural to assume $\xi = 0$. Even if this was so at tree level, the non-zero KM phase δ_{KM}, which is needed to explain the observed CPV in K and B meson decays, would induce a non-zero ξ through radiative corrections.

From eq. (64) we read off that at zero temperature the neutral components of the Higgs doublet fields have, in the electric charge conserving ground state, the expectation values

$$<0|\phi_1^0|0> = v_1 e^{\xi_1}/\sqrt{2}, \qquad <0|\phi_2^0|0> = v_2 e^{i\xi_2}/\sqrt{2}, \quad (66)$$

where $v = \sqrt{v_1^2 + v_2^2} = 246$ GeV, and $\xi_2 - \xi_1 = \xi$ is the physical CPV phase.

The spectrum of physical Higgs boson states of the two-doublet models consists of a charged Higgs boson and its antiparticle, $H^\pm$, and three neutral states. As far as CPV is concerned, $H^\pm$ carries the KM phase. This particle affects the (CPV) phenomenology of flavor-changing $|\Delta F| = 2$ neutral meson mixing and $|\Delta F| = 1$ weak decays of mesons and baryons. (Experimental data on $b \to s + \gamma$ imply that this particle must be quite heavy, $m_{H^+} > 210$ GeV.)

Let's briefly discuss some implications of Higgs sector CPV for present-day physics. If ξ were zero, the set of neutral Higgs boson states would consist of two scalar (CP=1) and one pseudoscalar (CP= -1) state. If $\xi \neq 0$ these states mix. As a consequence the 3 mass eigenstates, $|\varphi_{1,2,3}>$, no longer have a definite CP parity. That is, they couple both to scalar and to pseudoscalar quark and lepton currents. In terms of Weyl fields the corresponding Lagrangian reads

$$\mathcal{L}_\varphi = -\sum_\psi c_\psi \frac{m_\psi}{v} \bar\psi_L \psi_R \varphi + h.c.. \quad (67)$$

The sum over the Higgs fields $i = 1, 2, 3$ is implicit, ψ denotes a quark or lepton field, m_ψ is the mass of the associated particle, and the dimensionless reduced Yukawa couplings $c_\psi = a_\psi + ib_\psi$ (a_ψ, b_ψ real) depend on the parameters of the Higgs potential and on the type of model.

The Yukawa interaction (67) leads to CPV in *flavour-diagonal* reactions for quarks and for leptons ψ. The induced CP effects are proportional to some power $(m_\psi)^p$. For example, consider the reaction $\psi\bar{\psi} \to \psi\bar{\psi}$. The exchange of a φ boson at tree level induces an effective CPV interaction of the form $(\bar{\psi}\psi)(\bar{\psi}i\gamma_5\psi)$ with a coupling strength proportional to m_ψ^2/m_φ^2. The search for non-zero electric dipole moments (EDM) of the electron and the neutron has traditionally been a sensitive experimental method to trace non-SM CP violation [46]. If a light φ boson exists ($m_\varphi \sim 100$ GeV) and the CPV phase ξ is of order 1 the Yukawa interaction (67) can induce electron and neutron EDMs of the same order of magnitude as their present experimental upper bounds.

What happens at the EW phase transition in the early universe? We assume that the parameters of the 2HDM are such that the transition is strongly first order. Moreover, in order to simplify the discussion we assume that the passage from the symmetric to the broken phase occurs in one step, at some temperature T_c. Somewhat below T_c bubbles filled with Higgs fields start to nucleate and expand. That is, the Higgs VEVs are space and time dependent. Let's consider, for simplicity, only one of the bubbles and assume its expansion to be spherically symmetric. When the bubble has grown to some finite size we can use the following one-dimensional description. Consider the rest frame of the bubble wall. The wall is taken to be planar and the expansion of the bubble is taken along the z axis. The wall, i.e., the phase boundary has some finite thickness l_{wall}, extending from $z = 0$ to $z = z_0$. The symmetric phase lies to the right of this boundary, $z > z_0$ while the broken phase lies to the left, $z < 0$. Thus the neutral Higgs fields have VEVs whose magnitudes and phases vary with z:

$$<0|\phi_1^0|0>_T = \frac{\rho_1(z)}{\sqrt{2}} e^{i\theta(z)}, \qquad <0|\phi_2^0|0>_T = \frac{\rho_2(z)}{\sqrt{2}} e^{i\omega(z)}. \qquad (68)$$

In the symmetric phase, $z \gg z_0$, both VEVs vanish, whereas in the broken phase the VEVs should be close to their zero temperature values:

$$\rho_i(z) \simeq v_i, \qquad \theta(z) \simeq \xi_1, \qquad \omega(z) \simeq \xi_2, \qquad (69)$$

if $z \ll 0$. The variation of the moduli and phases with z can be determined by solving the field equations of motion that involve the finite-temperature effective potential of the model.

As to the couplings of the Higgs fields to fermions, we assume here and in the following subsection, for definiteness, that all quarks and leptons couple to Φ_1 only. Then the Yukawa coupling of a quark or lepton field $\psi = q, \ell$ to the neutral Higgs field is given by

$$\mathcal{L}_1 = -h_\psi \bar{\psi}_L \psi_R \phi_1^0 + \text{h.c.}$$
$$= -m_\psi(z) \bar{\psi}_L \psi_R - m_\psi^*(z) \bar{\psi}_R \psi_L + \ldots, \qquad (70)$$

where

$$m_\psi(z) = h_\psi \frac{\rho_1(z)}{\sqrt{2}} e^{i\theta(z)} \qquad (71)$$

is a complex-valued mass and the ellipses in (70) indicate the coupling of the quantum field, i.e., the coupling of a neutral Higgs particle to ψ. Thus the interaction of a fermion field $\psi(x)$ with the CP-violating Higgs bubble, treated as an external, classical background field, is summarized by the Lagrangian

$$\mathcal{L}_\psi = \bar{\psi}_L i \gamma^\mu \partial_\mu \psi_L + \bar{\psi}_R i \gamma^\mu \partial_\mu \psi_R - m_\psi(x) \bar{\psi}_L \psi_R - m_\psi^*(x) \bar{\psi}_R \psi_L . \qquad (72)$$

In section 5.4 we shall also use the plasma frame which is implicitly defined by requiring the form of the particle distributions to be the thermal ones. In this frame the Higgs VEVs are space- and time-dependent. The wall expands with a velocity v_wall. The interaction (72) is CP-violating because x-dependent phase $\theta(x)$ of $m(x)$. Obviously, the field $\theta(x)$ cannot be removed from $\mathcal{L}_\psi$ by redefining the fields $\psi_{L,R}(x)$. We shall investigate its consequences for baryogenesis in the next subsection.

CP Violation in the MSSM In the minimal supersymmetric extension (MSSM) of the Standard Model [47] CP-violating phases can appear, apart from the complex Yukawa interactions of the quarks yielding a non-zero KM phase δ_KM, in the so-called μ term in the superpotential (i), and in soft supersymmetry breaking terms (ii) - (iv). The requirement of gauge invariance and hermiticity of the Lagrangian allows for the following new sources of CP violation:

i) A complex mass parameter $\mu_c \equiv \mu \exp(i\varphi_\mu)$, μ real, describing the mixing of the two Higgs chiral superfields in the superpotential.

ii) A complex squared mass parameter m_{12}^2 describing the mixing of the two Higgs doublets[7] and contributes to the Higgs potential

$$V(\Phi_1, \Phi_1) \supset \mu_c \Phi_1^\dagger \cdot \Phi_2 + \text{h.c.} , \qquad (73)$$

iii) Complex Majorana masses M_i in the gaugino mass terms ($\epsilon \equiv i\sigma_2$),

$$-\sum_i M_i (\lambda_i^T \epsilon \lambda_i)/2 + \text{h.c.}, \qquad (74)$$

---

[7] In order to facilitate the comparison with the non-supersymmetric models, the non-SUSY convention for the Higgs doublets is employed here; i.e., the same hypercharge assignment is made for both SU(2) Higgs doublets, $\Phi_i = (\phi_i^+, \phi_i^0)^T$, (i=1,2) .

where $i = 1, 2, 3$ refers to the $U(1)_Y$, $SU(2)_L$ gauginos, and gluinos, respectively. A standard assumption is that the M_i have a common phase.

iv) Complex trilinear scalar couplings of the scalar quarks and scalar leptons, respectively, to the Higgs doublets Φ_1, Φ_2. These couplings form complex 3×3 matrices A_ψ in generation space. Motivated by supergravity models it is often assumed that the matrices A_ψ are proportional to the Yukawa coupling matrices h_ψ:

$$A_\psi = A h_\psi, \qquad \psi = u, d, \ell, \tag{75}$$

where A is a complex mass parameter.

Thus the parameter set μ_c, m_{12}^2, M_i, and A involves 4 complex phases. Exploiting two (softly broken) global $U(1)$ symmetries of the MSSM Lagrangian, two of these phases can be removed by re-phasing of the fields. A common choice, we shall also use, is a phase convention for the fields such that the gaugino masses M_i and the mass parameter m_{12}^2 are real. Then the observable CP phases in the MSSM (besides the KM phase) are $\varphi_\mu = \arg(\mu_c)$ and $\varphi_A = \arg(A)$. The experimental upper bounds on the electric dipole moments d_e, d_n of the electron and the neutron put, however, rather tight constraints on these CP phases, in particular on φ_μ. Even if there are correlations between these phases such that there are cancellations among the contributions to d_e and to d_n, Ref. [48] finds (see also [49,50]) that φ_μ is constrained by the data to be smaller than $|\varphi_\mu| \lesssim 0.03$. A way out of this constraint would be heavy first and second generation sleptons and squarks with masses of order 1 TeV.

What about Higgs sector CPV? In the MSSM the tree-level Higgs potential V_{tree} is CP-invariant. Supersymmetry does not allow for independent quartic couplings in V_{tree}. They are proportional to linear combinations of the $SU(2)_L$ and $U(1)_Y$ gauge couplings squared. At one-loop order the interactions of the Higgs fields $\Phi_{1,2}$ with charginos, neutralinos, (s)tops, etc. generate quartic Higgs self-interactions of the form

$$V_{\text{eff}} \supset \lambda_1 (\Phi_1^\dagger \Phi_2)^2 + \lambda_2 (\Phi_1^\dagger \Phi_2)(\Phi_1^\dagger \Phi_1) + \lambda_3 (\Phi_1^\dagger \Phi_2)(\Phi_2^\dagger \Phi_2) + \text{h.c.}, \tag{76}$$

in the effective potential. The CP phases φ_μ and φ_A induce complex $\lambda_{1,2,3}$. Thus, explicit CP violation in the Higgs sector occurs at the quantum level which leads to Yukawa interactions of the neutral Higgs bosons being of the form (67).

In the context of baryogenesis a potentially more interesting possibility is spontaneous CP violation at high temperatures $T \lesssim T_{\text{EW}}$. This kind of CP violation could not be traced any more in the laboratory! Ref. [51] pointed out that, irrespective of whether or not φ_μ and φ_A are sizeable, the MSSM effective potential receives, at high temperatures $T \lesssim T_{\text{EW}}$, quite large one-loop corrections of the form (76). As a consequence, the neutral Higgs fields can develop complex VEVs of the form (68) with a large CP-odd classical field. This would signify spontaneous CPV at finite temperatures, even if φ_μ and φ_A would be very small or even zero. However, ref. [52] finds that experimental constraints on the parameters of the MSSM and the requirement of the phase transition to be strongly first order preclude this possibility in the case of the MSSM.

Let's now come to those CP-violating interactions of the MSSM which are of relevance at the EW phase transition and involve φ_μ and φ_A at the tree level. As discussed above, there is a small, phenomenologically acceptable range of light Higgs and light stop mass parameters which allows for a strong first order transition. The Higgs VEVs are of the form

$$<0|\phi_1^0|0>_T = \rho_1(z), \qquad <0|\phi_2^0|0>_T = \rho_2(z), \qquad (77)$$

where ρ_i are real and for convenience, a normalization convention different from the one in (68) is used here.

These VEVs determine the interaction of the bubble wall with those MSSM particles that couple to the Higgs fields already at the classical level. Inspecting where the CP-violating phases φ_μ and φ_A are located in $\mathcal{L}_{MSSM}$ (we use the convention of the gaugino masses M_i being real) it becomes clear that the relevant interactions of the classical Higgs background fields are those with charginos, neutralinos, and sfermions, in particular top squarks. Contrary to the case of the 2HDM discussed above the interactions of quarks and leptons with a bubble wall do not – at the classical level – violate CP invariance if (77) applies.

Inserting (77) into the respective terms of the MSSM Lagrangian we obtain the Lagrangians describing the particle propagation in the presence of a Higgs bubble [53,55,56]. For the charged gauginos and Higgsinos in the gauge eigenstate basis we get

$$\mathcal{L}_c = \chi_R^\dagger \sigma_\mu \partial^\mu \chi_R + \chi_L^\dagger \bar{\sigma}_\mu \partial^\mu \chi_L$$
$$+ \chi_R^\dagger \mathcal{M}_c \chi_L + \chi_L^\dagger \mathcal{M}_c^\dagger \chi_R \qquad (78)$$

where $\sigma^\mu = (I, \sigma_i)$, $\bar{\sigma}^\mu = (I, -\sigma_i)$, and we have put

$$\chi_R^\dagger = (\widetilde{W}^+, \widetilde{H}_2^+), \qquad \chi_L = (\widetilde{W}^-, \widetilde{H}_1^-)^T, \qquad (79)$$

where $\widetilde{W}(x)$, $\widetilde{H}_{1,2}(x)$ are 2-component Weyl fields for the charged gauginos and Higgsinos, respectively. The chargino mass matrix is given by

$$\mathcal{M}_c(x) = \begin{pmatrix} M_2 & g_w \rho_2(x) \\ g_w \rho_1(x) & \mu_c \end{pmatrix}, \qquad (80)$$

where μ_c is the complex Higgsino mass parameter defined above.

For the scalar stop fields $\tilde{t}_R(x)$, $\tilde{t}_L(x)$ we obtain in the gauge eigenstate basis

$$\mathcal{L}_{\tilde{t}} = (\partial_\mu \tilde{t}_L^\dagger) \partial^\mu \tilde{t}_L + (\partial_\mu \tilde{t}_R^\dagger) \partial^\mu \tilde{t}_R - (\tilde{t}_L^\dagger, \tilde{t}_R^\dagger) \mathcal{M}_{\tilde{t}} \begin{pmatrix} \tilde{t}_L \\ \tilde{t}_R \end{pmatrix}, \qquad (81)$$

with

$$\mathcal{M}_{\tilde{t}}(x) = \begin{pmatrix} m_L^2 + h_t^2\, H_2^2(z) & h_t\left(A_t \rho_2(x) - \mu_c^* \rho_1(x)\right) \\ h_t\left(A_t^* \rho_2(x) - \mu_c \rho_1(x)\right) & m_R^2 + h_t^2\, \rho_2^2(x) \end{pmatrix}, \quad (82)$$

where $m_{L,R}^2$ are SUSY breaking squared mass parameters, h_t is the top-quark Yukawa coupling, and A_t is the left-right stop mixing parameter.

In the mass matrices (80) and (82) the CP-violating phases combine with the spatially varying VEVs and will give rise to x-dependent CP-violating phases when the mass matrices are diagonalized, analogously to the case of the 2HDM above. This causes CP-violating particle currents which we shall discuss further in the next subsection.

5.4 Electroweak Baryogenesis

As outlined above this scenario works only in extensions of the SM. The required departure from thermal equilibrium[8] is provided by the expansion of the Higgs bubbles, the true vacuum. When the bubble walls pass through a point in space, the classical Higgs fields change rapidly in the vicinity of such a point, see Fig. 11, as do the other fields that couple to those fields. As far as different mechanisms are concerned, the following distinction is made in the literature:

• *Nonlocal Baryogenesis* [60], also called "charge transport mechanism", refers to the case where particles and antiparticles have CP non-conserving interactions with a bubble wall. This causes an asymmetry in a quantum number other than B number which is carried by (anti)particle currents into the unbroken phase. There this asymmetry is converted by the $(B+L)$-violating sphaleron processes into an asymmetry in baryon number. Some instant later the wall sweeps over the region where $\Delta B \neq 0$, filling space with Higgs fields that obey (62). Thus the B-violating back-reactions are blocked and the asymmetry in baryon number persists. The mechanism is illustrated in Fig. 12.

• *Local Baryogenesis* [58,59] refers to case where the both the CP-violating and B-violating processes occur at or near the bubble walls.

In general, one may expect that both mechanisms were at work and $\Delta B \neq 0$ was produced by their joint effort. Which one of the mechanisms is more effective depends on the shape and velocity of the bubbles; i.e., on the underlying model of particle physics and its parameters.

In the following we discuss only the nonlocal baryogenesis mechanism. First, the case of Higgs sector CP violation is treated in some detail. For definiteness, we choose a 2-Higgs doublet extension of the SM with CP violation as decribed above. Then (72) applies. Because $|m_\psi(z)|$ becomes, at $T=0$, the mass of the fermion ψ, top quarks and, as far as leptons are concerned, τ leptons have the strongest interactions with the wall.

---

[8] The departure from thermal equilibrium could have been caused also by TeV scale topological defects that can arise in SM extensions [57].

Fig. 11. Sketch of the variation of the modulus and the CP-violating phase angle of a non-SM Higgs VEV, in the wall frame, at the boundary between the broken and the symmetric phase.

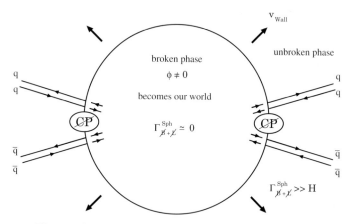

Fig. 12. Sketch of nonlocal electroweak baryogenesis.

We consider for simplicity only the so-called thin wall regime which applies if the mean free path of a fermion, l_ψ, is larger than the thickness l_wall. Then the quarks and leptons can be treated as free particles, interacting only in a small region with a non-trivial Higgs background field, see Fig. 11. Multiple scattering within the wall may be neglected. The expansion of the wall is supposed to be spherically symmetric and the 1-dimensional description as given in section 5.3.1 applies. Fig. 13 shows left-handed and right-handed quarks[9] q_L and q_R incident from the unbroken phase, which hit the moving wall and are reflected by the Higgs bubble into right-handed and left-handed quarks, respectively.

In the frame where the wall is at rest, the fermion interactions with the bubble wall are described by the Dirac equation following from (72):

$$(i\gamma^\mu \partial_\mu - m(z)P_R - m^*(z)P_L)\psi(z,t) = 0, \qquad (83)$$

where $P_{R,L} = (1 \pm \gamma_5)/2$ and ψ is a c-number Dirac spinor. Solving this equation with the appropriate boundary conditions yields the (anti)quark wave functions of either chirality [61,62].

---

[9] In this subsection the symbols q_L, $\bar{q}_L$, etc. do *not* denote fields but particle states.

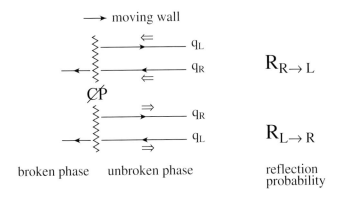

Fig. 13. Reflection of left- and right-handed quarks at a radially expanding Higgs bubble. The transmission of (anti)quarks from the broken into the symmetric phase is not depicted.

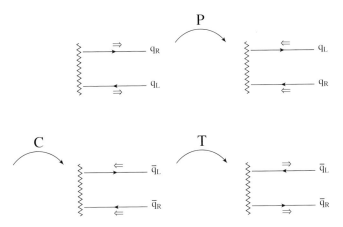

Fig. 14. The reflection $q_L \to q_R$ and the P-, CP-, and CPT-transformed process.

Instead of performing this calculation let's make a few general considerations. Let's have a look at the scattering process depicted in Fig. 14, where, in the symmmetric phase $z > z_0$, a left-handed quark q_L (having momentum $k_z < 0$) is reflected at the wall into a right-handed q_R. Notice that conservation of electric charge guarantees that a quark is reflected into a quark and not an antiquark. Angular momentum conservation tells us that q_L is reflected as q_R and vice versa. Also shown are the situations after a parity transformation (followed by a rotation around the wall axis in the paper plane orthogonal to the z axis by an angle π), and subsequent charge conjugation C, and time reversal (T) transformations. The analogous figure can be drawn for antiquark reflection. These figures immediately tell us that if CP were conserved then

$$\mathcal{R}_{L \to R} = \mathcal{R}_{\bar{R} \to \bar{L}}, \qquad \mathcal{R}_{R \to L} = \mathcal{R}_{\bar{L} \to \bar{R}} \qquad (84)$$

would hold. (The subscripts $\bar{R}$, $\bar{L}$ denote right-handed and left-handed antiquarks, respectively.) CPT invariance, which is respected by the particle physics models we consider, implies

$$\mathcal{R}_{L \to R} = \mathcal{R}_{\bar{L} \to \bar{R}}, \qquad \mathcal{R}_{R \to L} = \mathcal{R}_{\bar{R} \to \bar{L}}. \qquad (85)$$

The charge transport mechanism [61] works as follows. At some initial time we have equal numbers of quarks and antiquarks in the unbroken phase, in particular equal numbers of q_L and $\bar{q}_R$ and q_R and $\bar{q}_L$, respectively, which hit the expanding bubble wall. Reflection converts $q_L \to q_R$, $\bar{q}_R \to \bar{q}_L$, $q_R \to q_L$, and $\bar{q}_L \to \bar{q}_R$ and the particles move back to the region where the Higgs fields are zero. Because the interaction with the bubble wall is assumed to be CP-violating, the relations (84) for the reflection probabilities no longer hold. Actually, for the CP asymmetry

$$\Delta\mathcal{R}_{CP} \equiv \mathcal{R}_{\bar{L} \to \bar{R}} - \mathcal{R}_{R \to L} = \mathcal{R}_{L \to R} - \mathcal{R}_{\bar{R} \to \bar{L}} \qquad (86)$$

to be non-zero it is essential that $m_q(z)$ has a z dependent phase. The reflection coefficients are built up by the coherent superposition of the amplitudes for (anti)quarks to reflect at some point z in the bubble. When the phases vary with z the reflection probabilities $\mathcal{R}_{\bar{L} \to \bar{R}}$ and $\mathcal{R}_{R \to L}$ differ from each other. If the phase of $m_q(z)$ were constant these probabilities would be equal. (Keep in mind that we work at the level of 1-particle quantum mechanics.) An explicit computation yields [63]

$$\Delta\mathcal{R}_{CP}(k_z) \propto \int_{-\infty}^{+\infty} dz \, \cos(2k_z z) \text{Im}[m_q(z) M_q^*], \qquad (87)$$

where $M_q = m_q(z = -\infty)$ is the mass of the quark in the broken phase, and $\arg(M_q) = \xi_1$ – see eq. (69). This equation corroborates the above statement; if $m_q(z)$ had a constant phase, the asymmetry would be zero. Notice that at this stage the net baryon number is still zero. This is because the difference J_q^L of the fluxes of $\bar{q}_R$ and q_L, injected from the wall back into the symmetric phase, is equal[10] to J_q^R which we define as the difference of the fluxes of q_R and $\bar{q}_L$, as should be clear from (85). However, the $(B + L)$-violating weak sphaleron interactions, which are unsuppressed in the symmetric phase away from the wall, act only on the (massless) left-handed quarks and right-handed antiquarks. For instance, the reaction (49) decreases the baryon number by 3 units, while the corresponding reaction with right-handed antiquarks in the initial state increases B by the same amount. Thus if the functional form of the CP-violating part $\text{Im}[m_q(z) M_q^*]$ of the background Higgs field is such that

---

[10] Interactions with the other plasma particles are neglected.

$J_q^L > 0$ then, after the anomalous weak interactions took place, there are more left-handed quarks than right-handed antiquarks. The fluxes of the reflected $\bar{q}_L$ and q_R are not affected by the anomalous weak sphaleron interactions. Adding it all up we see that some place away from the wall a net baryon number $\Delta B > 0$ is produced. Some instant later the expanding bubble sweeps over that region and the associated non-zero Higgs fields strongly suppress the $(B + L)$-violating back reactions that would wash out ΔB. Thus the non-zero B number produced before is frozen in.

We must also take into account that (anti)particles in the broken phase can be transmitted into the symmetric phase and contribute to the (anti)particle fluxes discussed above. Using CPT invariance and unitarity, we find that the probabilities for transmission and the above reflection probabilities are related:

$$\mathcal{T}_{L \to L} = 1 - \mathcal{R}_{R \to L} = 1 - \mathcal{R}_{\bar{R} \to \bar{L}} = \mathcal{T}_{\bar{L} \to \bar{L}}, \qquad (88)$$

$$\mathcal{T}_{R \to R} = 1 - \mathcal{R}_{L \to R} = 1 - \mathcal{R}_{\bar{L} \to \bar{R}} = \mathcal{T}_{\bar{R} \to \bar{R}}. \qquad (89)$$

We can now write down a formula for the current J_q^L, which we define as the difference of the fluxes of $\bar{q}_R$ and q_L, injected from the wall into the symmetric phase. The contribution from the reflected particles involves the term $\Delta \mathcal{R}_{CP} f_s$ where f_s is the free-particle Fermi-Dirac phase-space distribution of the (anti)quarks in the region $z > z_0$ that move to the left, i.e., towards the wall. The contribution from the (anti)quarks which have returned from the broken phase involves $(\mathcal{T}_{\bar{R} \to \bar{R}} - \mathcal{T}_{L \to L}) f_b = -\Delta \mathcal{R}_{CP} f_b$, where f_b is the phase-space distribution of the transmitted (anti)quarks that move to the right. The reference frame is the wall frame. Notice that f_s and f_b differ because the wall moves with a velocity $v_{\text{wall}} \neq 0$ – in our convention from left to right. The current J_q^L is given by

$$J_q^L = \int_{k_z < 0} \frac{\mathrm{d}^3 k}{(2\pi)^3} \frac{|k_z|}{E} (f_s - f_b) \Delta \mathcal{R}_{CP}, \qquad (90)$$

where $|k_z|/E$ is the group velocity. The current is non-zero because two of the three Sakharov conditions, CP violation and departure from thermal equilibrium, are met. The current would vanish if the wall were at rest in the plasma frame – which leads to thermal equilibrium –, because then $(f_s - f_b) = 0$.

The current $J^L = \sum_\psi J_\psi^L$ is the source for baryogenesis some distance away from the wall as sketched above. We skip the analysis of diffusion and of the conditions under which local thermal equilibrium is maintained in front of the bubble wall [61,63,64]. This determines the densities of the left-handed quarks and right-handed antiquarks and their associated chemical potentials. The rate of baryon production per unit volume is determined by the equation [63]

$$\frac{\mathrm{d} n_B}{\mathrm{d} t} = -n_F \frac{\hat{\Gamma}_{\text{sph}}}{2T} \sum_{\text{generations}} (3\hat{\mu}_{U_L} + 3\hat{\mu}_{D_L} + \hat{\mu}_{\ell_L} + \hat{\mu}_{\nu_L}), \qquad (91)$$

where $n_F = 3$, $U = u, c, t$, $D = d, s, b$, $\hat{\Gamma}_{\rm sph}$ is the sphaleron rate per unit volume, which in the unbroken phase is given by eq. (53). Here the $\hat{\mu}_i = \mu_i - \bar{\mu}_i = 2\mu_i$ denote the difference between the respective particle and antiparticle chemical potentials. For a non-interacting gas of massless fermions i, the relation between $\hat{\mu}_i$ and the asymmetry in the corresponding particle and antiparticle number densities is $n_i - \bar{n}_i \simeq g\hat{\mu}_i T^2/12$, where $g = 1$ for a left-handed lepton and $g = 3$ for a left-handed quark because of three colors. In the symmetric phase (91) then reads

$$\frac{dn_B}{dt} = -6n_F \frac{\hat{\Gamma}_{\rm sph}}{T^3}(3B_L + L_L), \qquad (92)$$

where B_L and L_L denote the total left-handed baryon and lepton number densities, respectively. The factor of 3 comes from the definition of baryon number, which assigns baryon number 1/3 to a quark. This equation tells us what we already concluded qualitatively above: baryon rather than antibaryon production requires a negative left-handed fermion number density, i.e., a positive flux J_q^L. The total flux $\sum_\psi J_\psi^L$ determines the left-handed fermion number density. Then eq. (92) yields n_B and, using $s = 2\pi^2 g_{*s} T^3/45$ with $g_{*s} \simeq 110$ (see section 2.2), a prediction for the baryon-to-entropy ratio is obtained.

So much to the main aspects of the mechanism. There are, however, a number of issues that complicate this scenario. Decoherence effects during reflection should be studied. The propagation of fermions is affected by the ambient high temperature plasma leading to modifications of their vacuum dispersion relations. The shape and velocity of the wall is a critical issue. We refer to the quoted literature for a discussion of these and other points.

Because Higgs sector CP violation as discussed above is strongest for top quarks, one might expect that these quarks make the dominant contribution to the right hand side of (92). However, several effects tend to decrease their contribution relative to those of τ leptons. As top quarks interact much more strongly than τ leptons they have a shorter mean free path. This means that for typical wall thicknesses the thin-wall approximation does not hold for t quarks. Further the injected left-handed top current J_t^L is affected by QCD sphaleron fields which induce processes – unsuppressed at high T – where the chiralities of the quarks are flipped [66,67]. This damps the t quark contribution to B_L. Refs. [63,64] come to the conclusion that in this type of particle physics models the contribution of τ leptons to the left-handed fermion number density is the most important one. Ref. [64] finds that this induces a baryon-to-entropy ratio of about

$$\frac{n_B}{s} \simeq 10^{-12} \frac{\Delta\theta}{v_{\rm wall}}, \qquad (93)$$

where $v_{\rm wall}$ is the velocity of the wall and $\Delta\theta \simeq \theta(z = -\infty) - \theta(z = +\infty)$. Barring the possibility of spontaneous CP violation at non-zero temperatures in the 2-Higgs doublet models, $\Delta\theta$ should be roughly of the order of the CP-violating phase ξ in the 2-doublet potential (64). Using that primordial nucleosynthesis allows $n_B/s \simeq \eta/7 \simeq (2-8) \times 10^{-11}$ (cf. (26)) one gets the parameter constraint $\Delta\theta/v_{\text{wall}} \sim 40$. Even large CP violation, $\Delta\theta$ of order 1, would require small wall velocities, which might not be supported by investigations of the dynamics of the phase transition. Nevertheless, the 2-Higgs doublet models predict roughly the correct order of magnitude. In view of the complexity of this baryogenesis scenario, there are possibly additional, hitherto unnoticed effects that may influence n_B/s. For a treatment of the case when the bubble walls are thick, in the sense that fermions interact with the plasma many times as the wall sweeps through, see [65].

Only a few words on electroweak baryogenesis in the minimal supersymmetric standard model, see e.g. [53–56]. The essentials of the scenario are analogous to the 2HDM case, with CP-violating sources as described in section 5.3.2, the main source for baryogenesis being the phase φ_μ of the complex Higgsino mixing parameter μ_c. A number of authors conclude that the dominant baryogenesis source comes from the Higgsino sector, which produces a non-zero flux of left-handed quark chirality. The results for n_B/s may be presented in the form

$$\frac{n_B}{s} = 4 \times 10^{-11} a \sin\varphi_\mu . \qquad (94)$$

There is a considerable spread in the predicted values of a, respectively in the resulting estimates of the necessary magnitude of $\sin\varphi_\mu$. While refs. [54,56] find that a small CP phase $\varphi_\mu \gtrsim 0.04$ would suffice to obtain the correct order of magnitude of n_B/s (which requires, however, small wall velocities), ref. [55] concludes that $\sin\varphi_\mu$ must be of order 1. Large values of φ_μ, however, tend to be in conflict with the constraints from the experimental upper bounds on the electric dipole moments of the electron and neutron, see section 5.3.2. Electroweak baryogenesis in a next-to-minimal SUSY model was investigated in [68].

5.5 Role of the KM Phase

We haven't yet discussed which role is played in baryogenesis scenarios by the SM source of CP violation, the KM phase δ_{KM}. This question was put out of the limelight after it had become clear that the SM alone cannot explain the BAU, for reasons outlined above. Therefore, SM extensions must be invoked, and such extensions usually entail in a natural way new sources of CP violation which can be quite effective, as far as their role in baryogenesis scenarios is concerned, as we have seen – see also the next section. Nevertheless, this is a very relevant issue.

Recall the following well-known features of KM CP violation. All CP-violating effects, which are generated by the KM phase in the charged weak quark current couplings to W bosons, are proportional to the invariant [69,70]:

$$J_{CP} = \prod_{\substack{i>j \\ u,c,t}} (m_i^2 - m_j^2) \prod_{\substack{i>j \\ d,s,b}} (m_i^2 - m_j^2) \ \text{Im} \, Q \,, \qquad (95)$$

where i,j = 1,2,3 are generation indices, m_u, etc. denote the respective quark masses, and $\text{Im} \, Q$ is the imaginary part of a product of 4 CKM matrix elements, which is invariant under phase changes of the quark fields. There are a number of equivalent choices for $\text{Im} \, Q$. A standard choice is

$$\text{Im} \, Q = \text{Im}(V_{ud} V_{cb} V_{ub}^* V_{cd}^*) \,. \qquad (96)$$

Inserting the moduli of the measured CKM matrix elements yields $|\text{Im} \, Q|$ smaller than 2×10^{-5}, even if KM CP violation is maximal; i.e., $\delta_{\text{KM}} = \pi/2$ in the KM parameterization of the CKM matrix. We may write $\text{Im} \, Q \simeq 2 \times 10^{-5} \sin \delta_{\text{KM}}$. As far as the SM at temperatures $T \neq 0$ is concerned, the CP symmetry can be broken only in regions of space where the gauge symmetry is also broken, or at the boundaries of such regions, because $J_{CP} \neq 0$ requires non-degenerate quark masses. Imagine the EW transition would be first order due to a 2-Higgs doublet extension of the SM with *no* CP violation in the Higgs sector. The question is then: is the KM source of CP violation strong enough to create a sufficiently large asymmetry $\Delta \mathcal{R}_{CP}$ in the probabilities for reflection of (anti)quarks at the expanding wall as discussed above? It is clear that $\Delta \mathcal{R}_{CP}$ must be proportional to a dimensionless quantity of the form J_{CP}/D, where D has mass dimension 12. Reflection of quarks and antiquarks at a bubble wall is not CKM-suppressed; hence D does not contain small CKM matrix elements. If one recalls that in the symmetric phase the quark masses and thus J_{CP} vanish, it seems reasonable to treat the quark masses (perhaps not the top quark mass) as a perturbation. In the massless limit the mass scale of the theory at the EW transition is then given by the critical temperature $T_c \sim 100$ GeV. Thus one gets for the dimensionless measure of CP violation:

$$d_{CP} \equiv \frac{J_{CP}}{T_c^{12}} \sim 10^{-19} \qquad (97)$$

as an estimate of $\Delta \mathcal{R}_{CP}$. Clearly this number is orders of magnitude too small to account for the observed n_B/s. CP violation à la KM is therefore classified, by consensus of opinion, as being irrelevant for baryogenesis. It was argued, however, that there may exist significant enhancement effects [71], and there has been a considerable debate over this issue [71–73].

6   Out-of-Equilibrium Decay of Super-Heavy Particle(s)

Historically the first type of baryogenesis scenario which was developed in detail is the so-called out-of-equilibrium decay of a super-heavy particle [74,75],

[16,76]. The basic idea is that at a very early stage of the expanding universe, a super-heavy particle species X existed, the reaction rate of which became smaller than the expansion rate H of the universe at temperatures $T \gg T_{\text{EW}}$. Therefore, these particles decoupled from the thermal bath and became over-abundant. The decays of X and of the antiparticles $\bar{X}$ are supposed to be CP- and B-violating, such that a net baryon number $\Delta B \neq 0$ is produced when the $X, \bar{X}$ have decayed. This scenario has its natural setting in the framework of grand unified theories. A brief outline is given in the next subsection. A viable variant is baryogenesis via leptogenesis through the lepton-number violating decays of (a) heavy Majorana neutrino(s) [77]. This will be discussed in Subsect. 6.2.

6.1 GUT Baryogenesis

The "out-of-equilibrium decay" scenario is natural in the context of grand unified theories. Grand unification aims at unifying the strong and electroweak interactions at some high energy scale. It works, for instance, in the context of supersymmetry, where the effective couplings of the strong, weak, and electromagnetic interactions become equal at an energy scale $M_{\text{GUT}} \simeq 10^{16}$ GeV [47]. A matter multiplet forming a representation of the GUT gauge group G contains both quarks and leptons. Gauge bosons mediate transitions between the members of this multiplet, and – for many gauge groups – some of the gauge bosons induce B-violating processes. Also C violation and non-standard CP violation occurs naturally. As to the latter: the gauge group G must be broken at the GUT scale $M_{\text{GUT}} \simeq 10^{16}$ GeV to some smaller symmetry group $G' \supseteq \text{SU}(3)_c \times \text{SU}(2)_L \times \text{U}(1)_Y$. This is accomplished by scalar Higgs multiplets. As a consequence, GUTs contain in general super-heavy Higgs bosons with B-violating and CP-violating Yukawa couplings to quarks and leptons.

The simplest example of a GUT is based on the gauge group $G = SU(5)$. It is obsolete because the model is in conflict with the stability of the proton. Irrespective of this obstruction the minimal version of this model is of no use for implementing the scenario which we discuss below, because the interactions of minimal $SU(5)$ conserve $B - L$. A popular gauge group is $SO(10)$ which allows to construct models that avoid both obstacles [47].

Rather than going into the details of a specific GUT let us illustrate the baryogenesis mechanism with a well-known toy model [1]. Consider a super-heavy leptoquark gauge boson X which is supposed to have quark-quark and quark-lepton decay channels, $X \to qq, \ell \bar{q}$. In the table the branching ratios of these decays, of the decays of the antiparticle $\bar{X}$, and the baryon numbers B of the final states are tabulated.

final state f	branching ratio	B
$X \to qq$	r	$2/3$
$X \to \ell\bar{q}$	$1 - r$	$-1/3$
$\bar{X} \to \bar{q}\bar{q}$	$\bar{r}$	$-2/3$
$\bar{X} \to \bar{\ell}q$	$1 - \bar{r}$	$1/3$

The baryon number produced in the decays of X and $\bar{X}$ is:

$$B_X = \frac{2}{3}r - \frac{1}{3}(1-r),$$
$$B_{\bar{X}} = -\frac{2}{3}\bar{r} + \frac{1}{3}(1-\bar{r}), \quad (98)$$

and the net baryon number produced is

$$\Delta B_X \equiv B_X + B_{\bar{X}} = r - \bar{r}$$
$$= \frac{\Gamma(X \to f_1)}{\Gamma_{\rm tot}(X)} - \frac{\Gamma(\bar{X} \to \bar{f}_1)}{\Gamma_{\rm tot}(\bar{X})} = \frac{\Gamma(X \to f_1) - \Gamma(\bar{X} \to \bar{f}_1)}{\Gamma_{\rm tot}}, \quad (99)$$

where $f_1 = qq$ and we have used that $\Gamma_{\rm tot}(X) = \Gamma_{\rm tot}(\bar{X})$, which follows from CPT invariance. Obviously if C or CP were conserved then $\Delta B_X = 0$. Suppose the quarks and leptons q, ℓ couple to a spin-zero boson χ with Yukawa couplings that contain a non-removable CP-violating phase. It is natural to assume that C is already violated in the tree-level interactions of X to fermions. The CP-violating interactions affect the $X, \bar{X}$ decay amplitudes beyond the tree level, as shown in Fig. 15. The decay amplitude for $X \to qq$ is, up to spinors and a polarization vector describing the external particles,

$$A(X \to qq) = A_0 + A_1 = A_0 + B e^{i\delta_{CP}}, \quad (100)$$

where the tree amplitude A_0 is real. $\Delta B_X \neq 0$ requires, in addition to CP violation, also a non-zero final-state interaction phase. Therefore, the masses of the X boson and of the fermions must be such that the intermediate fermions in the 1-loop contribution to the amplitude can be on their respective mass shells and re-scatter to produce the final state. This causes a complex $B = |B|\exp(i\omega)$. The decay amplitude for $\bar{X} \to \bar{q}\bar{q}$ is

$$A(\bar{X} \to \bar{q}\bar{q}) = A_0 + |B|e^{i\omega}e^{-i\delta_{CP}}. \quad (101)$$

Using (100), (101) one obtains that

$$\Delta B_X \propto \frac{|AB|\sin\omega \sin\delta_{CP}}{\Gamma_{\rm tot}}, \quad (102)$$

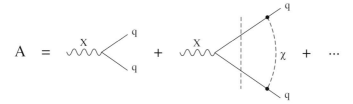

Fig. 15. Amplitude for the decay $X \to qq$ to one-loop approximation. The vertical dashed line indicates the absorptive part of the 1-loop contribution which enters ΔB_X.

and the constant of proportionality includes factors from phase space integration.[11] In addition, the baryon number ΔB_χ produced in the decays of the $\chi, \bar{\chi}$ bosons must also be computed. Let's assume that $\Delta B_X + \Delta B_\chi \neq 0$.

As long as the interactions of these bosons, which include decays, inverse decays, annihilation, the B-violating reactions $qq \to \ell \bar{q}$, $\bar{q}\bar{q} \to \bar{\ell} q$, etc. (remember the discussion in section 3.2) are fast compared to the expansion rate H, the $X, \bar{X}, \chi, \bar{\chi}$ have thermal distributions and the average baryon number of the plasma remains zero. Therefore the interactions of these bosons must be weak enough that they can fall out of equilibrium. This is a delicate issue, because these particles carry gauge charges and can couple quite strongly to the plasma of the early universe.

Let's outline the scenario for the $X, \bar{X}$. (It applies also to the scalar particles.) At temperatures $T \gg m_X$, where m_X is the mass of the X boson, the $X, \bar{X}$ particles have relativistic velocities and are assumed to be in thermal equilibrium. Then $n_X = n_{\bar{X}} \sim n_\gamma$ holds for their number densities. At lower temperatures, the $X, \bar{X}$ bosons become non-relativistic and, as long as they remain in thermal equilibrium, their densities get Boltzmann-suppressed with decreasing temperature, $n_{\bar{X}} = n_X \sim (m_X T)^{3/2} \exp(-m_X/T)$. Because $\Gamma_{\mathrm{ann}} \propto n_X$, the total rate $\Gamma_X \sim \alpha m_X$ of X and $\bar{X}$ decay is the relevant number to compare with H. If $\Gamma_X < H$, an excess of $X, \bar{X}$ with respect to their equilibrium numbers will develop. The $X, \bar{X}$ drift along in the expanding universe for a little while and decay. Notice that the inverse decays, $f \to X$, $\bar{f} \to \bar{X}$, by which bosons are created again by quark-quark annihilation, etc. are blocked, because the fraction of these fermions with sufficient energy to produce a super-heavy boson is Boltzmann suppressed for $T < m_X$. At the time of their decay, $t \sim \Gamma_X^{-1}$, there is quite an overabundance: $n_X = n_{\bar{X}} \sim n_\gamma(T_{\mathrm{decay}})$. Using that the entropy density $s \sim g_* n_\gamma$

---

[11] In charged B meson decays a CP asymmetry $A_{CP} = [\Gamma(B^+ \to f) - \Gamma(B^- \to \bar{f})]/[\Gamma(B^+ \to f) + \Gamma(B^- \to \bar{f})]$ arises in completely analogous fashion. $A_{CP} \neq 0$ requires CP violation and final state interactions.

(see sect. 2.2) one gets for the produced baryon asymmetry:

$$\frac{n_B}{s} \sim \frac{\Delta B\, n_\gamma}{g_* n_\gamma} \sim \frac{\Delta B}{g_*}, \tag{103}$$

where $\Delta B \simeq \Delta B_X$ is the baryon number produced per boson decay. For a (GUT) extension of the SM one may expect that g_* is somewhere between 10^2 and 10^3. Thus only a tiny CP asymmetry $\Delta B \sim 10^{-8} - 10^{-7}$ is required to obtain $n_B/s \sim 10^{-10}$. Of course, these crude estimates must be made quantitative by computing the relevant reaction rates using a specific particle physics model, and tracking the time evolution of the particle densities by solving the Boltzmann equations. A detailed exposition is given in [1].

The above condition for the decoupling of the X particles from the thermal bath, $\Gamma_X \sim \alpha m_X < H$, translates into a condition on the mass of the spin 1 gauge boson: $m_X > \alpha g_*^{-1/2} m_{\rm Pl} \sim 10^{16}$ GeV, where $\alpha = g_{\rm gauge}^2/(4\pi) \sim 10^{-2}$. For super-heavy scalar bosons with B-violating decays a mass bound obtains which is lower (cf., e.g., [6]).

There are several pitfalls that constrain this type of baryogenesis mechanism. First, remember that a scenario that tries to explain the BAU by a mechanism that operates above the temperature $T_{\rm EW} \sim 100$ GeV must involve interactions that violate $B - L$. In the context of grand unified theories, models based on the gauge group $SO(10)$ lead to $(B - L)$ non-conservation. These models have several attractive features, in particular with respect to the scenario discussed in the next subsection.

GUT baryogenesis may be in conflict with inflation. This is a serious problem. An essential assumption in the above scenario was that at very high temperatures T above m_X the $X, \bar{X}$ particles were in thermal equilibrium and were as abundant as photons. If these particles are the super-heavy gauge or Higgs bosons of a GUT this assumption may be wrong. It might be that the temperature of the quasi-adiabatically expanding plasma of particles in the early universe was always smaller than $M_{\rm GUT}$. There are a number of reasons to believe that the energy of the very early universe was dominated by vacuum energy, which led to exponential expansion of the cosmos. This is the basic assumption of the inflationary model(s). These models solve a number of fundamental cosmological problems, including the monopole problem. A number of GUTs predict super-heavy, stable magnetic monopoles. Their contribution to the energy density of the universe would over-close the cosmos – but that is not observed. Inflation would sweep away these monopoles, along with other particles, leaving an empty space. At the end of inflation the vacuum energy is converted through quantum fluctuations into pairs of relativistic particles and antiparticles which then thermalize. This process is called reheating and it can be characterized by an energy scale called the reheat temperature T_r. If the reheating process is fast, i.e., if T_r is above $M_{\rm GUT}$, the monopoles are re-created. On the other hand if reheating occurred slowly such that T_r is well below $M_{\rm GUT}$ then the re-production of $X\bar{X}$ super-heavy gauge and Higgs bosons – which were to initiate baryogenesis as described above – appears to be inhibited, or should at least be suppressed. See [6] for an overview on ways to circumvent this and associated problems.

6.2 Baryogenesis Through Leptogenesis

This mechanism is a special case of the "out-of-equilibrium decay" scenario. It assumes the existence of a heavy Majorana neutrino species in the early universe above $T_{\rm EW}$ with a particle mass, typically, of the order of $M \sim 10^{12}$ GeV – or, in fact, the more realistic case of three heavy Majorana neutrino species with non-degenerate masses. These particles interact only weakly with the other particle species in the early universe and fall out of equilibrium at some temperature $T \sim M \gg T_{\rm EW}$. It is essential that some of the interactions of the underlying particle physics model do not conserve $B - L$. The heavy Majorana neutrinos decay, for instance into ordinary leptons and Higgs bosons which are the most important channels, thereby generating a non-zero lepton number. Lepton-number violating scattering processes must not wash out this asymmetry. Then the $(B - L)$-conserving SM sphaleron reactions, which occur rapidly enough above $T_{\rm EW}$, convert this lepton asymmetry into a baryon asymmetry .

The scenario was suggested in [77], and it has been subsequently developed further – see [78–83] and the reviews [7]. The attractiveness of this scenario stems from the observed atmospheric and solar neutrino solar neutrino deficits which point to oscillations of the light neutrinos. It is well-known that these data can be explained by small differences in the masses of the electron, muon, and tau neutrinos. The value of $\Delta m_{23}^2 = m_3^2 - m_2^2$ extracted from the data indicates that the mass of the heaviest of the three light neutrinos is of the order of 10^{-2} eV. Such small masses can be explained in a satisfactory way by the so-called seesaw mechanism [84]. This mechanism requires (i) the neutrinos to be Majorana fermions and (ii) three very heavy right-handed neutrinos which are singlets with respect to $SU(2)_L \times U(1)_Y$ – see Appendix B.

Within the framework of GUTs, popular models are based on the gauge group $SO(10)$ which contain in their particle spectra ultra-heavy right-handed Majorana neutrinos with lepton-number violating decays. We consider here only a minimal, non-GUT model. Take the electroweak standard model and add three right-handed $SU(2)_L \times U(1)_Y$ singlet fields $\nu_{\alpha R}$ ($\alpha = 1, 2, 3$) with a Majorana mass term for these fields involving mass parameters much larger than $v = 246$ GeV. The general Yukawa interaction for the charged leptons and neutrinos is then given by eq. (141) of appendix B with $\Phi \equiv \Phi_1 = \Phi_2$, where $\Phi = (\phi^+, \phi^0)^T$ is the SM $SU(2)_L$ doublet field. As described in appendix B we have in the mass basis three very light, practically left-handed Majorana neutrinos, which we identify with the neutrinos we know, and three very heavy, right-handed Majorana neutrinos N_i. Let's switch back to the early universe when the N_i were still around. The interaction (141) implies that the N_i have lepton-number violating decays at tree-level, $N_i \to \ell\phi$ and $N_i \to \bar{\ell}\phi^*$, where ℓ, ϕ denotes either a negatively charged lepton and a ϕ^+ (which later ends up as the longitudinal component of the W^+ boson) or a light neutrino and a ϕ^0. C and CP violation cause a difference in these two rates – see below. As long as the N_i are in thermal equilibrium CPT invariance and the unitarity of the S matrix (cf. section 3.2) guarantee that the average lepton number remains zero. (The N_i are to be described as on-shell resonances in corresponding $2 \leftrightarrow 2$ processes.) When the N_i have fallen out of equilibrium, there is still the danger of lepton-number violating wash-out processes, for instance $|\Delta L| = 2$ reactions mediated by N_i exchange. The requirement $\Gamma_{|\Delta L|=2}(T) < H(T)$ for temperatures T smaller than the leptogenesis temperature, e.g. $T \lesssim 10^{10}$ GeV, implies an upper bound on the masses of the light neutrinos [7].

We assume the N_i to be non-degenerate and put the labels such that $M_3 > M_2 > M_1$ holds for the masses. The decay width of N_i at tree level in its rest frame, see Fig. 16, is easily computed using (141):

$$\Gamma_i \equiv \Gamma(N_i \to \ell\phi) + \Gamma(N_i \to \bar{\ell}\phi^*)$$
$$= \frac{(M_D^\dagger M_D)_{ii}}{4\pi v^2} M_i, \qquad (104)$$

where M_D is the Dirac mass matrix (142). For leptogenesis to work the decays of the N_i must be slow as compared to H. The condition $\Gamma_i < H(T = M_i)$ is fulfilled only if the masses of the light neutrinos are small, roughly $m_{\nu_i} < 10^{-3}$ eV [85], which is compatible with observations. There is then an excess of the heavy neutrinos with respect to their rapidly decreasing equilibrium distributions $n_{\text{eq}} \sim \exp(-M/T)$.

Eventually the N_i decay and lepton number is produced. It is due to the CP-asymmetry in the decay rates which is generated by the interference of the tree amplitude with the absorptive part of the 1-loop amplitude depicted in Fig. 16, analogous to eq. (102). If $M_1 \ll M_2, M_3$ one obtains for the decay of N_1:

$$\epsilon_1 \equiv \frac{\Gamma(N_1 \to \ell\phi) - \Gamma(N_1 \to \bar{\ell}\phi^*)}{\Gamma(N_1 \to \ell\phi) + \Gamma(N_1 \to \bar{\ell}\phi^*)}$$
$$\simeq -\frac{3}{4\pi v^2} \frac{1}{(M_D^\dagger M_D)_{11}} \sum_{j=2,3} \text{Im}[(M_D^\dagger M_D)_{1j}^2] \frac{M_1}{M_j}. \qquad (105)$$

The asymmetries ϵ_i are determined by the moduli and the CP-violating phases of the elements of the matrix $M_D^\dagger M_D$. The moduli are related to the light neutrino masses, while the CP-violating phases are in general unrelated to the CP-violating phases of the mixing matrix in the leptonic charged current-interactions involving the light neutrinos – see appendix B.

The asymmetry (105) corresponds to an asymmetry in the density of leptons versus antileptons, $n_L \equiv n_\ell - n_{\bar{\ell}} \neq 0$. A crude estimate of the lepton-

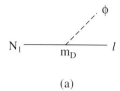

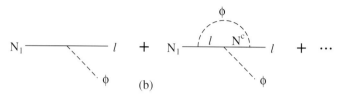

Fig. 16. Two-body decay of a super-heavy Majorana neutrino $N_1 \to \ell\phi$: Born amplitude (a) and a 1-loop contribution (b). Self-energy contributions are not depicted.

number-to-entropy ratio $Y_L = n_L/s$ gives

$$Y_L \sim \epsilon_1 \frac{n_N}{s} \sim \frac{\epsilon_1}{g_*}, \tag{106}$$

where $g_* \sim 100$ in the SM. Due to wash-out processes like those mentioned above this ratio is, in fact, smaller than (106). In order to determine the suppression factor κ, the Boltzmann equations for the time evolution of the particle number densities must be solved [78,80,7]. A typical solution for n_{N_1} is sketched in Fig. 17. Refs. [80,7] find $\kappa \sim 10^{-1} - 10^{-3}$, depending on the particle physics model.

The asymmetry in lepton number feeds the $(B - L)$-conserving weak sphaleron reactions, which occur rapidly enough above $T_{\rm EW}$, and produce an asymmetry in baryon number $Y_B = n_B/s$. There is a relation between Y_B and the corresponding asymmetries Y_{B-L} and Y_L. For a given particle physics model this relation depends on the processes which are in thermal equilibrium, and it is given by [86,87]:

$$Y_B = C Y_{B-L} = \frac{C}{C-1} Y_L. \tag{107}$$

The particle reactions which are fast enough as compared with H yield relations among the various chemical potentials, and these relations determine the number C. For the minimal model considered above, one has $C = 28/79$ in the high temperature phase if all but the $|\Delta L| = 2$ reactions are in thermal equilibrium [86]. (In general C depends on the ratio v_T/T, where v_T is the Higgs VEV which develops in the broken phase [88,89].) Using (106), (107)

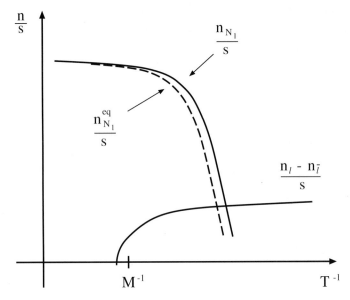

Fig. 17. Evolution of the ratio n_{N_1}/s as the universe cools off. Departure from thermal equilibrium occurs at $T \lesssim M_{N_1}$ and a leptonic asymmetry is generated [7].

the generated baryon-to-entropy ratio is estimated to be

$$Y_B \sim -Y_L = -\kappa \frac{\epsilon_1}{g_*} \,. \tag{108}$$

Using $g_* \sim 100$ and a dilution factor $\kappa \sim 10^{-2}$, we see that only a very small lepton-asymmetry $\epsilon_1 \sim 10^{-6}$ is needed. In fact, lepton-number violation must not be too strong, in order that the whole scenario works.

Detailed studies of leptogenesis have been made, for a number of SM extensions and using *Ansätze* for the neutrino mass matrices that fit well to the observations concerning the solar and atmospheric neutrino deficits [7]. The conclusion is that $Y_B \sim 10^{-10}$ is naturally explained by the decay of heavy Majorana neutrinos, the lightest of which having a mass $M_1 \sim 10^{10}$ GeV, and the required pattern of the 3 light neutrino masses is consistent with observations.

7 Summary

In these lectures I have outlined two popular theories of baryogenesis, which presently seem to be the most plausible ones: electroweak baryogenesis and out-of-equilibrium decay scenarios, in particular baryogenesis via leptogenesis by the decays of ultra-heavy Majorana neutrinos. A number of other, quite ingenious mechanisms for generating the BAU were conceived. Their discussion is, however, beyond the scope of these notes and the reader is referred to the quoted reviews.

Electroweak baryogenesis (EWBG) will be testable in the not too distant future. The clarification of the origin of electroweak symmetry breaking will be a central physics issue at the Tevatron and at future colliders, and the outcome will be crucial for the EWBG scenario. An important result was already obtained: Theoretical investigations of the SM electroweak phase transition and the experimental lower bound from the LEP experiments on the mass of the SM Higgs boson, $m_H^{\text{SM}} > 114$ GeV, led to the conclusion that the standard model of particle physics fails to explain the BAU. EWBG is still viable in extensions of the SM, the most popular of which is the minimal supersymmetric extension. However, the requirement of the electroweak phase transition to be strongly first order translates into tight upper bounds on the mass of the lightest Higgs boson, $m_H < 115$ GeV, and on the mass of the lighter of the two stop particles, $m_{\tilde{t}_1} < 170$ GeV, of the MSSM. Another important ingredient to EWBG is non-standard CP violation. This motivates the search for T-violating effects in experiments with atoms and molecules, and neutrons. Non-SM CP violation can also be traced in B meson decays or in high-energy reactions including the production and decays of top quarks and Higgs bosons, if Higgs particles will be discovered.

GUT-type baryogenesis scenarios cannot be falsified by laboratory experiments, but they would, of course, get spectacular empirical support if proton decay would be found, etc. That's what makes leptogenesis by the decays of ultra-heavy Majorana neutrinos attractive: it has, albeit indirect, support from the observed atmospheric and solar neutrino deficits. Theoretical investigations have shown that the scenario is consistent. As far as unknown parameters are concerned, the degree of arbitrariness is constrained: in order to obtain the correct order of magnitude of the BAU the masses of the light neutrinos must lie in range which is consistent with the interpretation of the solar and atmospheric neutrino data. The scenario would get a further push if the light neutrinos would turn out to be Majorana particles. Future particle physics experiments and/or astrophysical observations will bring us closer to understanding what is responsible for the matter-antimatter asymmetry of the universe.

Acknowledgments

I wish to thank the organizers of this school, in particular M. Beyer, for having arranged this pleasant meeting. I am grateful to A. Brandenburg for comments on the manuscript and I am indebted to T. Leineweber for the production of the postscript figures.

Appendix A

Let $q(\mathbf{x},t)$ be the Dirac field operator that describes a quark of flavor $q = u,...,t$, $q^\dagger(\mathbf{x},t)$ denotes its Hermitean adjoint, and $\bar{q} = q^\dagger \gamma^0$. The baryon number operator (36) is

$$\hat{B} = \frac{1}{3} \sum_q \int \mathrm{d}^3 x : q^\dagger(\mathbf{x},t) q(\mathbf{x},t) :, \tag{109}$$

and the colons denote normal ordering. Let C, P denote the unitary and T the anti-unitary operator which implement the charge conjugation, parity, and time reversal transformations, respectively, in the space of states. Their action on the quark fields is, adopting standard phase conventions,

$$Pq(\mathbf{x},t)P^{-1} = \gamma^0 q(-\mathbf{x},t), \tag{110}$$
$$Pq^\dagger(\mathbf{x},t)P^{-1} = q^\dagger(-\mathbf{x},t)\gamma^0, \tag{111}$$
$$Cq(\mathbf{x},t)C^{-1} = i\gamma^2 q^\dagger(\mathbf{x},t), \tag{112}$$
$$Cq^\dagger(\mathbf{x},t)C^{-1} = iq(\mathbf{x},t)\gamma^2, \tag{113}$$
$$Tq(\mathbf{x},t)T^{-1} = -\mathrm{i}\, q(\mathbf{x},-t)\gamma_5\gamma^0\gamma^2, \tag{114}$$
$$Tq^\dagger(\mathbf{x},t)T^{-1} = -\mathrm{i}\gamma^2\gamma^0\gamma_5 q^\dagger(\mathbf{x},-t), \tag{115}$$

where γ^0, γ^2, and $\gamma_5 = i\gamma^0\gamma^1\gamma^2\gamma^3$ denote Dirac matrices. Then

$$P : q^\dagger(\mathbf{x},t)q(\mathbf{x},t) : P^{-1} = : q^\dagger(-\mathbf{x},t)q(-\mathbf{x},t) :, \tag{116}$$
$$C : q^\dagger(\mathbf{x},t)q(\mathbf{x},t) : C^{-1} = : q(\mathbf{x},t)q^\dagger(\mathbf{x},t) := -: q^\dagger(\mathbf{x},t)q(\mathbf{x},t) :, \tag{117}$$
$$T : q^\dagger(\mathbf{x},t)q(\mathbf{x},t) : T^{-1} = : q^\dagger(\mathbf{x},-t)q(\mathbf{x},-t) : . \tag{118}$$

With these relations we immediately obtain:

$$P\hat{B}P^{-1} = \hat{B}, \tag{119}$$
$$C\hat{B}C^{-1} = -\hat{B}. \tag{120}$$

As shown in section 4 the baryon number operator is time-dependent due to non-perturbative effects. Using translation invariance we have $\hat{B}(t) = \mathrm{e}^{\mathrm{i}Ht}\hat{B}(0)\mathrm{e}^{-\mathrm{i}Ht}$, where H is the Hamiltonian of the system. The operator $\hat{B}(0)$ is even with respect to T and odd with respect to $\Theta \equiv CPT$:

$$\Theta \hat{B}(0) \Theta^{-1} = -\hat{B}(0). \tag{121}$$

Appendix B

Here we discuss the general structure of $SU(2)_L \times U(1)_Y$ invariant Yukawa interactions in the lepton sector if neutrinos are Majorana particles. Let's first collect some basic formulae for Majorana fields. Consider a Dirac field

$$\psi(x) = \begin{pmatrix} \xi(x) \\ \eta(x) \end{pmatrix}, \tag{122}$$

where ξ, η are 2-component spinor fields. In the chiral representation of the γ matrices, using the convention where $\gamma_5 = \text{diag}(I_2, -I_2)$, we have $\xi = \psi_R, \eta = \psi_L$, where ψ_R, ψ_L are the right-handed and left-handed Weyl fields. In the chiral representation the charge conjugated spinor field ψ^c reads

$$\psi^c \equiv i\gamma^2 \psi^\dagger = \begin{pmatrix} i\sigma_2 \eta^\dagger \\ -i\sigma_2 \xi^\dagger \end{pmatrix}, \tag{123}$$

and σ_2 is the second Pauli matrix. Let's use the Weyl fields in 4-component form, $\psi_R = (\xi, 0)^T, \psi_L = (0, \eta)^T$, and determine, using (123), their charge-conjugates:

$$\psi_L^c \equiv (\psi_L)^c = \begin{pmatrix} i\sigma_2 \eta^\dagger \\ 0 \end{pmatrix}, \tag{124}$$

$$\psi_R^c \equiv (\psi_R)^c = \begin{pmatrix} 0 \\ -i\sigma_2 \xi^\dagger \end{pmatrix}. \tag{125}$$

From this equation we can also read off the relation between the 2-component Weyl fields and their charge conjugates. Eq. (125) tells us that $\psi_L^c (\psi_R^c)$ is a right-handed (left-handed) Weyl field. Thus the Weyl field operator $\psi_L(\psi_R)$ annihilates a Dirac fermion state $|\psi>$ having L (R) chirality and creates an antifermion state $|\bar\psi>$ with R (L) chirality, while $\psi_L^c(\psi_R^c)$ annihilates $|\bar\psi>$ having R (L) chirality and creates a state $|\psi>$ with L (R) chirality. Moreover, we immediately obtain that

$$\overline{\psi_L^c} \equiv (\psi_L^c)^\dagger \gamma^0 = (0, i\eta^T \sigma_2), \tag{126}$$

$$\overline{\psi_R^c} \equiv (\psi_R^c)^\dagger \gamma^0 = (-i\xi^T \sigma_2, 0). \tag{127}$$

As to neutrinos, there are two options concerning their nature (which must eventually be resolved experimentally): either Dirac or Majorana fermion. The latter means, loosely speaking, that a neutrino would be its own antiparticle. Actually, for a Majorana fermion the distinction between particle and antiparticle looses its meaning because there is no longer a conserved quantum number that would discriminate between them (see below). A Majorana field is defined by the condition

$$\psi^c \stackrel{!}{=} r\psi, \tag{128}$$

where $|r| = 1$ is a phase chosen by convention. For $r = +1$ the four-component field $\psi_1 = (i\sigma_2\eta^\dagger, \eta)^T$ is a solution of this equation. In terms of Weyl fields this solution reads

$$\psi_1 = \psi_L + \psi_L^c. \tag{129}$$

The other solution of eq. (128) with $r = 1$ is

$$\psi_2 = \psi_R + \psi_R^c. \tag{130}$$

Next we consider the Majorana mass terms. For Majorana particles described by ψ_1 and ψ_2 with masses m_1 and m_2, respectively, we can write down the following Majorana mass terms

$$\mathcal{L}_M^{(1)} = -\frac{m_1}{2}\bar{\psi}_1\psi_1 = -\frac{m_1}{2}\overline{\psi_L^c}\psi_L + \text{h.c.}, \tag{131}$$

$$\mathcal{L}_M^{(2)} = -\frac{m_2}{2}\bar{\psi}_2\psi_2 = -\frac{m_2}{2}\overline{\psi_R^c}\psi_R + \text{h.c.}, \tag{132}$$

where we have used that $\bar{\psi}_A\psi_A = \overline{\psi_A^c}\psi_A^c = 0$ for A=L,R. These mass terms violate the "ψ-number" by 2 units, $|\Delta L_\psi| = 2$. For instance $<\bar{\psi}_R|\overline{\psi_L^c}\psi_L|\psi_L> \neq 0$; i.e., the first term in $\mathcal{L}_M^{(1)}$ flips a left-handed $|\psi_L>$ into a right-handed $|\bar{\psi}_R>$. Recalling the connection between symmetries and conservation laws we see that this non-conservation of ψ-number is related to the fact that $\mathcal{L}_M^{(1,2)}$ are not invariant under the global U(1) transformation $\psi_{L,R} \to e^{i\omega}\psi_{L,R}$, $\bar{\psi}_{L,R} \to e^{-i\omega}\bar{\psi}_{L,R}$.

The general mass term for neutrino fields ν_L and ν_R contains both Majorana and Dirac terms with complex mass parameters. The 1-flavor case reads

$$-\mathcal{L}_{D+M} = \frac{m_L}{2}\overline{\nu_L^c}\nu_L + \frac{m_R}{2}\overline{\nu_R^c}\nu_R + m_D\bar{\nu}_R\nu_L + \text{h.c.} \tag{133}$$

$$= \frac{1}{2}(\bar{\psi}_1, \bar{\psi}_2)\begin{pmatrix} m_L & m_D \\ m_D & m_R \end{pmatrix}\begin{pmatrix} \psi_1 \\ \psi_2 \end{pmatrix}, \tag{134}$$

where

$$\psi_1 = \nu_L + \nu_L^c, \tag{135}$$

$$\psi_2 = \nu_R + \nu_R^c \tag{136}$$

are Majorana fields. The mass parameters in (134) are taken to be real. Let's diagonalize the mass matrix for the case $m_R \gg m_D \gg m_L$. Putting $m_L = 0$ we have in the mass basis

$$-\mathcal{L}_{D+M} = \frac{m_\nu}{2}\bar{\nu}\nu + \frac{m_N}{2}\bar{N}N, \tag{137}$$

where

$$\nu \simeq \psi_1, \quad N \simeq \psi_2, \tag{138}$$

and
$$-m_\nu \simeq \frac{m_D^2}{m_R} \ll m_D, \tag{139}$$
$$m_N \simeq m_R. \tag{140}$$

The eigenvalue m_ν can be made positive by an appropriate change of phase of the field ν. For $m_R \gg m_D$ the neutrino mass eigenstates consist of a very light left-handed state $|\nu>$ and a very heavy right-handed state $|N>$. Eq. (139) constitutes the seesaw mechanism [84] for generating a very small mass for a left-handed neutrino from $m_D = \mathcal{O}(h_\ell v)$ and a large m_R.

Finally we consider the case of 3 lepton generations. Denoting the $\mathrm{SU}(2)_L$ doublets $\ell \equiv (\nu_{\alpha L}, \ell_{\alpha L})^T$, and the $\mathrm{SU}(2)_L$ singlets $e_R \equiv \ell_{\alpha R}$, the $\mathrm{SU}(2)_L \times \mathrm{U}(1)_Y$ singlets $\nu_R \equiv \nu_{\alpha R}$, where $\alpha = e, \mu, \tau$ labels the lepton generations in the weak basis and $\tilde{\Phi}_r \equiv i\sigma_2 \Phi_r^\dagger$, $r = 1, 2$, where Φ_r are Higgs doublet fields, the general $\mathrm{SU}(2)_L \times \mathrm{U}(1)_Y$ invariant Yukawa interactions in the lepton sector read
$$-\mathcal{L}_Y = \bar{\ell}_L \Phi_1 h_e e_R + \bar{\ell}_L \tilde{\Phi}_2 h_\nu \nu_R + \frac{1}{2}\overline{\nu_R^c} M_R \nu_R + h.c.. \tag{141}$$

Here h_e, h_ν denote the complex, 3×3 Yukawa coupling matrices, and M_R is the 3×3 mass matrix for the right-handed neutrino fields which may be taken to be diagonal without loss of generality. (M_R can be generated by a large VEV of a gauge singlet Higgs field.) Spontaneous symmetry breaking at the electroweak phase transition, $<0|\Phi_r|0>_T = v_{rT}/\sqrt{2}$, leads to Dirac mass matrices for the charged leptons and neutrinos,
$$M_e = h_e \frac{v_{1T}}{\sqrt{2}}, \quad M_D = h_\nu \frac{v_{2T}}{\sqrt{2}}. \tag{142}$$

Let us change from the weak basis to the mass basis by performing appropriate unitary transformations in flavor space. Using that the matrix elements of M_R are much larger than those of M_D one obtains [7]
$$\nu_i \simeq (K^\dagger)_{i\alpha}\nu_{\alpha L} + \nu_{\alpha L}^c K_{\alpha i}, \tag{143}$$
$$N_i \simeq \nu_{\alpha R} + \nu_{\alpha R}^c, \tag{144}$$

with the diagonal mass matrices
$$M_\nu = -K^\dagger M_D M_R^{-1} M_D^T K^* + \mathcal{O}(M_R^{-3}), \tag{145}$$
$$M_N = M_R + \mathcal{O}(M_R^{-1}), \tag{146}$$

where $i = 1, 2, 3$ labels the fields in the mass basis and K is the unitary 3×3 matrix which describes the mixing of the lepton flavors in the charged current interactions
$$\mathcal{L}_{cc}^{lept} = -\frac{g_w}{\sqrt{2}} \bar{\ell}_{iL} \gamma^\mu K_{ij} \nu_j W_\mu^- + h.c.. \tag{147}$$

We can decompose the Dirac mass matrix M_D into the form

$$M_D = VRU^\dagger, \qquad (148)$$

where U, V are unitary matrices and $R = diag(r_1, r_2, r_3)$. From (145) it follows that the moduli and phases of the matrix elements of K, which are relevant for present-day neutrino physics – e.g., for neutrino oscillations or for the search for neutrinoless double beta decay $Z \to (Z+2) + 2e^-$ – depend on the mass ratios m_j/m_i of the light, left-handed neutrinos, and on the angles and phases of U and V. On the other hand the matrix $M_D^\dagger M_D$, on which the quantities responsible for leptogenesis, in particular the CP asymmetry depend (see section 6.2), is given by

$$M_D^\dagger M_D = UR^2 U^\dagger. \qquad (149)$$

Hence for leptogenesis only the CP-violating phases of U are relevant! Therefore, in this scenario there is in general no connection between possible CP-violating effects that could be traced in the laboratory and the CP-violating phases which are responsible for the generation of the BAU.

References

1. E. W. Kolb and M. S. Turner, *The Early Universe*, Addison-Wesley Publishing Company, Reading (1993).
2. A. D. Dolgov, Phys. Rept. **222** (1992) 309.
3. A. G. Cohen, D. B. Kaplan and A. E. Nelson, Ann. Rev. Nucl. Part. Sci. **43** (1993) 27 [arXiv:hep-ph/9302210].
4. V. A. Rubakov and M. E. Shaposhnikov, Usp. Fiz. Nauk **166** (1996) 493 [Phys. Usp. **39** (1996) 461] [arXiv:hep-ph/9603208].
5. M. Trodden, Rev. Mod. Phys. **71** (1999) 1463 [arXiv:hep-ph/9803479].
6. A. Riotto and M. Trodden, Ann. Rev. Nucl. Part. Sci. **49** (1999) 35 [arXiv:hep-ph/9901362].
7. W. Buchmüller and M. Plümacher, [arXiv:hep-ph/0007176];
   W. Buchmüller, [arXiv:hep-ph/0101102].
8. D. E. Groom et al. [Particle Data Group Collaboration], Eur. Phys. J. C **15** (2000) 1.
9. E. W. Kolb and M. S. Turner, in ref. [8].
10. K. A. Olive, in ref. [8].
11. T. Saeki et al. [BESS Collaboration], Phys. Lett. B **422** (1998) 319 [arXiv:astro-ph/9710228];
    T. Sanuki et al.[BESS Collaboration], Astrophys. J. **545** (2000) 1135 [arXiv:astro-ph/0002481].
12. J. Alcaraz et al. [AMS Collaboration], Phys. Lett. B **461** (1999) 387 [arXiv:hep-ex/0002048].
13. A. G. Cohen, A. De Rujula and S. L. Glashow, Astrophys. J. **495** (1998) 539 [arXiv:astro-ph/9707087].
14. P. de Bernardis, talk given at the DESY Theory Workshop on *Gravity and Particle Physics*, Hamburg, October 2001.

15. A. D. Sakharov, JETP Letters **5** (1967) 24.
16. D. Toussaint, S. B. Treiman, F. Wilczek and A. Zee, Phys. Rev. D **19** (1979) 1036.
17. K. Kleinknecht, lectures given at this school.
18. A. Ali, lectures given at this school.
19. R. Waldi, lectures given at this school.
20. S. L. Adler, Phys. Rev. **177** (1969) 2426.
21. J. S. Bell and R. Jackiw, Nuovo Cim. A **60** (1969) 47.
22. G. 't Hooft, Phys. Rev. Lett. **37** (1976) 8.
23. A. A. Belavin, A. M. Polyakov, A. S. Shvarts and Y. S. Tyupkin, Phys. Lett. B **59** (1975) 85.
24. G. 't Hooft, Phys. Rev. D **14** (1976) 3432 [Erratum-ibid. D **18** (1976) 2199].
25. V. A. Kuzmin, V. A. Rubakov and M. E. Shaposhnikov, Phys. Lett. B **155** (1985) 36.
26. P. Arnold and L. D. McLerran, Phys. Rev. D **36** (1987) 581.
27. N. S. Manton, Phys. Rev. D **28** (1983) 2019.
28. F. R. Klinkhamer and N. S. Manton, Phys. Rev. D **30** (1984) 2212.
29. G. D. Moore, Phys. Rev. D **59** (1999) 014503 [arXiv:hep-ph/9805264].
30. D. Bödeker, G. D. Moore and K. Rummukainen, Phys. Rev. D **61** (2000) 056003 [arXiv:hep-ph/9907545].
31. G. D. Moore, Phys. Rev. D **62** (2000) 085011 [arXiv:hep-ph/0001216].
32. D. A. Kirzhnits and A. D. Linde, Phys. Lett. B **42** (1972) 471.
33. LEP Collaborations, LEPEWG, SLDEWG, [arXiv:hep-ex/0112021].
34. W. Buchmüller, Z. Fodor and A. Hebecker, Nucl. Phys. B **447** (1995) 317 [arXiv:hep-ph/9502321].
35. Z. Fodor, J. Hein, K. Jansen, A. Jaster and I. Montvay, Nucl. Phys. B **439** (1995) 147 [arXiv:hep-lat/9409017].
36. K. Kajantie, M. Laine, K. Rummukainen and M. E. Shaposhnikov, Nucl. Phys. B **466** (1996) 189 [arXiv:hep-lat/9510020].
37. W. Buchmüller and O. Philipsen, Phys. Lett. B **397** (1997) 112 [arXiv:hep-ph/9612286].
38. K. Rummukainen, M. Tsypin, K. Kajantie, M. Laine and M. E. Shaposhnikov, Nucl. Phys. B **532** (1998) 283 [arXiv:hep-lat/9805013].
39. Z. Fodor, Nucl. Phys. Proc. Suppl. **83** (2000) 121 [arXiv:hep-lat/9909162].
40. M. Laine, [arXiv:hep-ph/0010275].
41. J. M. Cline, Pramana **54** (2000) 1 [Pramana **55** (2000) 33] [arXiv:hep-ph/0003029].
42. D. Land and E. D. Carlson, Phys. Lett. B **292** (1992) 107 [arXiv:hep-ph/9208227].
43. A. Hammerschmitt, J. Kripfganz and M. G. Schmidt, Z. Phys. C **64** (1994) 105 [arXiv:hep-ph/9404272].
44. J. M. Cline and P. A. Lemieux, Phys. Rev. D **55** (1997) 3873 [arXiv:hep-ph/9609240].
45. W. Bernreuther, [arXiv:hep-ph/9808453], in: *Lecture Notes in Physics*, Vol. 521, L. Mathelitsch, W. Plessas (Eds.), Springer-Verlag (1999).
46. C. R. Gould, lectures given at this school.
47. J. F. Giudice, lectures given at this school.
48. A. Bartl, T. Gajdosik, W. Porod, P. Stockinger and H. Stremnitzer, Phys. Rev. D **60** (1999) 073003 [arXiv:hep-ph/9903402].

49. T. Ibrahim and P. Nath, Phys. Rev. D **57** (1998) 478 [Erratum-ibid. D **58** (1998) 019901] [arXiv:hep-ph/9708456].
50. M. Brhlik, G. J. Good and G. L. Kane, Phys. Rev. D **59** (1999) 115004 [arXiv:hep-ph/9810457].
51. D. Comelli and M. Pietroni, Phys. Lett. B **306** (1993) 67 [arXiv:hep-ph/9302207];
D. Comelli, M. Pietroni and A. Riotto, Nucl. Phys. B **412** (1994) 441 [arXiv:hep-ph/9304267].
52. S. J. Huber, P. John, M. Laine and M. G. Schmidt, Phys. Lett. B **475** (2000) 104 [arXiv:hep-ph/9912278].
53. P. Huet and A. E. Nelson, Phys. Rev. D **53** (1996) 4578 [arXiv:hep-ph/9506477].
54. A. Riotto, Phys. Rev. D **58** (1998) 095009 [arXiv:hep-ph/9803357].
55. J. M. Cline, M. Joyce and K. Kainulainen, JHEP **0007** (2000) 018 [arXiv:hep-ph/0006119], Erratum arXiv:hep-ph/0110031.
56. M. Carena, J. M. Moreno, M. Quiros, M. Seco and C. E. Wagner, Nucl. Phys. B **599** (2001) 158 [arXiv:hep-ph/0011055].
57. T. Prokopec, R. H. Brandenberger, A. C. Davis and M. Trodden, Phys. Lett. B **384** (1996) 175 [arXiv:hep-ph/9511349];
R. H. Brandenberger and A. Riotto, Phys. Lett. B **445** (1999) 323 [arXiv:hep-ph/9801448].
58. L. D. McLerran, M. E. Shaposhnikov, N. Turok and M. B. Voloshin, Phys. Lett. B **256** (1991) 451.
59. N. Turok and J. Zadrozny, Nucl. Phys. B **358** (1991) 471.
60. A. G. Cohen, D. B. Kaplan and A. E. Nelson, Nucl. Phys. B **349** (1991) 727.
61. A. E. Nelson, D. B. Kaplan and A. G. Cohen, Nucl. Phys. B **373** (1992) 453.
62. K. Funakubo, A. Kakuto, S. Otsuki, K. Takenaga and F. Toyoda, Phys. Rev. D **50** (1994) 1105 [arXiv:hep-ph/9402204].
63. M. Joyce, T. Prokopec and N. Turok, Phys. Rev. D **53** (1996) 2930 [arXiv:hep-ph/9410281].
64. J. M. Cline, K. Kainulainen and A. P. Vischer, Phys. Rev. D **54** (1996) 2451 [arXiv:hep-ph/9506284].
65. M. Joyce, T. Prokopec and N. Turok, Phys. Rev. D **53** (1996) 2958 [arXiv:hep-ph/9410282].
66. R. N. Mohapatra and X. M. Zhang, Phys. Rev. D **45** (1992) 2699.
67. G. F. Giudice and M. E. Shaposhnikov, Phys. Lett. B **326** (1994) 118 [arXiv:hep-ph/9311367].
68. S. J. Huber and M. G. Schmidt, Nucl. Phys. B **606** (2001) 183 [arXiv:hep-ph/0003122].
69. C. Jarlskog, Phys. Rev. Lett. **55** (1985) 1039.
70. J. Bernabéu, G. C. Branco and M. Gronau, Phys. Lett. **B** 169 (1986) 243.
71. G. R. Farrar and M. E. Shaposhnikov, Phys. Rev. D **50** (1994) 774 [arXiv:hep-ph/9305275].
72. M. B. Gavela, M. Lozano, J. Orloff and O. Pene, Nucl. Phys. B **430** (1994) 345 [arXiv:hep-ph/9406288]; Nucl. Phys. B **430** (1994) 382 [arXiv:hep-ph/9406289].
73. P. Huet and E. Sather, Phys. Rev. D **51** (1995) 379 [arXiv:hep-ph/9404302].
74. M. Yoshimura, Phys. Rev. Lett. **41** (1978) 281 [Erratum-ibid. **42** (1978) 746].
75. S. Dimopoulos and L. Susskind, Phys. Rev. D **18** (1978) 4500.
76. S. Weinberg, Phys. Rev. Lett. **42** (1979) 850.
77. M. Fukugita and T. Yanagida, Phys. Lett. B **174** (1986) 45.

78. M. A. Luty, Phys. Rev. D **45** (1992) 455.
79. M. Flanz, E. A. Paschos and U. Sarkar, Phys. Lett. B **345** (1995) 248 [Erratum-ibid. B **382** (1995) 447] [arXiv:hep-ph/9411366];
    M. Flanz, E. A. Paschos, U. Sarkar and J. Weiss, Phys. Lett. B **389** (1996) 693 [arXiv:hep-ph/9607310].
80. M. Plümacher, Z. Phys. C **74** (1997) 549 [arXiv:hep-ph/9604229].
81. W. Buchmüller and M. Plümacher, Phys. Lett. B **431** (1998) 354 [arXiv:hep-ph/9710460].
82. L. Covi, E. Roulet and F. Vissani, Phys. Lett. B **384** (1996) 169 [arXiv:hep-ph/9605319];
    E. Roulet, L. Covi and F. Vissani, Phys. Lett. B **424** (1998) 101 [arXiv:hep-ph/9712468].
83. A. Pilaftsis, Int. J. Mod. Phys. A **14** (1999) 1811 [arXiv:hep-ph/9812256].
84. T. Yanagida, in *Workshop on Unified Theories*, KEK report 79-18 (1979) p. 95;
    M. Gell-Mann, P. Ramond and R. Slansky, in *Supergravity*, North-Holland, Amsterdam (1979), ed. by P. van Nieuvenhuizen and D. Freedman, p. 315.
85. W. Fischler, G. F. Giudice, R. G. Leigh and S. Paban, Phys. Lett. B **258** (1991) 45.
86. S. Y. Khlebnikov and M. E. Shaposhnikov, Nucl. Phys. B **308** (1988) 885.
87. J. A. Harvey and M. S. Turner, Phys. Rev. D **42** (1990) 3344.
88. S. Y. Khlebnikov and M. E. Shaposhnikov, Phys. Lett. B **387** (1996) 817 [arXiv:hep-ph/9607386].
89. M. Laine and M. E. Shaposhnikov, Phys. Rev. D **61** (2000) 117302 [arXiv:hep-ph/9911473].

Physics Beyond the Standard Model

Gian Francesco Giudice

Theoretical Physics Division, CERN
Geneva, Switzerland

Abstract. In these lectures I give a short review of the main theoretical ideas underlying the extensions of the Standard Model of elementary particle interactions. I will discuss grand unified theories, supersymmetry, technicolour, and theories with extra spatial dimensions.

1 Introduction

Studying physics beyond the Standard Model means looking for the conditions of the Universe in the first billionth of a second, when its temperature was above 10^{14} K. This clearly requires a gigantic intellectual leap in the investigation. It is even more striking that modern accelerators can reproduce particle collisions similar to those that continually occurred in the thermal bath in the very first instants of our Universe. We are now entering the age in which, with the joint effort of experiments and theory, we are likely to unravel the mystery of the fundamental principles of particle interactions lying beyond the Standard Model.

The Standard Model [1] describes the interactions of three generations of quarks and leptons defined by a non-Abelian gauge theory based on the group $SU(3) \times SU(2) \times U(1)$. The precision measurements at LEP have given an extraordinary confirmation of the validity of the Standard Model up to the electroweak energy scale (for reviews, see ref. [2]), and we have no firm experimental indications for failures of this theory at higher energies. Our belief that the Standard Model is a low-energy approximation of a new and fundamental theory is based only on theoretical, but well-motivated, arguments.

First of all, the electroweak symmetry breaking sector is not on firm experimental ground. The Higgs mechanism, which is invoked by the Standard Model to generate the Z^0 and $W^\pm$ masses, predicts the existence of a new scalar particle, still to be discovered. From the theoretical point of view, the Higgs mechanism suffers from the so-called "hierarchy" or "naturalness" problem which, as discussed in Sect. 3, leads us to believe that new physics must take place at the TeV energy scale.

Furthermore, the complexity of the fermionic and gauge structures makes the Standard Model look like an improbable fundamental theory. To put it in a less qualitative way, the Standard Model contains many free parameters (e.g. the three gauge coupling constants, the nine fermion masses and the four Cabibbo–Kobayashi–Maskawa mixing parameters); these correspond to important physical quantities, but cannot be computed in the context of the model. Simplifying the Standard Model structure and predicting its free parameters are therefore basic tasks of a successful theory.

In these lectures I review the main current ideas about theories beyond the Standard Model, keeping the discussion at a qualitative level and making no use of advanced mathematics. More comprehensive reviews can be found in refs. [3] (for GUTs), [4] (for supersymmetry) and [5] (for technicolour).

2 Grand Unified Theories

2.1 SU(5)

The first attempts to extend the structure of the Standard Model have led to the construction of Grand Unified Theories (GUTs) [6]. The basic idea is that gauge interactions are described by a single simple gauge group, which contains the Standard Model SU(3) × SU(2) × U(1) as a subgroup and as a low-energy manifestation. At first this may seem impossible, since a simple gauge group contains a single coupling constant g_X and the strong, weak and electromagnetic couplings have different numerical values. However it should be remembered that, in a quantum field theory, the coupling constants depend on the energy scale at which they are probed, as a consequence of the exchange of virtual particles surrounding the charge. The evolution of the gauge coupling constants as a function of the energy scale can be computed using renormalization group techniques and perturbation theory, and the relevant equations are described in Sect. 4.1. There, we will also find that, as we include the quantum effects of all Standard Model particles, the three gauge coupling constants approach one another as the energy scale is raised. For the moment, let us assume that the three gauge couplings meet at a single value for a specific energy scale (M_X) and study possible GUT candidates describing the physics above M_X with a single gauge coupling constant g_X.

The simplest example of a GUT is based on the group SU(5). Each fermion family is contained in a **10** + **5̄** representation of SU(5). This can be understood from the decomposition in terms of the Standard Model group:

$$\begin{array}{rcl}
\mathrm{SU}(5) & \to & \mathrm{SU}(3) \times \mathrm{SU}(2) \times \mathrm{U}(1) \\
\mathbf{10} & \to & (\bar{\mathbf{3}}, \mathbf{1}, -\tfrac{2}{3})_{u_R^c} + (\mathbf{3}, \mathbf{2}, \tfrac{1}{6})_{q_L} + (\mathbf{1}, \mathbf{1}, \mathbf{1})_{e_R^c} \\
\bar{\mathbf{5}} & \to & (\bar{\mathbf{3}}, \mathbf{1}, \tfrac{1}{3})_{d_R^c} + (\mathbf{1}, \mathbf{2}, -\tfrac{1}{2})_{\ell_L} \ .
\end{array} \quad (1)$$

Here the numbers inside the brackets respectively denote the SU(3) and SU(2) representations and the U(1) quantum numbers. Equation (1) shows that the degrees of freedom for all the (left-handed) fields in one Standard Model family are described by the two SU(5) fields **10** and **5̄**. In GUTs not only is the gauge group unified, going from SU(3) × SU(2) × U(1) to SU(5) in this specific example, but also the fermionic spectrum is simplified. As quarks in QCD come with different colours, in GUTs different quarks and leptons are just different aspects of the same particle. This also explains the simple integer relations among the electric charges of different quarks and leptons.

2.2 Experimental Tests of GUTs

Theoretical elegance is of course not a sufficient argument to convince us that GUTs have anything to do with Nature. We need to establish GUTs predictions which can be confronted with experimental data. The basic idea of GUTs, gauge coupling unification, provides such a prediction. Indeed at the GUT scale M_X we can compute the weak mixing angle:

$$\sin^2 \theta_W \equiv \frac{e^2}{g^2} = \frac{\text{Tr}(T_3^2)}{\text{Tr}(Q^2)} = \frac{3}{8} \ . \tag{2}$$

Here T_3 is the third isospin-component and Q is the electric charge. The trace in Eq. (2), taken over any SU(5) representation, follows from a correct normalization of the GUT generators. Before comparing Eq. (2) with experiment, one has to rescale it to the low energies where coupling constants are measured. We will do this in Sect. 4.1, and show that Eq. (2) gives a successful prediction for a class of theories which we have not yet introduced, supersymmetric GUTs. We anticipate here that, if gauge coupling unification has any chance to succeed, the unification scale M_X must be extremely large, of the order of $10^{15} \ldots 10^{16}$ GeV, which, in the thermal history of our Universe, brings us to consider events occurring in the first $10^{-35} \ldots 10^{-38}$ s.

Since we have promoted the gauge group to SU(5), we expect new gauge bosons and therefore new forces which may have experimental consequences. The decomposition of the SU(5) gauge bosons in terms of Standard Model ones is:

$$\begin{aligned}\text{SU}(5) &\to \text{SU}(3) \times \text{SU}(2) \times \text{U}(1) \\ \mathbf{24} &\to (\mathbf{8},\mathbf{1},0)_g + (\mathbf{1},\mathbf{3},0)_W + (\mathbf{1},\mathbf{1},0)_B + (\mathbf{3},\mathbf{2},-\tfrac{5}{6})_X + (\bar{\mathbf{3}},\mathbf{2},\tfrac{5}{6})_{\bar{X}}\end{aligned} \tag{3}$$

Together with the familiar degrees of freedom for the gluons (g) and the electroweak gauge bosons ($W^\pm, W^0, B$), we find new particles (X and $\bar{X}$) which carry both colour and weak quantum numbers. The gauge bosons X and $\bar{X}$ affect weak interactions, but modify standard processes only by an amount $(M_W/M_X)^2$, a fantastically small number, whose effect is completely undetectable even in the most precise measurements. Nevertheless, the X-mediated interactions may not be so invisible. Let us inspect the interactions between $X, \bar{X}$ and the fermionic currents, which are dictated by SU(5) gauge invariance:

$$\mathcal{L} = \frac{g_X}{\sqrt{2}} \left\{ X_\mu^\alpha \left[\bar{d}_{R\alpha} \gamma^\mu e_R^c + \bar{d}_{L\alpha} \gamma^\mu e_L^c + \varepsilon_{\alpha\beta\gamma} \bar{u}_L^{c\gamma} \gamma^\mu u_L^\beta \right] + \right. \\ \left. + \bar{X}_\mu^\alpha \left[-\bar{d}_{R\alpha} \gamma^\mu \nu_R^c - \bar{u}_{L\alpha} \gamma^\mu e_L^c + \varepsilon_{\alpha\beta\gamma} \bar{u}_L^{c\gamma} \gamma^\mu d_L^\beta \right] + \text{h.c.} \right\} \ . \tag{4}$$

Notice that one cannot assign a conserved baryon (B) and lepton (L) quantum number to X and $\bar{X}$; the new interactions violate both B and L. In the Standard Model B and L are accidental global symmetries, in the sense that they are just a consequence of gauge invariance and renormalizability. It is not surprising that B and L are then violated in extensions of the Standard Model, in particular in GUTs where quarks and leptons are different aspects of the same particle.

The experimental discovery of processes that violate B and L would be clear evidence for physics beyond the Standard Model. One of the most important of such processes is proton decay, which has the dramatic consequence that ordinary matter is not stable. It is easy to see from Eq. (4) that the X boson mediates the transition $uu \to e^+ \bar{d}$. When dressed between physical hadronic states, this transition is converted into the proton decay modes $p \to e^+\pi^0, e^+\rho^0, e^+\eta, e^+\pi^+\pi^-$, and so on. The calculation of the proton lifetime yields

$$\tau_p = (0.2 \ldots 8.0) \times 10^{31} \left(\frac{M_X}{10^{15} \text{ GeV}} \right)^4 \text{ yr} . \tag{5}$$

The uncertainties in the numerical coefficient in Eq. (5) come mainly from the difficulty in estimating the matrix elements relating quarks to hadrons. For reasonable GUT masses, $M_X \simeq 10^{15} \ldots 10^{16}$ GeV, Eq. (5) predicts a proton lifetime $10^{21} \ldots 10^{25}$ times larger than the age of the Universe. It is fascinating that experiments can probe such slow processes by studying very large samples of matter. The present experimental bound on the lifetime of the decay mode $p \to e^+\pi^0$, the dominant proton decay channel in SU(5), is [7]

$$\tau(p \to e^+\pi^0) > 1.6 \times 10^{33} \text{ yr} . \tag{6}$$

This bound already sets important constraints on possible GUT models.

GUTs also provide a framework in which the creation of a primordial baryon asymmetry can be understood and computed. Although this is not an experimental test, it is clearly a very attractive theoretical feature. Observations tell us that the present ratio of baryons to photons in the Universe is a very small number, $n_B/n_\gamma = (4 \ldots 7) \times 10^{-10}$. If n_B/n_γ is then extrapolated back in time following the thermal history of the Universe, one finds that the excess of baryons over antibaryons at the time of the big bang must have been $\Delta_B \equiv (n_B - n_{\bar{B}})/n_B \sim 3 \times 10^{-8}$. We find it disturbing to consider that the present observed Universe is determined by a peculiar initial condition prescribing that for each three hundred million baryons there are three hundred million minus one antibaryons.

The hypothesis of baryogenesis is that $\Delta_B = 0$ at the time of the big bang and that the small cosmic baryon asymmetry was dynamically created during the evolution of the Universe. The physics responsible for the creation of Δ_B must necessarily involve interactions which violate B. GUTs are therefore a natural framework for baryogenesis and it has been proved [8] that they have

2.3 SO(10) and Neutrino Masses

I have presented SU(5) as the simplest GUT, but models based on larger groups can also be constructed. Probably the most interesting of them [9] is based on the orthogonal group SO(10), which contains SU(5) as a subgroup. The 16-dimensional spinorial representation of SO(10) decomposes into **10** + **5̄** + **1** under SU(5). We recognize the fermion content of one Standard Model family. It is quite satisfactory that quarks and leptons with their different quantum number assignments can be described by a single SO(10) particle, for each generation.

In addition to the ordinary quarks and leptons contained in the **10** + **5̄** of SU(5), the spinorial representation of SO(10) contains also a gauge singlet. This can be interpreted as the right-handed component of the neutrino, allowing the possibility of Dirac neutrino masses. The neutrino mass term can now be written in the form

$$(\bar{\nu}_L \bar{\nu}_L^c) \mathcal{M} \begin{pmatrix} \nu_R^c \\ \nu_R \end{pmatrix} + \text{h.c.} , \tag{7}$$

where, for simplicity, we are considering only the one-generation case. The different entries of the neutrino mass matrix $\mathcal{M}$

$$\mathcal{M} = \begin{pmatrix} T & D \\ D^T & S \end{pmatrix} \tag{8}$$

can be understood in terms of symmetry principles. The term S transforms as a singlet under the Standard Model gauge group and therefore is naturally generated at the scale where the SO(10) symmetry is broken, $S \sim M_X$. The other two terms, T and D, transform respectively as a triplet and a doublet under the weak group SU(2); therefore they can be generated only after the Standard Model gauge group is broken. However, vacuum expectation values of triplet fields lead to an incorrect relation between the strengths of neutral and charged weak currents. We conclude therefore that $T \simeq 0$ and $D \simeq m_f$, where m_f is a typical fermion (quark or charged lepton) mass. After diagonalization of the matrix in Eq. (8), we find one heavy eigenstate with mass of order M_X and one (mainly left-handed) eigenstate with mass [10]:

$$m_\nu \simeq \frac{m_f^2}{M_X} = 10^{-6} \text{eV} \left(\frac{m_f}{\text{GeV}}\right)^2 \left(\frac{10^{15} \text{ GeV}}{M_X}\right) . \tag{9}$$

In the context of the SO(10) GUT, not only do we expect neutrinos to be massive, but we also understand in terms of symmetries why their masses must be much smaller than the typical scale of the other fermion masses.

3 The Hierarchy Problem

The hierarchy (or naturalness) problem [11] is considered to be one of the most serious theoretical drawbacks of the Standard Model and most of the attempts to build theories beyond the Standard Model have concentrated on its solution. It springs from the difficulty in field theory in keeping fundamental scalar particles much lighter than Λ_{max}, the maximum energy scale up to which the theory remains valid.

It is intuitive to require that if a particle mass is much smaller than Λ_{max}, there should exist a (possibly approximate) symmetry under which the mass term is forbidden. We know an example of such a symmetry for spin-one particles. The photon is, theoretically speaking, naturally massless since the gauge symmetry $A_\mu \to A_\mu + \partial_\mu \lambda$ forbids the occurrence of the photon mass term $m^2 A_\mu A^\mu$. Similarly, we can identify a symmetry which protects the mass of a fermionic particle. A chiral symmetry, under which the left-handed and right-handed fermionic components transform differently $\psi_L \to e^{i\alpha}\psi_L, \psi_R \to e^{i\beta}\psi_R, \alpha \neq \beta$, forbids the mass term $m\bar\psi_L \psi_R$ + h.c. Scalar particles can be naturally light if they are Goldstone bosons of some broken global symmetry since their non-linear transformation property $\varphi \to \varphi + a$ forbids the mass term $m^2 \varphi^2$.

In the case of the Higgs particle , required in the Standard Model by the electroweak symmetry breaking mechanism, we cannot rely on any of the above-mentioned symmetries. In the absence of any symmetry principle, we expect the Higgs potential mass parameter m_H^2 to be of the order of Λ_{max}^2. Even if we artificially set the classical value of m_H^2 to zero, it will be generated by quadratically divergent quantum corrections:

$$m_H^2 = \frac{\alpha}{\pi} \Lambda_{\text{max}}^2 , \qquad (10)$$

where α measures the effect of a typical coupling constant.

One may argue that in a renormalizable theory, the bare value of any parameter is an infinite (or, in other words, cut-off dependent) quantity, without a precise physical meaning. Since all divergences can be reabsorbed, one can just choose the renormalized quantity to be equal to any appropriate value. However, we believe that a complete description of particle interactions in a final theory will be free from divergences. From this point of view, the cancellation between a bare value and quadratically divergent quantum corrections looks like a conspiracy between the infra-red (below Λ_{max}) and the ultraviolet (above Λ_{max}) components of the theory. We do not accept such a conspiracy, but, on the other hand, we know that the parameter m_H^2 sets the scale for electroweak symmetry breaking and it is therefore directly related to m_W^2. We thus require that the quantum corrections in Eq. (10) do not exceed m_W^2. This implies an upper bound on Λ_{max}:

$$\Lambda_{\text{max}} \lesssim \sqrt{\frac{\pi}{\alpha}} M_W \simeq \text{TeV} . \qquad (11)$$

We can conclude that the Standard Model has a natural upper bound at the TeV scale, where new physics should appear and modify the ultraviolet behaviour of the theory.

The hierarchy problem becomes most apparent when one considers GUTs. Here the Higgs potential of the model contains two different mass parameters: one is of order M_X and sets the scale for the breaking of the unified group; the other is of order M_W and sets the scale for the ordinary electroweak breaking. By explicit calculation, one can show [12] that these parameters mix at the quantum level and the hierarchy of the two mass scales can be maintained only at the price of fine-tuning the parameters by an amount $(M_X/M_W)^2$.

4 Supersymmetry

Supersymmetry [13], contrary to all other ordinary symmetries in field theory, transforms bosons to fermions and vice versa. This means that bosons and fermions sit in the same supersymmetric multiplet. In the simplest version of supersymmetry (the so-called $N = 1$ supersymmetry), each complex scalar has a Weyl fermion companion and each massless gauge boson also has a Weyl fermion companion; similarly the spin-2 graviton has a spin-3/2 companion, the gravitino. Invariance under supersymmetry implies that particles inside a supermultiplet are degenerate in mass. It is therefore evident that, in a supersymmetric theory, if a chiral symmetry forbids a fermion mass term, it forbids also the appearance of a scalar mass term, such as the notorious Higgs mass parameter. The hierarchy problem discussed in the previous section can now be solved. Indeed, it has been proved that a supersymmetric theory is free from quadratic divergences [14]. The contribution to m_H^2 proportional to $\Lambda_{\max}^2$ in Eq. (10) coming from a bosonic loop is exactly cancelled by a loop involving fermionic particles. Since the dependence on $\Lambda_{\max}^2$ has now disappeared, we can extend the scale of validity of the theory without provoking any hierarchy problem.

It should also be mentioned that when supersymmetry is promoted to a local symmetry, which means that the transformation parameter depends on space-time, then the theory automatically includes gravity and is called supergravity. Because of this characteristic, supersymmetry is believed to be a necessary ingredient for the complete unification of forces.

Here we are interested in the minimal extension of the Standard Model compatible with supersymmetry. Each Standard Model particle is accompanied by a supersymmetric partner: scalar particles (squarks and sleptons) are the partners of quarks and leptons, and fermion particles (e.g. gluinos) are the partners of the Standard Model bosons (e.g. gluons). Supersymmetry also requires two Higgs doublets, as opposed to the single Higgs doublet of the Standard Model, and their fermionic partners mix with the fermionic partners of the electroweak gauge bosons to produce particles with one unit of electric charge (charginos) or no electric charge (neutralinos).

Supersymmetry ensures that the couplings of all these new particles are strictly related to ordinary couplings. For instance, the couplings of squarks to one or two gluons, of gluinos to gluons, of squarks and gluinos to quarks are solely determined by α_s, the QCD gauge coupling constant.

The supersymmetric generalization of the Standard Model is therefore a well-defined theory where all new interactions are described by the mathematical properties of the supersymmetric transformation. As such, however, the theory is not acceptable since it predicts a mass degeneracy between the ordinary and the supersymmetric particles; in Nature, therefore, supersymmetry is not an exact symmetry. In order to preserve the solution of the hierarchy problem we need to break supersymmetry while maintaining the good ultraviolet behaviour of the theory. It has been shown [15] that if only a certain set of supersymmetry-breaking terms with dimensionful couplings are introduced, then the quadratic divergences still cancel, but the mass degeneracy is removed. Let us generically call m_S the mass that sets the scale for the dimensionful couplings which softly break supersymmetry. This scale has a definite physical meaning, since all new supersymmetric particles acquire masses of order m_S. It is the energy scale at which supersymmetry has to be looked for in experiments.

By explicit calculation one finds that, in a softly broken supersymmetric theory, quadratic divergences cancel, but some finite terms of the kind $(\alpha/\pi)m_S^2$ remain. From Eq. (10) we recognize that m_S behaves as the cut-off of quadratic divergences in the Standard Model. This is not entirely surprising since, in the limit $m_S \to \infty$, all supersymmetric particles decouple and one should recover the ultraviolet behaviour of the Standard Model. Therefore we conclude that, in a softly broken supersymmetric theory, the cut-off of quadratic divergences has a physical meaning since it is related to m_S, the mass scale of the new particles. Moreover, following the same argument that led us to Eq. (11) we find that these new particles cannot be much heavier than the TeV scale, if supersymmetry solves the hierarchy problem. In Sect. 4.2, I will make this argument more quantitative.

Although technically successful, it may appear that the introduction of the soft supersymmetry-breaking terms is too arbitrary to be entirely satisfactory. But, on the contrary, it has a very appealing explanation [16]. Let us first promote supersymmetry to supergravity, possibly a necessary step towards complete unification of forces. Then assume that supergravity is either spontaneously or dynamically broken in a sector of the theory that does not directly couple to ordinary particles. In this case, gravity communicates the supersymmetry breaking, and the low-energy effective theory of the supersymmetric Standard Model contains exactly all the terms which break supersymmetry without introducing quadratic divergences.

From this point of view, the appearance of the soft-breaking terms can be understood in terms of well-defined dynamics. However, we do not yet know which mechanism breaks supersymmetry and therefore we are not able to compute the soft-breaking terms. This is unfortunate because these define the mass spectrum of the new particles. All we can do now is to keep them as free parameters and hope they will be determined by experimental measurements or calculated, if theoretical progress is made. In the minimal version of the theory, there are only four such parameters but, if some assumptions are relaxed, the number of free parameters can grow enormously.

4.1 Supersymmetric Unification

In the previous section, we have extended the Standard Model to include supersymmetry in order to solve the hierarchy problem. We can now incorporate within this model the ideas of grand unification, and construct a supersymmetric GUT [17].

As discussed in Sect. 2, the first test of a GUT is gauge coupling unification. At the one-loop approximation the evolution of the $SU(3) \times SU(2) \times U(1)$ gauge coupling constants with the energy scale Q^2 is given by

$$\frac{d\alpha_i}{dt} = -\frac{b_i}{4\pi}\alpha_i^2 \quad \Rightarrow \quad \alpha_i(t) = \frac{\alpha_i(0)}{1 + \frac{b_i}{4\pi}\alpha_i(0)t}, \quad i = 1, 2, 3, \quad (12)$$

where $t = \log(M_X^2/Q^2)$. The coefficients b_i take into account the numbers of degrees of freedom and the gauge quantum numbers of all particles involved in virtual exchanges. For the Standard Model, we find

$$b_3 = -7 + \frac{4}{3}(N_g - 3), \quad b_2 = -\frac{19}{6} + \frac{4}{3}(N_g - 3), \quad b_1 = \frac{41}{6} + \frac{20}{9}(N_g - 3), \quad (13)$$

where N_g is the number of generations. In the supersymmetric case all new particles influence the running of the gauge coupling constants and modify the b_i parameters,

$$b_3 = -3 + 2(N_g - 3), \quad b_2 = 1 + 2(N_g - 3), \quad b_1 = 11 + \frac{10}{3}(N_g - 3). \quad (14)$$

Assuming $N_g = 3$ and gauge coupling unification, i.e. $\alpha_3(0) = \alpha_2(0) = 5/3\alpha_1(0)$, we can compute the QCD coupling $\alpha_s(M_Z)(\equiv \alpha_3(M_Z))$ and $\sin^2\theta_W (\equiv [1+\alpha_2(M_Z)/\alpha_1(M_Z)]^{-1})$ as a function of M_X, taking $\alpha^{-1}(M_Z) = 127.9 \pm 0.1$ ($\alpha^{-1} \equiv \alpha_1^{-1} + \alpha_2^{-1}$). It turns out that unification of couplings is inconsistent with the Standard Model evolution for any value of M_X. This rules out any simple GUT which breaks directly into $SU(3) \times SU(2) \times U(1)$, with only ordinary matter content. Inclusion of additional light particles or intermediate steps of gauge symmetry breaking may reconcile the Standard Model with the idea of unification. Of course, in this case, any prediction from gauge coupling unification is necessarily lost. More interesting is the supersymmetric case, in which unification can be achieved in the minimal version of the model, with $M_X \simeq 10^{16}$ GeV.

4.2 Electroweak Symmetry Breaking

As a realistic theory of particle interactions, the supersymmetric model should describe the correct pattern of electroweak symmetry breaking. This is obtained by the Higgs mechanism. As already mentioned above, supersymmetry requires two Higgs doublets, as opposed to the single one of the Standard Model. Along the neutral components of the two Higgs fields, the scalar potential is:

$$V(H_1^0, H_2^0) = m_1^2 |H_1^0|^2 + m_2^2 |H_2^0|^2 - m_3^2(H_1^0 H_2^0 + \text{h.c.}) + \frac{g^2 + g'}{8}\left(|H_1^0|^2 - |H_2^0|^2\right)^2 \tag{15}$$

where g, g' are respectively the SU(2) and U(1) gauge coupling constants. The mass parameters m_1^2, m_2^2 and m_3^2 originate from soft-breaking terms and are therefore of the order of m_S. The stability of the potential for large values of fields along the direction $H_1^0 = H_2^0$ requires

$$m_1^2 + m_2^2 > 2|m_3^2| \,. \tag{16}$$

Since electroweak symmetry is broken, the origin $H_1^0 = H_2^0 = 0$ must correspond to an unstable configuration, which implies:

$$m_1^2 m_2^2 < m_3^4 \,. \tag{17}$$

It is often assumed that the soft-breaking terms satisfy some universality conditions around M_X. Notice that, should for instance $m_1^2 = m_2^2$, Eqs. (16) and (17) cannot be simultaneously satisfied and electroweak symmetry remains unbroken. Nevertheless, before drawing any conclusion, we have to include the renormalization effects of changing the scale from M_X to the electroweak scale M_W. These effects are important as they are proportional to a large logarithm, $\log(M_X^2/M_W^2)$, and they have been systematically computed up to two loops [19]. Generically, the effect of gauge interactions is to increase the masses as we evolve from M_X to M_W. Therefore, if all masses are equal at M_X, we expect gluinos to be heavier than charginos and neutralinos, and similarly squarks to be heavier than sleptons, because of the dominant QCD effects. On the other hand, Yukawa interactions decrease the masses in the renormalization from high to low energies. Therefore, the stops will be the lightest among squarks, since the top quark coupling gives the dominant Yukawa effect.

Let us now consider the evolution of the Higgs mass parameters. As they do not feel QCD forces at one loop, their gauge renormalization is not very significant. The Yukawa coupling effect is important for m_2^2, because H_2 is the Higgs field responsible for the top quark mass, but not for m_1^2. Therefore, as an effect of the heavy top quark, m_2^2 decreases and it is likely to be driven negative around the weak scale, while m_1^2 remains positive. For $m_1^2 > 0$ and $m_2^2 < 0$, Eqs. (16) and (17) can be easily satisfied and electroweak symmetry is broken [20].

In conclusion, the supersymmetric model is consistent with electroweak symmetry breaking and the mechanism involved is appealing in several ways. First of all, the breaking is driven by purely quantum effects, a theoretically attractive feature. Then it needs a heavy top quark, which agrees with the Tevatron discovery. Finally, we have found that the dynamics itself chooses to break down SU(2). In a supersymmetric theory, colour SU(3) could spontaneously break if squarks get a vacuum expectation value, but this does not happen since squark masses squared receive large positive radiative corrections.

The minimization of the Higgs potential in Eq. (15) gives:

$$\frac{M_Z^2}{2} \equiv \frac{g^2 + g'}{8} v^2 = \frac{m_1^2 - m_2^2 \tan^2 \beta}{\tan^2 \beta - 1} , \qquad (18)$$

$$\sin^2 \beta = \frac{2m_3^2}{m_1^2 + m_2^2} , \qquad (19)$$

where

$$\langle H_1^0 \rangle = \frac{v}{\sqrt{2}} \cos \beta , \quad \langle H_2^0 \rangle = \frac{v}{\sqrt{2}} \sin \beta . \qquad (20)$$

Equation (18) can be interpreted as a prediction of M_Z in terms of the soft supersymmetry-breaking parameters (a_i) which determine m_1^2, m_2^2, and m_3^2. Unfortunately, we are not able to compute supersymmetry breaking, and therefore we can only use Eq. (18) as a constraint which fixes one of the parameters a_i in terms of the others.

We can also use Eq. (18) to define a quantitative criterion for obtaining upper bounds on supersymmetric particle masses from the naturalness requirement [21]. It is intuitive that, as the supersymmetry-breaking scale m_S grows, Eq. (18) can hold only with an increasingly precise cancellation among the different terms. We therefore require, for each parameter a_i:

$$\left| \frac{a_i}{M_Z^2} \frac{\partial M_Z^2}{\partial a_i} \right| < \Delta , \qquad (21)$$

where M_Z^2 is given by Eq. (18) and Δ is the degree of fine-tuning. Equation (21) can now be translated into upper bounds on the supersymmetric particle masses. Independently of specific universality assumptions on supersymmetry-breaking terms, we find [22], for instance, that the chargino and the gluino are respectively lighter than 120 and 500 GeV, if fine tunings no greater than 10% ($\Delta = 10$) are required.

4.3 Higgs Sector

Supersymmetry requires two Higgs doublets and therefore an extended spectrum of physical Higgs particles. Out of the eight degrees of freedom of the two complex doublets, three are eaten in the Higgs mechanism and five correspond to physical particles. These form two real CP-even scalars (h, H), one real CP-odd scalar (A), and one complex scalar (H^+). As we have seen in the previous section, the Higgs potential contains three parameters (m_1^2, m_2^2, m_3^2) and one of them is fixed by the electroweak symmetry-breaking condition, Eq. (18). Therefore, all tree-level masses and gauge couplings of the five Higgs particles are completely described by only two free parameters.

Another important feature of the supersymmetric Higgs potential is that the quartic coupling is given in terms of gauge couplings, see Eq. (15). In the Standard Model case, the quartic Higgs coupling measures the Higgs mass. Therefore, it is not surprising to find that in supersymmetry the mass of the lightest Higgs is bounded from above:

$$m_h < M_Z |\cos 2\beta| \ . \tag{22}$$

Supersymmetry does not only provide a solution to the hierarchy problem by stabilizing the Higgs mass parameter, but also predicts the existence of a Higgs boson lighter than the Z^0.

Note that Eq. (22) holds only at the classical level. There are important radiative corrections to the lightest Higgs mass proportional to m_t^4 [23]:

$$\delta m_h^2 \simeq \frac{3}{\pi^2} \frac{m_t^4}{v^2} \log \frac{m_S}{v} \ . \tag{23}$$

The upper bound given in Eq. (22) is then modified. For extreme values of the parameters, m_h can be as heavy as 130 GeV, but it is generally lighter.

This meant an excellent opportunity for LEP2, where the Standard Model Higgs boson can be probed via the process $e^+e^- \to hZ^0$ in essentially the entire kinematical range $m_h < \sqrt{s} - M_Z$. In the supersymmetric case, the search is more involved, because of the extended Higgs sector. For $\tan\beta$ close to 1, the supersymmetric Higgs boson resembles the Standard Model counterpart and the LEP2 search is unchanged. For large values of $\tan\beta$, the cross-section for $e^+e^- \to hZ^0$ is reduced and can become unobservable at LEP2. However, at the same time, the CP-odd Higgs boson A becomes light and the cross-section for the process $e^+e^- \to hA$ is then sizeable. The two different Higgs production mechanisms are therefore complementary and allow the search for the supersymmetric Higgs boson at LEP2 for most of the parameters. Nevertheless, a complete exploration of the whole supersymmetric parameter space will be possible only at the LHC, at the beginning of the next millenium.

The discovery of a light Higgs boson is certainly not a proof of the existence of supersymmetry at low energies. However, in the Standard Model, vacuum stability imposes a *lower* bound on the Higgs mass as a function of the top quark mass [24]. The Higgs search can in general give good indications about the scale of new physics.

4.4 Supersymmetry and Experiments

If the Higgs search is certainly an important experimental test, evidence for low-energy supersymmetry will come only from the discovery of the partners of ordinary particles.

The most important feature of supersymmetry phenomenology is the existence of a discrete symmetry, called R parity, which distinguishes ordinary particles from their partners. This is not an accidental symmetry, in the sense that it is not an automatic consequence of supersymmetry and gauge invariance. Nevertheless, it is usually assumed, or else dangerous B- or L-violating interactions are introduced. It can be understood as a consequence of gauge symmetry in GUT models which contain left-right symmetric groups. If R parity is indeed conserved only an even number of supersymmetric partners can appear in each interaction. As a consequence, supersymmetric particles are produced in pairs and the lightest supersymmetric particle is stable.

In most of the models, this stable particle turns out to be the lightest neutralino (χ^0). This is fortunate for the model, since the present density of electric- or colour-charged heavy particles is very strongly limited by searches for exotic atoms [26]. A stable neutral particle is not only allowed by present searches but also welcome since it can explain the presence of dark matter in the Universe (see ref. [27]). From the point of view of collider experiments, χ^0 will behave as a heavy neutrino which escapes the detector, leaving an unbalanced momentum and missing energy in the observed event. The distinguishing signature of supersymmetry is therefore an excess of missing energy and momentum. For example, in e^+e^- colliders, charginos and sleptons are pair-produced with typical electroweak cross-sections and then decay, giving rise to events such as:

$$e^+e^- \to \chi^+\chi^- \to \text{isolated leptons and/or jets} + \not{E} ,$$
$$e^+e^- \to \tilde{\ell}^+\tilde{\ell}^- \to \text{isolated leptons} + \not{E} . \qquad (24)$$

Using these processes, LEP1, working at the Z^0 peak, was able to rule out the existence of these particles with masses less than $M_Z/2$ [7]. LEP2 has covered most of the available kinematical range, and excluded χ^+ and $\tilde{\ell}^+$ with masses almost up to $\sqrt{s}/2$. This is certainly a very critical region since, as we have seen in Sect. 4.2 the 10% fine-tuning limits place the weakly-interacting supersymmetric particles at the border of the LEP2 discovery reach.

Strongly-interacting particles, such as squarks and gluinos, can be best studied at hadron colliders where they are produced with large cross-sections. The signature is again missing transverse energy carried by the neutralinos produced in the decays of squarks and gluinos. Tevatron experiments have set limits on the masses of these particles of about 150...200 GeV, depending on the particular model assumptions. At the LHC squarks and gluinos can be searched even for masses of several TeV, well above the 10% fine-tuning limits.

It is worth pointing out that although e^+e^- colliders are the ideal machines for a systematic search of new weakly-interacting particles, charginos and neutralinos may also be discovered at hadron colliders, for instance in the process:

$$p\bar{p} \to \chi_1^\pm \chi_2^0 , \quad \chi_1^\pm \to \ell^\pm \nu \chi_1^0 , \quad \chi_2^0 \to \ell^+ \ell^- \chi_1^0 . \qquad (25)$$

The signal of three leptons and missing transverse energy in the final state has almost no Standard Model background, when sufficient lepton isolation requirements are imposed. However, it is difficult to obtain lower bounds on the new particle masses, because the leptonic branching ratios of charginos and neutralinos depend strongly on the model parameters.

In conclusion, this generation of colliders is testing the theoretically best-motivated region of parameters in the supersymmetric model. We can be confident that, after the LHC has run, either low-energy supersymmetry will have been discovered or it must be discarded, since its main motivation is no longer valid.

4.5 The Flavour Problem

The Standard Model Lagrangian for gauge interactions is invariant under a global U(3,5) symmetry, with each U(1) acting on the generation indices of the five irreducible fermionic representations of the gauge group $(q_L, u_R^c, d_R^c, \ell_L, e_R^c)_i$, $i = 1, 2, 3$. This symmetry, called flavour (or generation) symmetry, implies that gauge interactions do not distinguish among the three generations of quarks and leptons. In the real world, this symmetry must be broken, as quarks and leptons of different generations have different masses. However, the breaking must be such as to maintain an approximate cancellation of Flavour-Changing Neutral Currents (FCNC). This is called the flavour problem.

In the Standard Model the flavour problem is solved in a simple and rather elegant way. The flavour symmetry is broken only by the Yukawa interactions between the Higgs field and the fermions. After electroweak symmetry breaking, these interactions give rise to the various masses of the three generations of quarks and leptons. The attractive feature of this mechanism is that all FCNC exactly vanish at tree level [28]. This is a specific property of the Standard Model with minimal Higgs structure and it is not automatic in models with an enlarged Higgs sector. Small contributions to FCNC are generated at loop level and generally agree with experimental observations. Athough this mechanism provides a great success of the Standard Model, it prevents us from computing any of the quark or lepton masses, as these are introduced in terms of some free parameters.

In supersymmetry, the solution of the flavour problem is more arduous. The soft-breaking terms generally violate the flavour symmetry and give too large contributions to the FCNC. This can be understood by recalling that, in a softly-broken supersymmetric theory, the mass matrices for quarks and squarks are independent and therefore cannot be simultaneously diagonalized by an equal rotation of the quark and squark fields. Thus neutral currents involving gluino–quark–squark vertices can mediate significant transitions among the different generations. Only if squarks and gluinos were heavier than 10...100 TeV could generic soft-breaking terms be consistent with observations of FCNC processes. Since, as previously discussed, the very motivation for low-energy supersymmetry implies that squarks and gluinos must be lighter than 500...1000 GeV, we have to postulate that the supersymmetry-breaking terms have some specific property.

The first possibility is that the supersymmetry-breaking terms respect the flavour symmetry in the limit of vanishing Yukawa couplings. This possibility is often advocated in models based on supergravity, on the basis of the hypothesis that all gravitationally-induced interactions are flavour-invariant. However, this hypothesis has been shown to be incorrect both in supergravity models with generic Kähler metrics [29] and in models derived from superstrings [30]. Nevertheless, this is an interesting possibility, since it significantly reduces the number of free parameters in the supersymmetry-breaking terms and allows sharp predictions testable at future colliders.

The other possibility is that the supersymmetry-breaking terms violate the flavour symmetry but are approximately aligned with the corresponding flavour violation in the fermionic sector (e.g. with the Yukawa couplings). This can be the result of some new symmetry [31] or some dynamical mechanism [32].

It is likely that the solution of the flavour problem is linked with the mechanism of supersymmetry breaking and therefore it will only be unravelled after significant theoretical developments. Now we can only speculate that an understanding of the flavour problem may help us to calculate the amount of flavour breaking and ultimately all quark and lepton masses.

4.6 Recent Developments in Supersymmetry

Let us now turn to discussing some more recent theoretical developments in supersymmetric model building. Most of the activity has focused on understanding the structure of the soft supersymmetry-breaking terms, especially in view of the flavour problem I have just illustrated. The question of the origin of the supersymmetry-breaking terms is indeed a crucial one, because the soft terms represent the connection between theory (i.e. the mechanism of supersymmetry breaking) and experiment (i.e. the mass spectrum of the new particles).

For many years the paradigm has been that the soft terms are produced by the gravitational couplings between a hidden sector where supersymmetry is originally broken and an observable sector containing the ordinary degrees of freedom [33]. The scale of supersymmetry breaking is determined to lie at an intermediate scale $\sqrt{F} \sim 10^{11}$ GeV by requiring that the observable supersymmetric particle masses $\tilde{m}$ are close to the weak scale:

$$\tilde{m} = \frac{F}{M_{\rm Pl}} \sim {\rm TeV}. \qquad (26)$$

This mechanism is elegant and theoretically appealing, as gravity is directly participating in electroweak physics. However, in this framework, the soft terms are renormalizable parameters of the effective theory defined at energies below the Planck mass $M_{\rm Pl}$. As such, at the quantum level, they receive corrections that depend on the properties of the underlying theory in the far ultraviolet. Therefore the soft terms cannot be computed, as long as we do not know the ultimate theory including a full description of quantum gravity. This could just be a limitation due to our lack of knowledge but, from a pragmatic point of view, it introduces two main problems. The first one is the lack of theoretical predictivity. This is indeed an acute problem since, even with the minimal field content, the low-energy supersymmetric model contains more than 110 free renormalizable parameters, crippling our ability to give solid guidelines to experimental searches. Secondly, the sensitivity of the soft terms to ultraviolet physics implies that their flavour structure will retain the effects of any (unknown) flavour violation at very high energies [34]. In particular, flavour universality of the soft terms will be spoilt by new interactions, which include, for instance, effects from GUTs [35] or from the dynamics at the (unknown) scale Λ_F responsible for the origin of the flavour-violating Yukawa couplings. Even if we hypothetically took mass-degenerate squarks at $M_{\rm Pl}$, high-energy flavour violations would induce large squark splittings, not correlated to the Yukawa couplings, at low energy. This situation is experimentally ruled out because the flavour violations in squarks and sleptons induce, through loop diagrams, unacceptably large contributions to Δm_K, ϵ_K, Δm_B, $b \to s\gamma$, $\mu \to e\gamma$, and other flavour processes.

To solve the flavour problem in the context of gravity-mediated supersymmetry breaking one needs to have full control of the dynamics even beyond $M_{\rm Pl}$. It is possible that its solution lies in the properties of quantum gravity and its flavour symmetries. However, recently there have been theoretical attempts to pursue alternative solutions, finding mechanisms aimed at eliminating the ultraviolet sensitivity of the soft terms altogether. If such a program succeeds, there are two immediate advantages. First of all, one has control over the flavour violations in the soft terms. Moreover, in this case, the soft terms are necessarily computable (i.e. their quantum corrections are finite in the effective theory below $M_{\rm Pl}$) and therefore one can make definite mass predictions relevant to experimental searches.

The best known class of theories in which the soft terms are insensitive to the far ultraviolet is given by gauge-mediated models [36]. Here the original supersymmetry breaking is felt at tree level only by some new particles of mass M (the messengers) and then communicated to the observable sector by loop diagrams involving ordinary Standard Model gauge interactions. Quantum corrections to the soft terms vanish for momenta larger than M. Any dynamics occurring at energy scales above M do not affect the soft terms. If we assume that M lies below any new flavour dynamics, then the Yukawa couplings provide the only source of flavour violations and we recover a supersymmetric extension of the GIM mechanism. Flavour violations in low-energy hadronic and leptonic processes are fully under control.

In gauge mediation, the soft terms are finite and computable. In the simplest version of the model, the gaugino, squark, and slepton masses are given by

$$\tilde{m}_{g_i} = \frac{\alpha_i F}{4\pi M}, \qquad (27)$$

$$\tilde{m}_f^2 = 2 \sum_{i=1}^{3} C_f^i \left(\frac{\alpha_i F}{4\pi M} \right)^2. \qquad (28)$$

Here α_i are the Standard Model gauge coupling constants and C_f^i are the corresponding quadratic Casimir coefficients.

Recently a different approach to obtain ultraviolet insensitivity of the soft terms has been pursued. The central observation is that, in the presence of supersymmetry breaking, gravity generates soft terms even if there are no direct couplings between the hidden and observable sectors [37,38]. This is an effect of the superconformal anomaly and it gives rise to soft terms that are suppressed by loop factors. If tree-level soft terms exist, then the anomaly-induced terms are subdominant. However, in some cases, they can provide the leading contribution. For gauginos, this occurs when the theory does not contain any gauge-singlet superfield that breaks supersymmetry (as for theories with dynamical supersymmetry breaking) [38]. Indeed, in the absence of gauge singlets X, one cannot generate the gaugino masses $\tilde{m}_g$ from the usual operator

$$\int d^2\theta \frac{X}{M_{\rm Pl}} {\rm Tr} W^\alpha W_\alpha + {\rm h.c.}, \qquad (29)$$

and therefore one has to rely on higher-dimensional operators, which give at most $\tilde{m}_g \sim F^{3/2}/M_{\rm Pl}^2 \sim$ keV. For scalars, the absence of tree-level contributions to their soft masses can be obtained with specific structures of Kähler potentials. These structures occur when the supersymmetry-breaking and observable sectors reside on different branes embedded into a higher-dimensional space and separated by a sufficiently large distance [37].

Let us assume that the soft terms, for the reasons explained above (or for any other unknown reason), are dominated by the anomaly contribution. In this case, the gaugino masses are given by [37,38]

$$\tilde{m}_g = \frac{\beta_g}{g} m_{3/2}, \qquad (30)$$

where $m_{3/2}$ is the gravitino mass (a measure of the supersymmetry-breaking scale) and β is the corresponding gauge-coupling beta function. More explicitly, for the gauginos relative to the three factors of the Standard Model gauge group, Eq. (30) gives

$$M_3 = -\frac{3\alpha_s}{4\pi} m_{3/2}$$
$$M_2 = \frac{\alpha}{4\pi \sin^2 \theta_W} m_{3/2} \simeq -0.1 M_3$$
$$M_1 = \frac{11\alpha}{4\pi \cos^2 \theta_W} m_{3/2} \simeq -0.3 M_3. \tag{31}$$

This is to be compared with the usual gaugino mass relations under GUT assumptions, $\tilde{m}_g = (g^2/g^2_{\rm GUT})\tilde{m}_g(M_{\rm GUT})$, which give

$$M_2 = 0.30 \; M_3$$
$$M_1 = 0.17 \; M_3. \tag{32}$$

The anomaly-mediated mass relation in Eq. (30) is particularly interesting because it depends only on low-energy coupling constants and it makes no reference on high-energy boundary conditions (GUT, messengers, ...). Indeed the form of Eq. (30) is invariant under renormalization group transformations. This entails a large degree of predictivity, since all soft terms can be computed from known low-energy Standard Model parameters and a single mass scale, $m_{3/2}$. Also, it leads to robust predictions, since the renormalization group invariance guarantees complete insensitivity of the soft terms from ultraviolet physics. As demonstrated with specific examples in ref. [38], heavy states do not affect the soft terms, since their contributions to the β functions and to threshold corrections exactly compensate each other. This means that the gaugino mass predictions in Eqs. (31) are valid irrespective of the GUT gauge group in which the Standard Model may or may not be embedded[1]. Therefore, although the soft terms are generated at very high-energy scales, their renormalization group trajectories are determined in such a way that the low-energy values of the soft terms are specified only by low-energy parameters. Whatever the dynamics that breaks flavour symmetry at high energies may be, the low-energy soft terms will respect a super-GIM mechanism.

In spite of its great theoretical appeal, a supersymmetric model with anomaly-mediated mass spectrum is not phenomenologically acceptable. The problem lies in the form of the scalar masses [37]

$$\tilde{m}^2 = -\frac{1}{4}\left(\frac{\partial \gamma}{\partial g}\beta_g + \frac{\partial \gamma}{\partial y}\beta_y\right) m_{3/2}^2. \tag{33}$$

---

[1] However, exceptions to ultraviolet insensitivity appear in the presence of gauge-singlet superfields [39].

Here β_g and β_y are the beta functions for the gauge and Yukawa coupling y, and γ is the anomalous dimension of the corresponding superfield. In the supersymmetric model SU(3) is asymptotically free and has a negative β function, but SU(2) and U(1) have a positive β function. Therefore, Eq. (33) predicts positive squark squared masses, but negative slepton squared masses. This would induce a spontaneous breaking of QED.

Several possible solutions have been suggested in order to cure this problem [37,39,40]. All of these solutions of course require new positive contributions to the slepton masses. These new terms necessarily spoil the most attractive feature of anomaly mediation, i.e. the renormalization group invariance of the soft terms and the consequent ultraviolet insensitivity. This is the most disappointing aspect of this scenario. At present, it is too early to assess if some of the appealing features of anomaly mediation have any relevance in the description of the elementary particle world.

4.7 More on Experimental Consequences

The realization that there are several possible schemes of supersymmetry-breaking communication has profound experimental implications, not only because of the different patterns of the superpartner mass spectrum, but also because each scheme has very distinctive signatures at high-energy collisions. Therefore, the search for supersymmetry requires different experimental analyses aimed at identifying quite different signals.

The stereotype missing-energy signature of supersymmetry described in Sect. 4.4 is specific to gravity-mediated scenarios, in which the produced supersymmetric particles cascade decay into the invisible lightest neutralino.

In gauge-mediated scenarios, the gravitino is the lightest supersymmetric particle, because its mass is determined by gravitational interactions instead of gauge interactions as in the case of the other superpartners. The experimental signals are then determined by the nature of the next-to-lightest supersymmetric particle (either a neutralino or a stau, depending on model-dependent parameters) and the scale of supersymmetry breaking F (which determines the lifetime of the next-to-lightest supersymmetric particle). For $\sqrt{F} \lesssim 10^6$ GeV, the next-to-lightest supersymmetric particles promptly decay into their Standard Model partners and gravitinos, leaving topologies containing tau leptons and missing energy (in the case of the stau) or photons and missing energy (in the case of the neutralino). On the other hand, for $\sqrt{F} \gtrsim 10^6$ GeV, the next-to-lightest supersymmetric particle is quasi-stable, since its decay length is typically longer than the detector size. The experimental signature is given by missing energy in the case of the neutralino, while in the case of the stau there is a more unconventional signal coming from a heavy charged particle crossing the apparatus, leaving anomalous ionization tracks.

The gaugino mass relations in Eqs. (31), characteristic of anomaly mediation, lead to quite peculiar experimental signals. Indeed, Eqs. (31) predict

$M_2 < M_1$ (in contrast to the usual case of Eqs. (32), in which $M_1 < M_2$). This and the electroweak-breaking conditions imply that, in realistic models, the SU(2) W-ino triplet is almost degenerate in mass. The mass splitting inside the triplet is dominated by loop effects and the charged particle is heavier than the neutral one, with $m_{\chi^\pm} - m_{\chi^0}$ in the range between the pion mass and about 1 GeV [41,42]. The (mainly W-ino) neutralino is the lightest supersymmetric particle, and the first chargino decays into a neutralino and a relatively soft pion $\chi^\pm \to m_{\chi^0} \pi^\pm$. The experimental difficulty lies in triggering such events, although kinks in the vertex detector could be revealed at the analysis stage. Different strategies consist in tagging high-energy jets or photons [41] or focus on the production and decay of other supersymmetric particles [42].

From this brief discussion, it should be clear that very different experimental strategies and analyses are necessary to look for the diversified ways in which supersymmetry could reveal itself in high-energy collisions.

As we have previously discussed, the various schemes of supersymmetry-breaking communication differ in the way they address the flavour problem. Therefore it is not surprising that experiments searching for rare flavour-violating or CP-violating processes are of great value for discriminating between the different supersymmetric scenarios. We can distinguish between two classes of supersymmetry-breaking models: *i)* those (like gauge mediation or anomaly mediation with a universal extra contribution to scalar masses) which satisfy a super-GIM mechanism, and flavour or CP violation is originating only from CKM angles and phases; *ii)* those (like gravity mediation) which rely on some flavour symmetry valid at some very large scale in the proximity of M_{Pl}, and necessarily contain some new sources of flavour and CP violation in the supersymmetry-breaking parameters.

In models belonging to class *i)*, we can expect only rather moderate deviations from the Standard Model predictions in flavour processes. The only exceptions could come from processes that are accidentally suppressed in the Standard Model and are more sensitive to new physics corrections (as in the case of the rare decay $B \to X_s \gamma$). On the other hand, in the models of class *ii)*, it appears almost unavoidable that new flavour-violating and CP-violating effects should lurk just behind the present experimental limits [35]. In this respect, the rôle of B factories will be crucial in helping theorists to sort out the way in which supersymmetry breaking is realized. Similarly, improvements in the sensitivity on lepton-family violating processes (like $\mu \to e\gamma$ and μ–e conversion in nuclei) and CP-violation (like electron and neutron electric dipole moments) will bring very valuable information.

5 Technicolour

We have seen how supersymmetry can cure the hierachy problem of the Standard Model by stabilizing the mass scale in the Higgs potential. Technicolour

[43] offers a different solution to the hierarchy problem, based on the idea of removing all fundamental scalar particles from the theory. The mass scale which sets the electroweak breaking is dynamically determined in a strongly interacting gauge theory with purely fermionic matter.

The presence of light scalars (mesons) in the hadronic spectrum does not pose a problem of hierarchy. The description of mesons as fundamental particles is valid only up to about $\Lambda_{\rm QCD}$. Above this scale, physics is described in terms of quarks and gluons, and hadrons have to be interpreted as composite particles. Technicolour aims to describe the Higgs boson as a composite particle, similarly to the case of mesons in QCD.

In order to illustrate the main idea of technicolour, let us consider as a toy model QCD with only two massless flavours ($m_u = m_d = 0$). In this limit, the theory has a chiral $\mathrm{SU}(2)_L \times \mathrm{SU}(2)_R$ invariance, in which the left-handed and right-handed components of the up and down quarks are rotated independently. As QCD becomes strongly-interacting at $Q^2 \lesssim \Lambda^2_{\rm QCD}$, the quark condensates are formed:

$$\langle u\bar{u} \rangle = \langle d\bar{d} \rangle = \mathcal{O}(\Lambda^3_{\rm QCD}) \,. \tag{34}$$

If the two condensates are equal, the chiral symmetry is broken to the vectorial part $\mathrm{SU}(2)_{L+R}$. Goldstone's theorem ensures the existence of three massless scalar particles in the spectrum, the pions $\pi^0, \pi^\pm$. In the real world, quark masses explicitly break chiral symmetry and give small masses to the pions. Also, if the strange quark is included, the chiral symmetry $\mathrm{SU}(3)_L \times \mathrm{SU}(3)_R$ is broken to $\mathrm{SU}(3)_{L+R}$, giving rise to the meson octet as approximate Goldstone bosons.

Let us turn on weak interactions in our toy model. Since the W boson couples to quarks, it also interacts with the pions. This coupling can be obtained from PCAC, which determines the matrix element of the broken current (j^a_μ) in terms of the pion decay constant f_π:

$$\langle 0 | j^a_\mu | \pi^b \rangle = f_\pi q_\mu \delta^{ab} \,. \tag{35}$$

Here a, b are SU(2) indices and q_μ is the pion four-momentum. From Eq. (35) and the coupling of the W boson to the weak current, we obtain the coupling between W^a_μ and π^b:

$$\frac{g}{2} f_\pi q_\mu \delta^{ab} \,. \tag{36}$$

Consider now the correction of one-pion exchange in the W propagator:

$$\frac{1}{q^2} + \frac{1}{q^2}\left(\frac{g}{2}f_\pi q^\mu\right)\frac{1}{q^2}\left(\frac{g}{2}f_\pi q_\mu\right)\frac{1}{q^2} \,. \tag{37}$$

The first term corresponds to the uncorrected massless W propagator and the second term corresponds to the exchange of a massless pion between two

W propagators with the coupling given in Eq. (36). We can insert an infinite number of pion exchanges, but it is not difficult to sum the whole series:

$$\frac{1}{q^2}\sum_{n=0}^{\infty}\left[\left(\frac{g}{2}f_\pi^2\right)\frac{1}{q^2}\right]^n = \frac{1}{q^2 - \left(\frac{g}{2}f_\pi\right)^2} \ . \tag{38}$$

Equation (38) shows that the effect of the pion exchange is to shift the pole value of the W propagator to

$$M_W = \frac{g}{2}f_\pi \ . \tag{39}$$

The W boson has acquired mass, which is not a surprising result if we think that we have promoted a global broken symmetry to a local invariance. The value for the W mass given by Eq. (39) is about 30 MeV, certainly too small to explain the experimental data.

We can use the result of this toy model and explain the physical value of M_W, if we introduce a new force, called technicolour. Technicolour behaves in a similar fashion to the ordinary colour forces but it becomes strong at a much larger scale $\Lambda_{\text{TC}} \simeq 500$ GeV. The simplest technicolour model is very easy to construct. Take a doublet of fermions with the same electroweak quantum numbers as the up and down quarks, assign to them a technicolour charge and call them techniquarks U and D. The condensates

$$\langle \bar{U}U\rangle = \langle \bar{D}D\rangle = \mathcal{O}(\Lambda_{TC}^3) \tag{40}$$

generate three composite Goldstone modes, which become the longitudinal degrees of freedom of the W and Z gauge bosons. We have then built a model of electroweak symmetry breaking with no fundamental Higgs boson. The experimental signature is the presence of strongly interacting dynamics at the TeV scale, which produces new resonances similar to those found in the hadronic spectrum at the GeV scale.

Although the mechanism for generating electroweak breaking in technicolour is very elegant, several difficulties have prevented the construction of a fully realistic model. The first problem is the communication of electroweak breaking to the quark and leptonic sectors of the theory. This can be done via new interactions, called extended technicolour (ETC) forces [44], which couple quarks to techniquarks. If the ETC symmetry is broken (possibly by some dynamical mechanism) at a scale M_{ETC} larger than Λ_{TC}, quarks and leptons receive masses of the order of

$$m_f \sim \frac{\langle F\bar{F}\rangle}{M_{\text{ETC}}^2} \sim \frac{\Lambda_{\text{TC}}^3}{M_{\text{ETC}}^2} \ , \tag{41}$$

where $\langle F\bar{F}\rangle$ is the corresponding technifermionic condensate. The trouble is that measurements of FCNC processes generally impose stringent lower bounds on $M_{\rm ETC}$, of the order of 100 TeV. This means that the ETC mechanism can generate the masses for the first generation of fermions, but has difficulties to explain the larger masses of the second and third generations. The task is particularly arduous for the top quark, since a dynamical mechanism which explains the large isospin breaking in the difference between m_t and m_b generally leads to large corrections to the ρ parameter, the ratio between the strengths of the neutral and charged weak currents. Finally, the effect of the strong technicolour dynamics always gives sizeable corrections to the electroweak precision data in LEP1, which have been shown to agree with the Standard Model with great accuracy [2].

The hope is that these problems can be cured in technicolour theories with dynamics substantially different from a scaled-up QCD. There has been some effort in this direction, trying to construct theories in which the ultraviolet behaviour of the technifermion self-energy enhances the quark mass contribution, while the infra-red behaviour determines the W mass. This may occur in theories with slowly running coupling constants (the so-called walking technicolour [45]) or in fixed-point gauge theories [46], although the non-perturbative nature of the problem prevents us from making reliable calculations.

6 Extra Dimensions

One of the greatest scientific successes of the last twenty years has been the precise verification of the Standard Model as the correct theory describing elementary particle interactions up to the weak scale. Following the idea of grand unified theories, we are used to extrapolating our knowledge to much smaller length scale, of the order of $M_{\rm GUT}^{-1} \sim 10^{-32}$ m. Moreover, string theory suggests a way to unify gauge and gravity forces at an even smaller distance scale, $M_S^{-1} \sim 2/(\sqrt{k\alpha_{\rm GUT}} M_{\rm Pl})$.

These are certainly courageous theoretical extrapolations, but nevertheless are not at present experimentally confirmed. In particular, gravity has been experimentally tested only up to scales of the order of $\lambda \sim$ mm $\sim (2 \times 10^{-4}$ eV$)^{-1}$, i.e. 30 orders of magnitude larger than M_S^{-1}! It is therefore legitimate to question the standard scenario, and wonder whether the gravitational coupling α_G could evolve, at energies above λ^{-1}, quite differently from our traditional expectations. In particular, one could imagine that the gravitational coupling becomes of the order of the gauge couplings already at the weak scale, eliminating the need for the large mass parameter $M_{\rm Pl}$ or, in other words, eliminating the notorious hierarchy problem.

Arkani-Hamed, Dimopoulos, and Dvali [47] have suggested a physical setting in which this radical point of view can actually be realized. Their construction assumes that our 4-dimensional world, in which ordinary particle processes occur, is actually embedded into a higher-dimensional space, in which only gravitons are free to roam. Let us define the total number of dimensions as $D = 4 + \delta$ and assume that the δ extra dimensions are compactified in a space with volume V_δ. It is a simple geometrical exercise to prove that the effective Newton constant in the 4-dimensional theory is related to the fundamental energy scale M_D of the full D-dimensional gravitational theory by the equation

$$G_N^{-1} \equiv M_{\text{Pl}}^2 = M_D^{2+\delta} V_\delta. \tag{42}$$

From this, we infer the typical radius of the compactified space

$$R \sim V_\delta^{1/\delta} \sim \frac{1}{M_D} \left(\frac{M_{\text{Pl}}}{M_D}\right)^{2/\delta}. \tag{43}$$

If we want to realize the scenario in which the fundamental gravitational mass parameter is roughly equal to the weak mass scale, we have to insist that $M_D \sim$ TeV, and therefore the typical size of the compactification radius is

$$\begin{aligned} R &= (5 \times 10^{-4} \text{ eV})^{-1} \sim 0.4 \text{ mm} && \text{for } \delta = 2, \\ R &= (20 \text{ keV})^{-1} \sim 10^{-5} \text{ } \mu\text{m} && \text{for } \delta = 4, \\ R &= (7 \text{ MeV})^{-1} \sim 30 \text{ fm} && \text{for } \delta = 6. \end{aligned} \tag{44}$$

For $\delta = 1$ the size of R is of astronomical length and therefore excluded by standard observations. The case $\delta = 2$ is marginally allowed and therefore interesting for experiments aiming at improving gravitational tests at small distances. As δ grows, R approaches the inverse of the fundamental mass scale M_D.

Before proceeding, we have to discuss whether the construction of ref. [47] can be realized in a physical system. Localizing fields on subspaces with lower dimensions can be achieved in a field theoretical context, but requires the introduction of certain scalar fields with particular potential; it is therefore possible but not straightforward. The great interest in the proposal of ref. [47] has been stirred by the observation that this situation is rather generic in the context of string theory. Indeed, Dirichlet branes (the space defined by the end-points of open strings [48]) are defects intrinsic to string theory on which the gauge theory is confined. The picture of ordinary particles (open strings) localized on the brane with gravity (closed strings) propagating in the bulk can be realized in string models [49]. This observation could actually help in bringing closer two lines of research in theoretical physics (one more phenomenologically oriented and one more formally oriented), which seemed to follow different paths in the last years. Indeed many theoretical speculations intended for Planck energy scales could now be relevant at the TeV scale, and therefore experimentally tested.

As evident from Eq. (43), in the higher-dimensional context, the weakness of gravity or, in other words, the smallness of the ratio G_N/G_F is related to the largeness of the number RM_D, which measures the compactified radius in its natural units. The hierarchy problem is not completely solved unless we understand why $R^{-1} \ll M_D \sim$ TeV. There have been several attempts to find dynamical explanations for the radius stabilization [50]. This problem may be connected with the cosmological constant puzzle.

Randall and Sundrum [51] have proposed an alternative higher-dimensional scenario in which the hierarchy problem is solved without the need for large ($R \gg M_D^{-1}$) extra dimensions. They consider a 5-dimensional non-factorizable geometry (i.e. the 4-dimensional metric is not independent of the extra coordinates) in which the line element is given by

$$\mathrm{d}s^2 = \mathrm{e}^{-2kr_c\Phi}\eta_{\mu\nu}\mathrm{d}x^\mu \mathrm{d}x^\nu + r_c^2 \mathrm{d}\Phi^2. \tag{45}$$

Here k is an energy scale of the order of the 5-dimensional Planck mass M_5 and Φ ($0 \le \Phi \le \pi$) is the coordinate of the compactified extra dimensions of size r_c. This metric is the solution of the Einstein equation in a model with two 3-branes (at $\Phi = 0$ and $\Phi = \pi$) with opposite tensions tuned to preserve 4-dimensional Poincaré invariance. In this situation, the 4-dimensional Planck mass is given by

$$M_{\mathrm{Pl}}^2 = \frac{M_5^3}{k}\left(1 - \mathrm{e}^{-2\pi kr_c}\right). \tag{46}$$

We will be interested in the limit $kr_c \gg 1$, in which the exponential factor in Eq. (46) is irrelevant, and we take $M_5 \sim k \sim \mathcal{O}(M_{\mathrm{Pl}})$. The exponential factor is however important for the mass parameters of the fields confined on the 3-brane at $\Phi = \pi$ representing our world. As apparent from Eq. (45), the exponential $\mathrm{e}^{-2kr_c\Phi}$ acts as a conformal factor in the 4-dimensional theory and therefore it is not surprising that the physical mass parameters on the brane are given by $m_0 \mathrm{e}^{-\pi kr_c}$, if $m_0 \sim \mathcal{O}(M_{\mathrm{Pl}})$ is the mass parameter in the 5-dimensional theory. For the moderate number $kr_c \simeq 12$, the large hierarchy between the weak and the gravitational mass can be reproduced.

The emerging physical picture is the following. Because of the non-factorizable form of the geometry, the gravitational field configuration is highly non-trivial. Gravitons are localized on one brane, while the Standard Model particles live on the other brane. The small overlap of the graviton wavefunction with our brane explains the weakness of gravity. No hierarchically small numbers are required because of the exponential suppression. The mass gaps and the mass scale in the effective interactions of the Kaluza-Klein gravitons are both of the order of the weak scale, since the weak scale is the only relevant mass in this physical picture.

This proposal has been further elaborated and an alternative scenario for a solution to the hierarchy problem has been suggested in ref. [52]. The crucial observation [53] is that, in the presence of non-factorizable metrics we can envision non-compact extra dimensions without conflicting with observations. The graviton is again localized, but its Kaluza-Klein spectrum has no mass gap. Nevertheless this is not problematic, because all excited Kaluza-Klein modes give corrections to the gravitational couplings of the order of e^2/M_{Pl}^2, where E is the typical process energy. It is now possible to consider a setup in which the Standard Model resides on one brane while gravity is localized on a different brane, and both branes have positive tensions. The separation between the two branes reproduces the hierarchy $M_W/M_{\rm Pl}$ and the fifth dimension is infinitely large.

6.1 Opening New Problems

The idea of having a unique fundamental mass scale, of the order of the TeV, for both weak and gravity interactions clearly requires a complete rethinking of much of the accepted understanding of the high-energy behaviour and of early cosmology.

First of all, one has to abandon a very successful feature of the traditional constructs: certain symmetry-breaking interactions are small because they arise from physics at very large scales. Usually one describes these symmetry-breaking effects with effective operators suppressed by an unspecified mass scale Λ, such as

$$
\begin{aligned}
\text{neutrino masses} &\to \frac{1}{\Lambda}\ell\ell HH \\
\text{proton decay} &\to \frac{1}{\Lambda^2}qqq\ell \\
\text{flavour violation} &\to \frac{1}{\Lambda^2}\bar{s}d\bar{s}d \\
\text{lepton family violation} &\to \frac{1}{\Lambda}\bar{\mu}\sigma_{\mu\nu}eF^{\mu\nu}.
\end{aligned}
\tag{47}
$$

The smallness of the observed violation of the corresponding exact or approximate symmetries implies that the scale Λ is much larger than the weak scale. In theories with quantum gravity at the TeV scale, we cannot rely on such an explanation. These theories therefore require new mechanisms to understand small parameters. One possibility is that small parameters are not the consequence of approximate symmetries, as in the examples above, but instead in what I will call "locality and geometry". As suggested in ref. [54], suppose that all unwanted symmetry-breaking effects can only occur locally on branes that are physically separated by a distance d from the 3-dimensional brane of our world. In this case, the effective couplings of the symmetry-breaking interactions will be suppressed by a factor $e^{-m/d}$, where m is the typical mass of the bulk particle that mediates the interaction from one brane to the other. Large suppression factors can be obtained with moderate ratios of m/d.

The same mechanism can be used to obtain the flavour structure of the Yukawa couplings [55]. One can also extend this picture and place the three quark and lepton families on different locations in the directions orthogonal to the ordinary 3-dimensional space [56]. Depending on the profile of the fermion wave-functions along the extra dimensions, large hierarchies in the

Yukawa couplings could be obtained from numbers of order 1, using the above-mentioned exponential factor. If this conjecture were true, we could even hope to unravel unsuspected properties of the flavour symmetries. The pattern of Yukawa couplings could look much simpler when viewed in terms of exponential factors or some other functional dependence.

Neutrino masses cannot be any longer explained by the see-saw mechanism and require some new higher-dimensional mechanism. One possibility is that right-handed neutrinos, in contrast with the other Standard Model particles, live in the full D-dimensional space [57]. The Yukawa interaction between left- and right-handed neutrinos is localized on the brane. Since the wave-function of ν_R is spread in the bulk space, the effective Yukawa coupling is suppressed by the square root of the compactified volume V_δ. The neutrino mass is then given by

$$m_\nu = \frac{\lambda \langle H \rangle}{\sqrt{V_\delta M_D^\delta}} \sim \lambda \langle H \rangle \frac{M_D}{M_{\rm Pl}} \sim 10^{-4} \text{ eV} \left(\frac{M_D}{\text{TeV}}\right), \qquad (48)$$

where we have assumed that the Yukawa coupling λ in the D-dimensional theory is of order 1. Notice that the resulting neutrino mass is of the Dirac type and it is in the correct ballpark to explain the atmospheric neutrino data.

Although it first appears that gauge-coupling unification is irremediably lost, it is nevertheless possible to conceive new higher-dimensional schemes in which the success of the supersymmetric prediction is recovered. One possibility [58] is to assume that Standard Model particles have Kaluza-Klein excitations (with masses larger than a few TeV). Their effects in the β functions change the logarithmic dependence on the energy into a power dependence and speed up the unification, which can now occur at energies not much larger than the weak scale. ¿From the field-theoretical point of view, one loses control of the theory, but nevertheless it is possible that an actual gauge-coupling unification is achieved in a string theory with TeV scale. Another possibility [59] is to use field variations in the large extra dimensions to achieve a logarithmic unification.

The early cosmology of theories with quantum gravity at the TeV scale will also look drastically different from what has been traditionally assumed. In the scenario of ref. [47], a problem arises. During the early phase of the Universe, energy can be emitted from the brane into the bulk in the form of gravitons. The gravitons propagate in the extra dimensions and can decay into ordinary particles only by interacting with the brane, and therefore with a rate suppressed by $1/M_{\rm Pl}^2$. Their contribution to the present energy density exceeds the critical value unless [47]

$$T_\star < \frac{M_D}{\text{TeV}} \; 10^{\frac{6\delta-15}{\delta+2}} \text{ MeV}. \qquad (49)$$

Here $T_\star$ is the maximum temperature to which we can simply extrapolate the thermal history of the Universe, assuming it is in a stage with completely stabilized R and with vanishing energy density in the compactified space. As a possible example of its origin, $T_\star$ could correspond to the reheating temperature after an inflationary epoch. The bound in Eq. (49) is very constraining. In particular, for $\delta = 2$, only values of M_D larger than about 6 TeV can lead to $T_\star > 1$ MeV and allow for standard nucleosynthesis. Moreover, even for larger values of δ, Eq. (49) is very problematic for any mechanism of baryogenesis [60].

The graviton emission is also dangerous in an astrophysical context. Extra-dimensional gravitons would speed up supernova cooling in contradiction with the neutrino observation from SN1987A, unless [61]

$$\begin{aligned} M_D &> 50 \text{ TeV} \quad \text{for} \quad \delta = 2, \\ M_D &> 4 \text{ TeV} \quad \text{for} \quad \delta = 3. \end{aligned} \quad (50)$$

An even stronger limit comes from distortion of the diffuse gamma-ray background [62],

$$\begin{aligned} M_D &> 110 \text{ TeV} \quad \text{for} \quad \delta = 2, \\ M_D &> 5 \text{ TeV} \quad \text{for} \quad \delta = 3. \end{aligned} \quad (51)$$

This bound is very constraining in the case of two extra dimensions, and it rapidly decreases with δ, because of the power-law suppression of graviton interactions. Notice that these limits are determined by the infrared behaviour of the gravitational theory. Therefore they do not apply to theories that have large Kaluza-Klein graviton gaps. They can also be evaded in the scenario of ref. [47], in the case of very particular compactified spaces which enhance the masses of the first Kaluza-Klein excitations.

6.2 Experimental Tests of Extra Dimensions

The idea that quantum gravity resides at the weak scale can be put under experimental scrutiny. We started our discussion on the motivations of extra dimensions by pointing out that gravity has been tested only to scales just below the millimetre. It is therefore clear that improvements in the experimental sensitivity will be of great importance. Indeed there are ongoing experiments [63] that aim at testing gravity up to distances of several tens of microns.

Unfortunately, the astrophysical bounds presented in Eq. (51) can be translated into a limit on the Compton wavelength of the first graviton Kaluza-Klein mode of 5×10^{-2} μm. The possibility of experimentally observing a deviation of gravity caused by higher-dimensional gravitons is then ruled out, at least in near-future experiments. Any modification of the compactified space capable of avoiding the astrophysical bound will also exclude a visible signal at short-distance gravitational experiments. Nevertheless, in many models realizing the idea of low-scale quantum gravity, there exist other light bulk particles, which could lead to observable signals [47]. A possible effect could also come from other light particles in scenarios with low-energy supersymmetry breaking [64].

High-energy collider experiments can directly probe the new dynamics of quantum gravity at the weak scale. At first, one may believe that the experimental signal should depend on the specific quantum gravitational theory, and therefore no solid prediction could be made. However, in a certain kinematical regime, it is possible to make rather model-independent estimates of graviton production in high-energy collisions. The strategy is to use an effective theory [65,66], valid below the fundamental mass scale M_D, where one can perform an expansion in E/M_D (here E is the typical process energy) and use our knowledge of the infrared properties of gravity.

In the scenario of ref. [47], gravitons are massless particles propagating in D dimensions. Therefore, the relation between their energy E and their momentum is $E^2 = \boldsymbol{p}^2 + p_{\text{extra}}^2$, where $\boldsymbol{p}$ describes the usual 3-dimensional components and p_{extra} is the momentum along the extra dimensions. This relation gives an intuitive explanation of how a D-dimensional particle can be described by a collection of 4-dimensional modes (called the Kaluza-Klein excitations) with mass $m = |p_{\text{extra}}|$.

We will be interested in the production of the Kaluza-Klein graviton modes in high-energy collisions. The single production of a graviton with non-vanishing $|p_{\text{extra}}|$ violates momentum conservation along the extra dimensions. This is not surprising, since the presence of the 3-brane breaks translational invariance in the directions orthogonal to the brane. It is like playing tennis against a wall: the momentum along the direction orthogonal to the wall is not conserved. Gravitons cannot be directly detected. Therefore the signal in collider experiments is missing energy and imbalance in final-state momenta, caused by the graviton escaping in the extra-dimension compactified space. Just for illustration, we can visualize elementary-particle interactions as the collisions of balls on a pool table. The balls can only move on a 2-dimensional surface (the brane), but as they knock each other they can emit a sound wave (the graviton), which travels in the air (the bulk). Because of this energy loss, an observer living on the surface of the table can infer the existence of the extra dimension by measuring the kinematics of the balls before and after the collision.

Each graviton Kaluza-Klein mode G_n has a production probability proportional to E^2/M_{Pl}^2, which gives rise to a cross section at hadron colliders of

$$\sigma(pp \to G_n \text{ jet}) \simeq \frac{\alpha_s}{\pi} G_N = 10^{-28} \text{ fb}. \qquad (52)$$

This is hopelessly small and it cannot be observed. However, experiments are sensitive to inclusive processes, in which we sum over all kinematically accessible Kaluza-Klein modes. Because of the large volume in the extra dimensions, the number of graviton Kaluza-Klein modes with mass less than a typical energy E is very large $\sim E^\delta M_{\text{Pl}}^2/M_D^{2+\delta}$. As a result, the dependence of the inclusive cross section on M_{Pl} cancels out, and we find

$$\sum_n \sigma(pp \to G_n \text{jet}) \simeq \frac{\alpha_s}{\pi} \frac{E^\delta}{M_D^{2+\delta}}. \qquad (53)$$

By studying final states with photons and missing energy, LEP has already set bounds on the fundamental quantum gravity scale M_D of about 1 TeV (for a number of extra dimensions $\delta = 2$). Future studies at the Tevatron, LHC, linear colliders or muon colliders can significantly extend the sensitivity region of M_D by analysing final states with jets and missing energy or photons and missing energy [65,66].

It should be stressed that in a complete quantum gravity theory there will certainly exist other experimental signals, quite different from the graviton signal considered above. However, these new signatures are model-dependent and cannot be predicted without a complete knowledge of the final theory. Therefore, the effective-theory signal discussed here, although it does not necessarily represent the discovery mode, is best suited for setting reliable bounds on M_D.

In general, one can parametrize new physics effects at the scale M_D with all possible effective interactions with couplings of order 1 in units of M_D. However, there is one particular operator that could play a special role,

$$\mathcal{T} \equiv T_{\mu\nu} T^{\mu\nu} - \frac{1}{\delta+2} T^\mu_\mu T^\nu_\nu. \qquad (54)$$

Here $T_{\mu\nu}$ is the energy–momentum tensor. The operator in Eq. (54) is induced by tree-level virtual graviton exchange and it will appear in the effective Lagrangian with a coefficient of order $1/M_D^4$. Unfortunately the precise form of this coefficient cannot be computed by using only the effective theory, because it depends on ultraviolet properties. Nevertheless, experimental searches on the existence of this operator are interesting because they represent a test on the spin-2 nature of the particle that mediates the effective interactions. The operator $\mathcal{T}$ gives rise to a variety of experimental signals, which include, in e^+e^- colliders, d-wave contributions to fermion pair production, $\gamma\gamma$ and multijet final states and, in hadron colliders, dilepton or $\gamma\gamma$ production [65,67]. All these signals are in principle related, because they originate from the same interaction.

The graviton-production signal is characteristic of theories with large extra dimensions $R \gg M_D^{-1}$. In models in which the graviton Kaluza-Klein gaps are of the order of M_D (as for instance in the scenario of ref. [51]), the interesting experimental signal is given by the production of the new gravitational excitations with weak-scale masses. Actually, it is possible that all Standard Model particles have Kaluza-Klein modes at the TeV scale [68]. This is the case, for instance, in the proposal of ref. [58] to achieve gauge-coupling unification at low-energy scales. This situation is not inconsistent with the large extra dimension scenario. The Standard Model could live in a D'-dimensional space with $4 < D' < D$ and with compactification radius $R' \sim$ TeV. Gravity propagates also in the extra $D - D'$ dimensions characterized by a radius $R \gg R'$. Precision electroweak measurements constrain at present R'^{-1} to be above about 3...4 TeV [69,70]. Nevertheless, LHC still has the chance of observing the first Kaluza-Klein excitations of Standard Model particles or, at least, of setting bounds on R'^{-1} of more than 6 TeV [70,71].

If indeed quantum gravity sets in at the electroweak scale, future collider experiments will directly test the structure of its unknown dynamics. For instance, if string theory becomes relevant at M_D [72], experiments could observe Regge recurrences with higher masses and spins. It is certain that, whatever the underlying weak-scale quantum gravity theory may be, collider experiments in the TeV range will be quite exciting.

7 Conclusions

We are now entering a phase in which searches for new physics are becoming the main experimental goal. The community in theoretical physics beyond the Standard Model is therefore facing a special responsibility. I believe that we are responding to this challenge, since in the last few years numerous new theoretical ideas have arisen to question some of the traditional beyond-the-Standard-Model assumptions. It is too early to make definite assessments, but it is very plausible to believe that some of these ideas may lead to a profound revision of our views on the underlying high-energy theory.

References

1. S.L. Glashow, *Nucl. Phys.* **22** (1961) 579;
   S. Weinberg, *Phys. Rev. Lett.* **19** (1967) 1264;
   A. Salam, in *Elementary Particle Theory*, ed. N. Svartholm (Almquist and Wiksells, Stockholm, 1969), p. 367.
2. P. Langacker, in *Precision Tests of the Standard Electroweak Model*, ed. by P. Langacker (World Scientific, Singapore, 1994);
   G. Altarelli, in *Proc. of the Tennessee Int. Symp. on Radiative Corrections*, Gatlinburg TN, June 1994.
3. P. Langacker, *Phys. Rep.* **72** (1981) 185;
   G.G. Ross, *Grand Unified Theories* (Benjamin/Cummings Publ. Co., Menlo Park, 1985).
4. J. Wess and J. Bagger, *Supersymmetry and Supergravity* (Princeton University Press, 1983);
   H.P. Nilles, *Phys. Rep.* **110** (1984) 1;
   R. Arnowitt, A. Chamseddine, and P. Nath, *Applied N=1 Supergravity* (World Scientific, Singapore, 1984);
   H.E. Haber and G. Kane, *Phys. Rep.* **117** (1985) 75;
   P. West, *Introduction to Supersymmetry and Supergravity* (World Scientific, Singapore, 1986);

R.N. Mohapatra, *Unification and Supersymmetry* (Springer-Verlag, Heidelberg, 1986);
R. Barbieri, *Riv. Nuovo Cim.* **11** (1988) 1.
5. E. Farhi and L. Susskind, *Phys. Rep.* **74** (1981) 277;
R. Kaul, *Rev. Mod. Phys.* **55** (1983) 449;
M. Chanowitz, *Annu. Rev. Nucl. Part. Sci.* **38** (1988) 323;
T. Appelquist, in *Proc. Mexican School of Particles and Fields*, Mexico City, December 1990.
6. H. Georgi and S.L. Glashow, *Phys. Rev. Lett.* **32** (1974) 438.
7. *Review of particle properties*, D.E. Groom *et al.*, The European Physical Journal **15** (2000) 1.
8. M. Yoshimura, *Phys. Rev. Lett.* **41** (1978) 281;
S. Dimopoulos and L. Susskind, *Phys. Rev.* **D18** (1978) 4500;
D. Toussaint, S.B. Treiman, F. Wilczek, and A. Zee, *Phys. Rev.* **D19** (1979) 1036;
S. Weinberg, *Phys. Rev. Lett.* **42** (1979) 850.
9. H. Georgi, in *Particles and Fields*, ed. by C.E. Carlson, (American Institute of Physics, New York, 1974);
H. Fritzsch and P. Minkowski, *Ann. Phys.* **93** (1975) 193.
10. M. Gell-Mann, P. Ramond, and R. Slansky, in *Supergravity*, ed. by D.Z. Freeman and P. van Nieuwenhuizen (North-Holland, Amsterdam, 1979).
11. K. Wilson, as quoted by L. Susskind, *Phys. Rev.* **D20** (1979) 2619;
G. 't Hooft, in *Recent Developments in Gauge Theories*, ed. by G. 't Hooft *et al.* (Plenum Press, 1980).
12. E. Gildener, *Phys. Rev.* **D14** (1976) 1667.
13. Y. Gol'fand and E. Likhtam, *JETP Lett.* **13** (1971) 323;
D. Volkov and V. Akulov, *Phys. Lett.* **B46** (1973) 109;
J. Wess and B. Zumino, *Nucl. Phys.* **B70** (1974) 39.
14. J. Wess and B. Zumino, *Phys. Lett.* **B49** (1974) 52;
J. Iliopoulos and B. Zumino, *Nucl. Phys.* **B76** (1974) 310;
S. Ferrara and O. Piguet, *Nucl. Phys.* **B93** (1975) 261;
M.T. Grisaru, W. Siegel, and M. Rocek, *Nucl. Phys.* **B159** (1979) 429.
15. L. Girardello and M.T. Grisaru, *Nucl. Phys.* **B194** (1982) 65.
16. A.H. Chamseddine, R. Arnowitt, and P. Nath, *Phys. Rev.* **D49** (1982) 970;
R. Barbieri, S. Ferrara, and C.A. Savoy, *Phys. Lett.* **B119** (1982) 343.
17. S. Dimopoulos and H. Georgi, *Nucl. Phys.* **B193** (1981) 150;
N. Sakai, *Z. Phys.* **C11** (1981) 153.
18. W. Marciano and A. Sirlin, in *Proc. Second Workshop on Grand Unification*, ed. by J. Leveille *et al.*, Ann Arbor, April 1981.
19. Y. Yamada, *Phys. Lett.* **B316** (1993) 109;
S.P. Martin and M.T. Vaughn, *Phys. Lett.* **B318** (1993) 331.
20. H.P. Nilles, *Phys. Lett.* **B115** (1982) 193;
L. Ibañez, *Phys. Lett.* **B118** (1982) 73.
21. J. Ellis, K. Enqvist, D.V. Nanopoulos, and F. Zwirner, *Mod. Phys. Lett.* **A1** (1986) 57;
R. Barbieri and G.F. Giudice, *Nucl. Phys.* **B306** (1988) 63.
22. S. Dimopoulos and G.F. Giudice, *Phys. Lett.* **B357** (1995) 573.
23. Y. Okada, M. Yamaguchi, and T. Yanagida, *Prog. Theor. Phys.* **85** (1991) 1;
H.E. Haber and R. Hempfling, *Phys. Rev. Lett.* **66** (1991) 1815;
J. Ellis, G. Ridolfi, and F. Zwirner, *Phys. Lett.* **B257** (1991) 83.

24. N. Cabibbo, L. Maiani, G. Parisi, and R. Petronzio, *Nucl. Phys.* **B158** (1979) 295.
25. J.R. Espinosa and M. Quiros, preprint CERN-TH/95-18;
    M. Quiros, preprint CERN-TH/95-197.
26. J. Ellis, J.S. Hagelin, D.V. Nanopoulos, K. Olive, and M. Srednicki, *Nucl. Phys.* **B238** (1984) 453.
27. M. Spiro, in these Proceedings.
28. S.L. Glashow, J. Iliopoulos, and L. Maiani, *Phys. Rev.* **D2** (1970) 1285.
29. S. Soni and A. Weldon, *Phys. Lett.* **B126** (1983) 215;
    L.J. Hall, J. Lykken, and S. Weinberg, *Phys. Rev.* **D27** (1983) 2359.
30. L. Ibañez and D. Lüst, *Nucl. Phys.* **B382** (1992) 305;
    V.S. Kaplunovsky and J. Louis, *Phys. Lett.* **B306** (1993) 269.
31. Y. Nir and N. Seiberg, *Phys. Lett.* **B309** (1993) 337.
32. S. Dimopoulos, G.F. Giudice, and N. Tetradis, *Nucl. Phys.* **B454** (1995) 59.
33. A.H. Chamseddine, R. Arnowitt and P. Nath, Phys. Rev. Lett. **49**, 970 (1982);
    R. Barbieri, S. Ferrara and C.A. Savoy, Phys. Lett. **B119**, 343 (1982).
34. L.J. Hall, V.A. Kostelecky and S. Raby, Nucl. Phys. **B267**, 415 (1986).
35. R. Barbieri and L.J. Hall, Phys. Lett. **B338**, 212 (1994); R. Barbieri, L. Hall and A. Strumia, Nucl. Phys. **B445**, 219 (1995).
36. M. Dine and A.E. Nelson, Phys. Rev. **D48**, 1277 (1993); M. Dine, A.E. Nelson and Y. Shirman, Phys. Rev. **D51**, 1362 (1995); M. Dine, A.E. Nelson, Y. Nir and Y. Shirman, Phys. Rev. **D53**, 2658 (1996); G.F. Giudice and R. Rattazzi, hep-ph/9801271.
37. L. Randall and R. Sundrum, hep-th/9810155.
38. G.F. Giudice, M.A. Luty, H. Murayama and R. Rattazzi, JHEP **12**, 027 (1998).
39. A. Pomarol and R. Rattazzi, JHEP **05**, 013 (1999).
40. Z. Chacko, M.A. Luty, I. Maksymyk and E. Ponton, hep-ph/9905390; E. Katz, Y. Shadmi and Y. Shirman, JHEP **08**, 015 (1999).
41. J.L. Feng, T. Moroi, L. Randall, M. Strassler and S. Su, Phys. Rev. Lett. **83**, 1731 (1999).
42. T. Gherghetta, G.F. Giudice and J.D. Wells, hep-ph/9904378.
43. S. Weinberg, *Phys. Rev.* **D19** (1978) 1277;
    L. Susskind, *Phys. Rev.* **D20** (1979) 2619.
44. S. Dimopoulos and L. Susskind, *Nucl. Phys.* **B155** (1979) 237;
    E. Eichten and K.D. Lane, *Phys. Lett.* **B90** (1980) 125.
45. B. Holdom, *Phys. Rev.* **D24** (1981) 1441 and *Phys. Lett.* **B150** (1985) 301;
    T. Appelquist, D. Karabali, and L.C.R. Wijewardhana, *Phys. Rev. Lett.* **57** (1986) 957;
    T. Appelquist and L.C.R. Wijewardhana, *Phys. Rev.* **D36** (1987) 568.
46. K. Yamawaki, M. Bando, and K. Matumoto, *Phys. Rev. Lett.* **56** (1986) 1335;
    G.F. Giudice and S. Raby, *Nucl. Phys.* **B368** (1992) 221.
47. N. Arkani-Hamed, S. Dimopoulos and G. Dvali, Phys. Lett. **B429**, 263 (1998) and Phys. Rev. **D59**, 086004 (1999).
48. J. Polchinski, hep-th/9611050; C.P. Bachas, hep-th/9806199.
49. P. Horava and E. Witten, Nucl. Phys. **B460**, 506 (1996) and Nucl. Phys. **B475**, 94 (1996); I. Antoniadis, N. Arkani-Hamed, S. Dimopoulos and G. Dvali, Phys. Lett. **B436**, 257 (1998); Z. Kakushadze and S.H. Tye, Nucl. Phys. **B548**, 180 (1999).
50. R. Sundrum, Phys. Rev. **D59**, 085010 (1999); N. Arkani-Hamed, S. Dimopoulos and J. March-Russell, hep-th/9809124.

51. L. Randall and R. Sundrum, Phys. Rev. Lett. **83**, 3370 (1999).
52. J. Lykken and L. Randall, hep-th/9908076.
53. L. Randall and R. Sundrum, hep-th/9906064.
54. N. Arkani-Hamed and S. Dimopoulos, hep-ph/9811353; Z. Berezhiani and G. Dvali, Phys. Lett. **B450**, 24 (1999).
55. N. Arkani-Hamed, L. Hall, D. Smith and N. Weiner, hep-ph/9909326.
56. N. Arkani-Hamed and M. Schmaltz, hep-ph/9903417.
57. K.R. Dienes, E. Dudas and T. Gherghetta, hep-ph/9811428; N. Arkani-Hamed, S. Dimopoulos, G. Dvali and J. March-Russell, hep-ph/9811448; A.E. Faraggi and M. Pospelov, Phys. Lett. **B458**, 237 (1999).
58. K.R. Dienes, E. Dudas and T. Gherghetta, Phys. Lett. **B436**, 55 (1998) and Nucl. Phys. **B537**, 47 (1999).
59. N. Arkani-Hamed, S. Dimopoulos and J. March-Russell, hep-th/9908146.
60. K. Benakli and S. Davidson, Phys. Rev. **D60**, 025004 (1999).
61. S. Cullen and M. Perelstein, Phys. Rev. Lett. **83**, 268 (1999).
62. L.J. Hall and D. Smith, Phys. Rev. **D60**, 085008 (1999).
63. J.C. Long, H.W. Chan and J.C. Price, Nucl. Phys. **B539**, 23 (1999).
64. S. Dimopoulos and G.F. Giudice, Phys. Lett. **B379**, 105 (1996).
65. G.F. Giudice, R. Rattazzi and J.D. Wells, Nucl. Phys. **B544**, 3 (1999).
66. E.A. Mirabelli, M. Perelstein and M.E. Peskin, Phys. Rev. Lett. **82**, 2236 (1999); T. Han, J.D. Lykken and R. Zhang, Phys. Rev. **D59**, 105006 (1999).
67. J.L. Hewett, Phys. Rev. Lett. **82**, 4765 (1999).
68. I. Antoniadis, Phys. Lett. **B246**, 377 (1990).
69. P. Nath and M. Yamaguchi, hep-ph/9902323; M. Masip and A. Pomarol, hep-ph/9902467; W.J. Marciano, hep-ph/9903451; A. Strumia, hep-ph/9906266.
70. T.G. Rizzo and J.D. Wells, hep-ph/9906234.
71. I. Antoniadis, K. Benakli and M. Quiros, Phys. Lett. **B331**, 313 (1994) and hep-ph/9905311.
72. J.D. Lykken, Phys. Rev. **D54**, 3693 (1996).

Subject Index

B meson 43
– oscillation 74, 202
– total decay rate 107
B_s meson
– oscillation 70, 76, 93
D meson
– oscillation 74
K meson
– CP violation 203
– decay asymmetry 196
– oscillation 73, 199
K system 31
– CP transformation 10
– CP violation 10, 27, 30
– decay matrix 32
– mass matrix 32
R parity 306
T, dimensionless time variable 63
Λ term 239
α, mixing parameter 60
δ_ϵ, CP violation parameter 61
ϵ, mixing parameter 60
η_m, mixing parameter 59
μ term 266
θ term 19
d_α, CP violation parameter 61
p, q, mixing parameters 59

Abundance
– elements 244
– light nuclei 245
Adler-Bell-Jackiw anomaly 250
Anapole moment 6, 211
Anomaly 310
Anomaly equations 250
Anticounter 35
Antigalaxies 244
Antimatter 97
– primordial 27, 244
Antinuclei 244
Antiparticle 45, 96, 97, 134
Arrow of time 246
Asymmetry
– baryon 237, 243, 244, 281
– lepton 281
– matter-antimatter 97, 237
Atom
– Cs 6, 209
– exotic 306
– Tl 209
Atomic nucleus
– ^{2}H 212
– ^{3}He 209
– ^{118}Sn 210
– ^{139}La 210
– ^{165}Ho 210, 212
– ^{18}F 211
– ^{199}Hg 18, 209, 212
– ^{57}Fe 16, 212
Axion 21
– invisible 21
– mass 21
– searches 21, 23

B+L number violation 251
Background radiation 27, 245
Baryogenesis 237, 297
– electroweak 255, 269, 275
– GUT 277
– local 269
– nonlocal 269
Baryon asymmetry 237
– primordial 297
– universe (BAU) 243, 244, 281
Baryon number violation 28, 246
– standard model 249

Baryon-to-entropy ratio 274, 284
beat frequency
– $B^0 \bar{B}^0$ 66
Beon
– see B meson 196
Big bang 245, 297
Boltzmann equation 244
Boundary conditions
– thermodynamic 259
Brane 317
Bubble nucleation 260

Cabibbo angle 12, 47
Charge
– generalized 7
– topological 252
Charge conjugation 7, 9, 96
– Dirac spinors 8
– eigenvalue 7
– eigenvector 7
– particle-antiparticle system 8
Charge parity 7
Charginos 300
Chemical potential 241
Chern-Simons number 252, 254
Chiral symmetry 314
CKM, Cabibbo-Kobayashi-Maskawa
  matrix 11, 45, 202, 249
– parameter α 51, 100, 133
– parameter β 51, 99, 129, 183, 197
– parameter γ 51, 132, 145
– phase 46, 52, 275
– unitarity 12, 46
Cold neutrons 210
Compactification radius 317
Compound nucleus 211
Conservation law 28
Correlation
– fivefold 210, 213
– triple 210, 212
Cosmological constant 239, 318
CP asymmetry
– B meson 197
CP symmetry
– conserved 101
CP transformation 9
– B system 98
– K system 10
– boson 9

– eigenstates 98
– particle-antiparticle system 10
CP violation
– B meson 202
– B system 102
– B_s system 117
– K meson 203
– K system 10, 27, 30, 102
– baryogenesis 237
– direct 31–33, 102, 103, 202
– – experiment 34, 37, 105
– discovery 30
– experiment 39
– interference mixed/unmixed 103, 110
– – experiment 183
– mixing 31
– non-standard 263, 277
– – Higgs sector 263
– – minimal supersymmetry 266
– oscillation 102, 106, 202
– – experiment 110
– phase 32, 202
– strong CP problem 19
– supersymmetry 313
CPT invariance 31, 57, 247
CPT symmetry 200
CPT theorem 19, 22, 29
– consequences 23, 57
– tests of 23
Cross section
– K^0 nucleon 131
– coefficients 225
– partial 225
– total 223

Dark energy 239
Dark matter 306
Decay asymmetry
– K meson 196
Decay rates
– time dependent 201
Decaying state
– differential equation 197
Density matrix 214, 216, 229
– statistical 217
Density operator 247
– statistical 217, 231
Detector

- KTeV 36
- NA48 34

Early universe 254, 262, 281
Eigenstates 10, 11
- B meson 201
- CP 29, 31, 98, 113, 129, 141
- mass 11, 58, 115
- T 14
- weak 10, 11
Einstein equations 238
- with extra dimensions 318
Electric dipole moment 4, 18, 20, 98, 207, 265
- atom 6, 17, 209
- neutron 6, 17, 207
- upper limit 18, 207
Electroweak symmetry breaking 299, 303
Energy density 241
Entropy density 241, 242
Equation of state 239
Equilibrium
- local thermal 243, 255
- thermal 243
Equilibrium thermodynamics 241
Extra dimensions 316
- experimental tests 321
- open problems 319

FCNC, Flavour-Changing Neutral Currents 307
Fermion masses 55
Fine-tuning 300, 304
Fivefold correlation 210, 213
Flavour oscillation
- mechanical analogon 65
Flavour problem 307
Flavour tagging 64
flavour tagging 81, 147
Forward scattering amplitude 210, 214
- decomposition 227
-- spin-1/2 230
- elastic 215
- expansion coefficients 228
Freeze out 244
Friedmann equation 239, 240

Gauge symmetry
- electroweak 249
- left-right symmetric 306
- SO(10) 277
- spontaneously broken 249
- SU(5) 277
Gauge theory
- non-Abelian 294
Gaugino 266
- mass generation 310
Gedanken-Experiment
- Sakharov conditions 248
GIM mechanism 310
Gluinos 300
Goldstone boson 299
Goldstone theorem 314
Gravitino 300
GUT, Grand unified theories 277, 295
- SO(10) 298
- SU(5) 295
- experimental tests 296
- gauge boson 296
- gauge coupling constant 295
- scale 295
- SO(10) 277
- SU(5) 277
- supersymmetric 302
- unification scale 296

Hamiltonian
- two-particle 57
Helmholtz free energy 257
Hierarchy problem 299
- technicolour 314
Higgs field 55, 249
- fermion coupling 55
- mass 259
-- radiative correction 305
- mechanism 303
- potential 259
- two doublet model 263, 303
Higgs particle 299
Higgs potential
- classical 259
- high temperatures 259
- parameters 305
- two doublet model 303
Higgsino 275
Hubble distance 238

Hubble parameter   239, 243

Ideal gas
– distribution   241
Inflation   241, 246, 280
Instantons   253
Interaction
– milliweak   30
– superweak   11, 30
– T-odd QCD   19
Interference technique   208
Isotropic pressure   241

Jarlskog parameter   49

Kähler metrics   308
Kaluza-Klein   318, 321
KM, Kobayashi-Maskawa
– mechanism   31, 237

Larmor frequency   208
Larmor precession   210
Leptogenesis   284
Lepton-number-to-entropy ratio   283

Mach's principle   9
Madison convention   232
Majorana fields   287
Majorana masses   266
Majorana neutrino   281
Mass matrix   55
Mass scale
– fundamental   319
Matter-antimatter asymmetry   97, 237
Microwave photon background   245
Mirror
– image   29
– reflection   4
– transition   198
Mixing
– K meson   200
– p-wave   211
– s- and d-wave   211
Molecules
– PbO   209
– TlF   209
Monopole problem   280

Naturalness problem   299

Neutralino   300, 306
Neutrino
– Dirac   249, 298
– Majorana   238, 249, 277, 281
– mass matrix   298
– massive   298, 320
– oscillation   281
– right-handed   298
– seesaw mechanism   281, 320
– solar   281
Neutron transmission   230
Noether's theorem   250
Nonequilibrium
– thermal   243, 246, 255
– thermodynamic   28
Nucleosynthesis
– primordial   245, 275
Number density   241

Operator
– antiunitary   22
Optical theorem
– generalized   214, 217
Oscillation (particle-antiparticle)   197
– B meson   74, 202
– B_s meson   76
– D meson   74
– K meson   73, 199
– formalism   57, 197
– mechanical analogon   65
Out-of-equilibrium   281

Parity   4, 96
– eigenvalues   5
– even, odd   17
– intrinsic   5
– spherical harmonics   6
– spinors   6
Parity violation   86
– neutron scattering   6
– proton proton scattering   6
Particle-antiparticle conjugation
– *see also* Charge conjugation   29
PCAC, Partially conserved axial vector current   314
Penguin diagram   100, 105, 135
Phase
– unobservable   52
Phase convention   232

Phase diagram 256
Phase transition 256
- electroweak 261, 262
- first order 260
- second order 261
Photon temperature 242
Planck mass 243
Planck scale 317
PNC, parity nonconserving 6
Polarization geometry correlation
- factors 225
- matrix 230
Polarization vector 230
Proton decay 297

QCD
- θ term 19
- anomaly 21
- instanton 21
- strong CP problem 19
- sum rules 20
- T-odd interaction 19
- vacuum 20
Quantum gravity 321
Quark masses
- generation 55
- two equal 50

Radiative correction
- Higgs mass 305
Radioactive ion beam 209
Rare deacy
- supersymmetry 313
Reflection probability 271, 273
Robertson-Walker metric 239

Sakharov conditions 246
Schiff shielding 209
Seesaw mechanism 281, 320
Solar neutrino problem 281
Space reflection 28
Sphaleron 254, 261, 274
Spinor transport 211
Standard model 294
- baryon number violation 249
- beyond 294
- extensions 262
Standard Model of Cosmology 238, 240, 245, 246, 249

Statistical tensor 217
Storage bottles 209
String theory 316
Strong CP problem 19
Super-heavy particles 276
Supergravity 267, 300
Superstrings 308
Supersymmetric partner 300
- mass spectrum 312
Supersymmetry 262, 300
- breaking 301
- breaking scale 308
- CP violation 313
- experimental tests 306
- Higgs sector 304
- minimal (MSSM) 262
- next-to-minimal 263
- rare decay 313
- soft breaking parameters 304
- soft breaking scale 301

T invariance 248
T-matrix 215
- partial wave expansion 219
- relation to S-matrix 222
Tagging, flavour 64
tagging, flavour 81, 147
Technicolour 313
- extended 315
- walking 316
Techniquarks 315
Temperature
- critial 254, 257, 262, 276
- photons 242
- universe 242
Time dependence
- particle decay 197
Time reversal 12, 29
- Newton equation 12
- electromagnetic transition multipoles 15
- final state interactions 15, 16
- invariance 14
- Kramer's theorem 15
- null test 227
- of Hermitian observable 14
- Schrödinger equation 12
- sperical harmonics 13
- tests

– – β decays   16
– – γ transition   16
– – detailed balance   17
– – edm electric dipole moment   17
– – neutron transmission   210
– – nuclear physics   206
– – principle of reciprocity   16
– – resonance   212
– – transmission   18
Time reversal violation
– P-even   212
– P-odd   211
Topological charge   252
Transformation
– antiunitary   12
– CP *see also* CP transformation   9
– discrete   29
– non-unitary   4, 13
Transition rates
– B- and L-violating   254
– time evolution   200
Triple correlation   210, 212

Uncertainty principle   208
Unitarity triangle   48
– area   50
Universality   309
Universe
– big bang   27
– dark matter   306
– expanding   27

Vacuum configurations   253
Vacuum expectation value   257
– Higgs field   55, 257, 267
Vacuum structure   252
Violation
– B, L   297
– baryon number   246
– flavor symmetry   308

Weak mixing angle   296
Weyl fields   287
Wu-Yang triangle   32

Druck: Strauss Offsetdruck, Mörlenbach
Verarbeitung: Schäffer, Grünstadt

Lecture Notes in Physics

For information about Vols. 1–553
please contact your bookseller or Springer-Verlag

Vol. 554: K. R. Mecke, D. Stoyan (Eds.), Statistical Physics and Spatial Statistics. The Art of Analyzing and Modeling Spatial Structures and Pattern Formation. Proceedings, 1999. XII, 415 pages. 2000.

Vol. 555: A. Maurel, P. Petitjeans (Eds.), Vortex Structure and Dynamics. Proceedings, 1999. XII, 319 pages. 2000.

Vol. 556: D. Page, J. G. Hirsch (Eds.), From the Sun to the Great Attractor. X, 330 pages. 2000.

Vol. 557: J. A. Freund, T. Pöschel (Eds.), Stochastic Processes in Physics, Chemistry, and Biology. X, 330 pages. 2000.

Vol. 558: P. Breitenlohner, D. Maison (Eds.), Quantum Field Theory. Proceedings, 1998. VIII, 323 pages. 2000

Vol. 559: H.-P. Breuer, F. Petruccione (Eds.), Relativistic Quantum Measurement and Decoherence. Proceedings, 1999. X, 140 pages. 2000.

Vol. 560: S. Abe, Y. Okamoto (Eds.), Nonextensive Statistical Mechanics and Its Applications. IX, 272 pages. 2001.

Vol. 561: H. J. Carmichael, R. J. Glauber, M. O. Scully (Eds.), Directions in Quantum Optics. XVII, 369 pages. 2001.

Vol. 562: C. Lämmerzahl, C. W. F. Everitt, F. W. Hehl (Eds.), Gyros, Clocks, Interferometers...: Testing Relativistic Gravity in Space. XVII,507 pages. 2001.

Vol. 563: F. C. Lázaro, M. J. Arévalo (Eds.), Binary Stars. Selected Topics on Observations and Physical Processes. 1999.IX, 327 pages. 2001.

Vol. 564: T. Pöschel, S. Luding (Eds.), Granular Gases. VIII, 457 pages. 2001.

Vol. 565: E. Beaurepaire, F. Scheurer, G. Krill, J.-P. Kappler (Eds.), Magnetism and Synchrotron Radiation. XIV, 388 pages. 2001.

Vol. 566: J. L. Lumley (Ed.), Fluid Mechanics and the Environment: Dynamical Approaches. VIII, 412 pages. 2001.

Vol. 567: D. Reguera, L. L. Bonilla, J. M. Rubí (Eds.), Coherent Structures in Complex Systems. IX, 465 pages. 2001.

Vol. 568: P. A. Vermeer, S. Diebels, W. Ehlers, H. J. Herrmann, S. Luding, E. Ramm (Eds.), Continuous and Discontinuous Modelling of Cohesive-Frictional Materials. XIV, 307 pages. 2001.

Vol. 569: M. Ziese, M. J. Thornton (Eds.), Spin Electronics. XVII, 493 pages. 2001.

Vol. 570: S. G. Karshenboim, F. S. Pavone, F. Bassani, M. Inguscio, T. W. Hänsch (Eds.), The Hydrogen Atom: Precision Physics of Simple Atomic Systems. XXIII, 293 pages. 2001.

Vol. 571: C. F. Barenghi, R. J. Donnelly, W. F. Vinen (Eds.), Quantized Vortex Dynamics and Superfluid Turbulence. XXII, 455 pages. 2001.

Vol. 572: H. Latal, W. Schweiger (Eds.), Methods of Quantization. XI, 224 pages. 2001.

Vol. 573: H. M. J. Boffin, D. Steeghs, J. Cuypers (Eds.), Astrotomography. XX, 434 pages. 2001.

Vol. 574: J. Bricmont, D. Dürr, M. C. Galavotti, G. Ghirardi, F. Petruccione, N. Zanghi (Eds.), Chance in Physics. XI, 288 pages. 2001.

Vol. 575: M. Orszag, J. C. Retamal (Eds.), Modern Challenges in Quantum Optics. XXIII, 405 pages. 2001.

Vol. 576: M. Lemoine, G. Sigl (Eds.), Physics and Astrophysics of Ultra-High-Energy Cosmic Rays. X, 327 pages. 2001.

Vol. 577: I. P. Williams, N. Thomas (Eds.), Solar and Extra-Solar Planetary Systems. XVIII, 255 pages. 2001.

Vol. 578: D. Blaschke, N. K. Glendenning, A. Sedrakian (Eds.), Physics of Neutron Star Interiors. XI, 509 pages. 2001.

Vol. 579: R. Haug, H. Schoeller (Eds.), Interacting Electrons in Nanostructures. X, 227 pages. 2001.

Vol. 580: K. Baberschke, M. Donath, W. Nolting (Eds.), Band-Ferromagnetism: Ground-State and Finite-Temperature Phenomena. IX, 394 pages. 2001.

Vol.581: J. M. Arias, M. Lozano (Eds.), An Advanced Course in Modern Nuclear Physics. XI, 346 pages. 2001.

Vol.582: N. J. Balmforth, A. Provenzale (Eds.), Geomorphological Fluid Mechanics. X, 579 pages. 2001.

Vol.583: W. Plessas, L. Mathelitsch (Eds.), Lectures on Quark Matter, XIII, 334 pages. 2002.

Vol.584: W. Köhler, S. Wiegand (Eds.), Thermal Nonequilibrium Phenomena in Fluid Mixtures. XVII, 470 pages. 2002.

Vol.585: M. Lässig, A. Valleriani (Eds.), Biological Evolution and Statistical Physics. XI, 337 pages. 2002.

Vol.586: Y. Auregan, A. Maurel, V. Pagneux, J.-F. Pinton (Eds.), Sound–Flow Interactions. XVI, 286 pages. 2002

Vol.588: Y. Watanabe, S. Heun, G. Salviati, N. Yamamoto (Eds.), Nanoscale Spectroscopy and Its Applications to Semiconductor Research. XV, 306 pages. 2002.

Vol.589: A. W. Guthmann, M. Georganopoulos, A. Marcowith, K. Manolakou (Eds.), Relativistic Flows in Astrophysics. XII, 241 pages. 2002

Vol.590: D. Benest, C. Froeschlé (Eds.), Singularities in Gravitational Systems. Applications to Chaotic Transport in the Solar System. XI, 215 pages. 2002

Vol.591: M. Beyer (Ed.), CP Violation in Particle, Nuclear and Astrophysics. XI, 334 pages. 2002

Vol.592: S. Cotsakis, L. Papantonopoulos (Eds.), Cosmological Crossroads. An Advanced Course in Mathematical, Physical and String Cosmology. XVI, 578 pages. 2002

Vol.593: D. Shi, B. Aktaş, L. Pust, F. Mikhailov (Eds.), Nanostructured Magnetic Materials and Their Applications. XII, 299 pages. 2002

Vol.595: C. Berthier, L. P. Lévy, G. Martinez (Eds.), High Magnetic Fields. Applications in Condensed Matter Physics and Spectroscopy. X, 497 pages. 2002

Monographs
For information about Vols. 1–30
please contact your bookseller or Springer-Verlag

Vol. m 31 (Corr. Second Printing): P. Busch, M. Grabowski, P.J. Lahti, Operational Quantum Physics. XII, 230 pages. 1997.

Vol. m 32: L. de Broglie, Diverses questions de mécanique et de thermodynamique classiques et relativistes. XII, 198 pages. 1995.

Vol. m 33: R. Alkofer, H. Reinhardt, Chiral Quark Dynamics. VIII, 115 pages. 1995.

Vol. m 34: R. Jost, Das Märchen vom Elfenbeinernen Turm. VIII, 286 pages. 1995.

Vol. m 35: E. Elizalde, Ten Physical Applications of Spectral Zeta Functions. XIV, 224 pages. 1995.

Vol. m 36: G. Dunne, Self-Dual Chern-Simons Theories. X, 217 pages. 1995.

Vol. m 37: S. Childress, A.D. Gilbert, Stretch, Twist, Fold: The Fast Dynamo. XI, 406 pages. 1995.

Vol. m 38: J. González, M. A. Martín-Delgado, G. Sierra, A. H. Vozmediano, Quantum Electron Liquids and High-Tc Superconductivity. X, 299 pages. 1995.

Vol. m 39: L. Pittner, Algebraic Foundations of Non-Com-mutative Differential Geometry and Quantum Groups. XII, 469 pages. 1996.

Vol. m 40: H.-J. Borchers, Translation Group and Particle Representations in Quantum Field Theory. VII, 131 pages. 1996.

Vol. m 41: B. K. Chakrabarti, A. Dutta, P. Sen, Quantum Ising Phases and Transitions in Transverse Ising Models. X, 204 pages. 1996.

Vol. m 42: P. Bouwknegt, J. McCarthy, K. Pilch, The W3 Algebra. Modules, Semi-infinite Cohomology and BV Algebras. XI, 204 pages. 1996.

Vol. m 43: M. Schottenloher, A Mathematical Introduction to Conformal Field Theory. VIII, 142 pages. 1997.

Vol. m 44: A. Bach, Indistinguishable Classical Particles. VIII, 157 pages. 1997.

Vol. m 45: M. Ferrari, V. T. Granik, A. Imam, J. C. Nadeau (Eds.), Advances in Doublet Mechanics. XVI, 214 pages. 1997.

Vol. m 46: M. Camenzind, Les noyaux actifs de galaxies. XVIII, 218 pages. 1997.

Vol. m 47: L. M. Zubov, Nonlinear Theory of Dislocations and Disclinations in Elastic Body. VI, 205 pages. 1997.

Vol. m 48: P. Kopietz, Bosonization of Interacting Fermions in Arbitrary Dimensions. XII, 259 pages. 1997.

Vol. m 49: M. Zak, J. B. Zbilut, R. E. Meyers, From Instability to Intelligence. Complexity and Predictability in Nonlinear Dynamics. XIV, 552 pages. 1997.

Vol. m 50: J. Ambjørn, M. Carfora, A. Marzuoli, The Geometry of Dynamical Triangulations. VI, 197 pages. 1997.

Vol. m 51: G. Landi, An Introduction to Noncommutative Spaces and Their Geometries. XI, 200 pages. 1997.

Vol. m 52: M. Hénon, Generating Families in the Restricted Three-Body Problem. XI, 278 pages. 1997.

Vol. m 53: M. Gad-el-Hak, A. Pollard, J.-P. Bonnet (Eds.), Flow Control. Fundamentals and Practices. XII, 527 pages. 1998.

Vol. m 54: Y. Suzuki, K. Varga, Stochastic Variational Approach to Quantum-Mechanical Few-Body Problems. XIV, 324 pages. 1998.

Vol. m 55: F. Busse, S. C. Müller, Evolution of Spontaneous Structures in Dissipative Continuous Systems. X, 559 pages. 1998.

Vol. m 56: R. Haussmann, Self-consistent Quantum Field Theory and Bosonization for Strongly Correlated Electron Systems. VIII, 173 pages. 1999.

Vol. m 57: G. Cicogna, G. Gaeta, Symmetry and Perturbation Theory in Nonlinear Dynamics. XI, 208 pages. 1999.

Vol. m 58: J. Daillant, A. Gibaud (Eds.), X-Ray and Neutron Reflectivity: Principles and Applications. XVIII, 331 pages. 1999.

Vol. m 59: M. Kriele, Spacetime. Foundations of General Relativity and Differential Geometry. XV, 432 pages. 1999.

Vol. m 60: J. T. Londergan, J. P. Carini, D. P. Murdock, Binding and Scattering in Two-Dimensional Systems. Applications to Quantum Wires, Waveguides and Photonic Crystals. X, 222 pages. 1999.

Vol. m 61: V. Perlick, Ray Optics, Fermat's Principle, and Applications to General Relativity. X, 220 pages. 2000.

Vol. m 62: J. Berger, J. Rubinstein, Connectivity and Superconductivity. XI, 246 pages. 2000.

Vol. m 63: R. J. Szabo, Ray Optics, Equivariant Cohomology and Localization of Path Integrals. XII, 315 pages. 2000.

Vol. m 64: I. G. Avramidi, Heat Kernel and Quantum Gravity. X, 143 pages. 2000.

Vol. m 65: M. Hénon, Generating Families in the Restricted Three-Body Problem. Quantitative Study of Bifurcations. XII, 301 pages. 2001.

Vol. m 66: F. Calogero, Classical Many-Body Problems Amenable to Exact Treatments. XIX, 749 pages. 2001.

Vol. m 67: A. S. Holevo, Statistical Structure of Quantum Theory. IX, 159 pages. 2001.

Vol. m 68: N. Polonsky, Supersymmetry: Structure and Phenomena. Extensions of the Standard Model. XV, 169 pages. 2001.

Vol. m 69: W. Staude, Laser-Strophometry. High-Resolution Techniques for Velocity Gradient Measurements in Fluid Flows. XV, 178 pages. 2001.

Vol. m 70: P. T. Chruściel, J. Jezierski, J. Kijowski, Hamiltonian Field Theory in the Radiating Regime. VI, 172 pages. 2002.

Vol. m 71: S. Odenbach, Magnetoviscous Effects in Ferrofluids. X, 151 pages. 2002.

Vol. m 72: J. G. Muga, R. Sala Mayato, I. L. Egusquiza (Eds.), Time in Quantum Mechanics. XII, 419 pages. 2002.